Applied Fluvial Geomorphology for River Engineering and Management

Edited by

COLIN R. THORNE
University of Nottingham, UK

RICHARD D. HEY
University of East Anglia, UK

MALCOLM D. NEWSON
University of Newcastle, UK

JOHN WILEY & SONS
Chichester · New York · Weinheim · Brisbane · Singapore · Toronto

Baffins Lane, Chichester,
West Sussex PO19 1UD, England

National 01243 779777
International (+44) 1243 779777
e-mail (for orders and customer service enquiries): cs-books@wiley.co.uk
Visit our Home Page on http://www.wiley.co.uk
or http://www.wiley.com

Other Wiley Editorial Offices

John Wiley & Sons, Inc., 605 Third Avenue,
New York, NY 10158-0012, USA

WILEY-VCH Verlag GmbH, Pappelallee 3,
D-69469 Weinheim, Germany

Jacaranda Wiley Ltd, 33 Park Road, Milton,
Queensland 4064, Australia

John Wiley & Sons (Canada) Ltd, 22 Worcester Road,
Rexdale, Ontario M9W 1L1, Canada

John Wiley & Sons (Asia) Pte Ltd, 2 Clementi Loop #02-01,
Jin Xing Distripark, Singapore 129809

Library of Congress Cataloging-in-Publication Data

Applied fluvial geomorphology for river engineering and management
edited by Colin R. Thorne, Richard D. Hey, Malcolm D. Newson.
p. cm.
Includes bibliographical references and index.
ISBN 0-471-96968-0 (alk. paper)
1. Watersheds. 2. Geomorphology. 3. River engineering.
I. Thorne, C. R. (Colin R.) II. Hey, R. D. (Richard David)
III. Newson, Malcolm David.
GB562.ACG 1997
627'. 12–dc21 97-14904
CIP

British Library Cataloguing in Publication Data

A catalogue record for this book is available from the British Library

ISBN 0 471 96968 0 (paperback)
0 471 97852 3 (hardback)

Typeset in 10/11½ Times by Techset Composition, Salisbury, Wiltshire
Printed and bound in Great Britain by Bookcraft (Bath) Ltd
This book is printed on acid-free paper responsibly manufactured from sustainable forestation, for which at least two trees are planted for each one used

Applied Fluvial Geomorphology for River Engineering and Management

Contents

List of Contributors

James Bathurst	Department of Civil Engineering, University of Newcastle, Newcastle upon Tyne, Northumberland, NE1 7RU, UK
Andrew Brookes	River Appraisals, Thames Region, Environment Agency, Kings Meadow Road, Reading, Berkshire, RG1 8DQ, UK
Paul Carling	Department of Geography, University of Lancaster, Bailrigg, Lancaster, Lancashire, LA1 4YB, UK
Richard Hey	School of Environmental Sciences, University of East Anglia, Norwich, Norfolk, NR4 7TJ, UK
Janet Hooke	Department of Geography, University of Portsmouth, Lion Place, Portsmouth, Hampshire, PO1 3HE, UK
Stuart Lane	Department of Geography, University of Cambridge, Downing Place, Cambridge, CB2 3EN, UK
Damian Lawler	School of Geography, University of Birmingham, Birmingham, Warwickshire, B15 2TT, UK
John Lewin	Institute of Geography and Earth Sciences, University College of Wales, Llandinam Building, Penglais, Aberystwyth, SY23 3DB, UK
Mark Macklin	School of Geography, University of Leeds, Leeds, Yorkshire, LS2 9JT, UK
Malcolm Newson	Department of Geography, University of Newcastle, Newcastle upon Tyne, Northumberland, NE1 8ST, UK
Geoff Petts	School of Geography, University of Birmingham, Birmingham, Warwickshire, B15 2TT, UK
Ian Reid	Department of Geography, University of Loughborough, Loughborough, Leicestershire, LE11 3TU, UK
Keith Richards	Department of Geography, University of Cambridge, Downing Plane, Cambridge, CB2 3EN, UK
David Sear	Department of Geography, University of Southampton, Southampton, Hampshire, SO17 1BJ, UK
Colin Thorne	Department of Geography, University of Nottingham, Nottingham, Nottinghamshire, NG7 2RD, UK
Des Walling	Department of Geography, University of Exeter, Rennes Drive, Exeter, Devon, EX4 4RJ, UK
Bruce Webb	Department of Geography, University of Exeter, Rennes Drive, Exeter, Devon, EX4 4RJ, UK
Alan Werritty	Department of Geography, University of Dundee, Perth Road, Dundee, DD1 4HN, UK

Acknowledgements

This book was produced under the auspices of the River Dynamics Group, an ad hoc group of British fluvial geomorphologists with a common interest in the application of geomorphology to river and environmental management. A series of meetings of the group was funded by a grant (contract number DAJA4591MO172) from the US Army Research and Standardisation Group (London). The support, encouragement and patience of Mr Jerry Comati is gratefully acknowledged. Draft chapters were proofed and corrected by Eileen Thorne and the final manuscript was assembled at the University of Nottingham by Amanda Rowley. Their hard work and diligence was crucial to production of this volume. However, any uncorrected errors or omissions remain the responsibility of the editors.

Section I

Introduction

SECTION CO-ORDINATOR: R. D. HEY

1 River Engineering and Management in the 21st Century

RICHARD D. HEY
School of Environmental Sciences, University of East Anglia, Norwich, UK

1.1 RIVER ENGINEERING AND SOCIETY

Although the earliest settlers avoided valley bottom lands because they were prone to flooding, harboured disease and were barriers to communication, mankind eventually recognised the value of rivers and began to exploit them.

The engineering profession has made a major contribution to the health and wellbeing of mankind by designing and constructing a range of river engineering works to control floods, provide water supplies, generate power, aid navigation and stabilise and train rivers. The earliest projects were minor in scale and had only local impact, whereas mechanisation and greater technical knowledge now enable major projects to be implemented which can have more widespread and significant impacts on the environment.

1.1.1 Water Supply

Prior to the Industrial Revolution direct abstractions from rivers or wells were local and small scale. However, rapid urbanisation soon led to increased risk of failure from river sources, over-exploitation of groundwater, and contamination of potable supplies. As a consequence dams were constructed in unpolluted headwater catchments to create reservoirs which artificially increased the storage capacity within the river basin and enhanced the reliable yield. These were often single-purpose direct supply reservoirs with water treated on site and piped to the demand centre to avoid contamination. For example, Birmingham's water supply was obtained by pipeline from the Elan Valley reservoirs in mid-Wales.

More recently, reservoirs, both surface and groundwater, have been used to regulate river flow. The river is used as a natural aqueduct which allows multiple use of water along its course. Flows in the River Severn, UK, are regulated through operation of Clywedog and Vrynwy reservoirs and from the Shropshire groundwater scheme.

1.1.2 Flood Alleviation and Land Drainage

Provided that flooding can be prevented or ameliorated, floodplains represent a valuable resource either for urban and industrial development because of the flat terrain, or for agriculture, due to the fertility of the land.

Applied Fluvial Geomorphology for River Engineering and Management, Edited by C. R. Thorne, R. D. Hey and M. D. Newson.

A range of engineering solutions have been developed to reduce flood risk. One approach is the construction of storage reservoirs to retain flood waters, with subsequent releases regulated to ensure that flows remain in-bank. Another is to modify the river channel to increase its flood capacity. Channel works include widening, dredging and straightening the river, often in association with heavy maintenance to remove bank and in-channel vegetation.

Often, the floodplain is protected by levees or flood embankments along the river's course to restrict the effect of flooding. Recently, new flood defence measures such as two-stage channels and diversions have become more popular owing to their potential for river habitat improvement.

1.1.3 Power Generation

For centuries rivers have been exploited to supply mechanical power for milling, weaving and other purposes. On upland rivers, contour canals were often constructed around hillsides to provide the head of water to drive undershot or overshot mill wheels. Alternatively, on lowland rivers, low head, run of river, structures were built to drive wheels for milling. The ponded reach was often dredged and widened so that the volume of water stored overnight was sufficient to power the mill the next day. On many rivers several mills were constructed, each just upstream from the backwater of its neighbour, creating a series of linear, stepped reservoirs. On the River Wensum in Norfolk, UK, seven mills operated along a 30 km length of river.

More recently, in many countries hydroelectric power generation has supplanted direct mechanic power production. Upland sites predominate for commercial generation because of the requirement for a large head difference. Storage dams are generally constructed to provide a reliable yield to drive the turbines. On low gradient rivers, where the base flow is large, it can be viable to generate power as part of a multipurpose scheme as, for example, at Gezoubha on the River Yangtze, China, in association with navigation improvement and flood control. A further scheme is currently being planned within the Three Gorges, immediately upstream.

1.1.4 Navigation

Many lowland rivers are maintained for navigation and are often linked to canal systems. To enable navigation by larger vessels, and to extend the upper navigable limit, weirs with locking systems have been constructed to raise depths in the ponded reach. Often, variable sluices are incorporated in the weirs as, for example, on the Thames, Rhine and Mississippi, which enable a standard high water level to be maintained across a wide range of flows.

River training works are also carried out to remove or prevent shoaling on non-ponded reaches. This can be by dredging or by the construction of spur dykes to redirect flows and scour shallow sections. Channel straightening, often in association with flood control measures, has also been carried out to reduce navigation distance. The Greenville Reach of the Mississippi River has been shortened from 809 to 547 km by this means (Winkley, 1982).

1.1.5 River Stabilisation and Training

Engineers have developed a range of structural procedures to stabilise and train sections of channel to prevent bed scour or shoaling, bank erosion and channel migration.

Rivers which are being choked by excess sediment supply, manifested by wide braided reaches with numerous unvegetated islands, bars and channels, can be stabilised by narrowing the river between artificial non-eroding banks. This attempts to increase the river's sediment transport capacity to match the volume of sediment supplied from upstream. Alternatively, on smaller shoaling rivers, sediment traps can be constructed to reduce sediment supply to the aggrading reach, or erosion prevention measures can be implemented upstream (Sear and Newson, 1991).

Meandering channels are stabilised by preventing bank erosion on the outer bank in the bend by various forms of bank revetments, such as blockstone or gabion baskets, or by softer bioengineering treatments involving, for example, willow planting, facines and geotextiles, particularly jute-based ones which eventually degrade.

Incising channels which possess excess stream power are generally controlled by decreasing their sediment transport capacity by installing check weirs or grade control structures to reduce channel gradients.

1.1.6 Society and River Environments

The foregoing sections briefly illustrate some of the ways in which engineers have responded to the demands of society to harness the resources of rivers for the benefit of mankind. Major development programmes, aimed at transforming and underpinning a region's or a country's economy, have often been based on integrated river basin development projects. In the USA the Tennessee Valley Project is an excellent example of such a development and there are numerous examples in developing countries: Damodar Valley, India; Volta River Project in Nigeria; Aswan Dam on the River Nile in Egypt.

While economic objectives are still of prime importance, society is becoming increasingly aware of the environmental costs of such development. Environmental impact legislation in the USA and the European Community requires that the environmental implications of river engineering works are evaluated and incorporated in the decision-making process. Clearly, the design of future river engineering works must aim to minimise their impact on the environment as well as meeting economic objectives. Society's valuation of rivers and riverscapes, especially in the light of their continued exploitation and despoliation, has been heightened over the last decade. International RAMSAR sites have been designated to protect wetland areas, while nationally key river reaches with high conservation value are being given protection. Public opinion is becoming disenchanted with the wholesale exploitation of river systems and is demanding that some rivers, at least, should be preserved in a natural state. For example public pressure in France overturned plans to control and regulate the River Loire, one of the last remaining wild and natural large rivers in Europe. Equally 'Wild and Scenic River' designation protects key rivers in the USA. Where new river engineering works are required it is essential that environmentally sensitive solutions are adopted. Equally, every opportunity should be taken to restore heavily engineered reaches to a natural state or, if this is not possible, to carry out some rehabilitation measures. This requires a good understanding of what that natural state is and how it can be achieved.

1.2 IMPACT OF ENGINEERING AND MANAGEMENT

Rivers naturally adjust their shape and dimensions in response to the imposed discharge and sediment load. The habitats that are created are colonised by invertebrates, flora and fisheries which are characteristic of that particular type of river: upland or lowland, sand or gravel, ephemeral or perennial. Although water quality can also affect biological response, it is the physical characteristics of the river that are the prime determinants, since unpolluted rivers have little conservation or fisheries value if they have limited morphological and hydraulic variability.

Any modification to a river as a result of engineering works, or to its flow and sediment transport regime through land-use change or river regulation, can cause instability. This can change channel characteristics and adversely affect conservation and fisheries value.

1.2.1 Water Supply

While direct abstraction from a river reduces flows downstream from the intake and increases the frequency of low flows, the scale of such operations is usually relatively minor, in terms of impact on channel forming flows, or local in extent. In many cases abstractions are offset by treated effluent return flows.

Reservoir developments have a much greater impact on the river since both flow and sediment regimes downstream from the dam are changed significantly. With direct supply reservoirs, average flows are reduced, low flows are more frequent and flood levels are reduced. Regulating reservoirs, in contrast, decrease the frequency of both low and flood flows. While changing the flow regime can affect the range of velocities and flow depths downstream from the dam, it is the change in the sediment transport regime which causes the major morphological problem. As sediment is trapped in the reservoir, water releases are either starved of sediment or even entirely sediment free. During high level releases and overtopping flood flows, erosion will occur downstream from the dam. The degree and extent of erosion depend on the pattern of releases and associated hydraulic conditions, the nature of the bed material, bank stability and tributary influences (Raynov et al., 1986; Thorne and Osman, 1988). For example, downstream from Hoover Dam on the Colorado, the river bed degraded by 7.5 m immediately below the dam within 13 years of dam closure and erosion affected 120 km of river during that period (Williams and Wolman, 1984). Erosion also causes bed armouring as fine material is removed from the bed. As the bed becomes immobile, sands and silts which infiltrate the gravel framework reduce permeability and render the gravels unsuitable for salmonid spawning (Milner et al., 1981).

Interbasin transfers are often an integral part of river regulation schemes. If abstractions and releases alter the magnitude and frequency of sediment transport events downstream from the river intake/outfall, then instability will result. In order to maintain the natural stability of the river, regulation should be designed not to alter the frequency of flows that transport bed material downstream from the intake/outfall (Hey, 1976).

1.2.2 Flood Alleviation and Land Drainage

Traditional engineering works for flood alleviation and land drainage involve various combinations of channel widening, dredging and straightening. Not only does this directly destroy instream and riparian habitats and flora to create uniform conditions which lack habitat diversity, but it also disrupts the sediment transport continuity through the engineered reach promoting erosion and deposition which can affect adjacent reaches (Brookes, 1987a,b, 1988). These effects are particularly marked on rivers that transport significant amounts of bed material load, as they are able to respond very quickly to imposed changes. On rivers where bed material loads are low, the response is less dramatic and the unnatural condition can be perpetuated by light maintenance (Hey et al., 1990). However, the river remains ecologically impoverished, visually unattractive and, by common agreement, unsatisfactory. The Mississippi River is a classic example of the problems that can result when rivers are straightened. The steepened section in the straightened reach increased the river's sediment transport capacity. This, in turn, caused erosion which then progressed upstream along the main river and its tributaries. As a result the sediment supply to the engineered reach increased and caused severe sedimentation problems. Heavy engineering works are required to maintain the river in its unnatural straightened state. These have involved the construction of major spur dyke fields to maintain an adequate navigation channel, many kilometres of revetment to prevent bank erosion and extensive operational dredging to remove bars (Winkley, 1982).

1.2.3 Power Production

Reservoir construction and operation for hydroelectric power production can have a major impact on river systems. As with water supply reservoirs, dam construction and reservoir operation can have a significant impact on the flow and sediment transport regime of the river downstream from the dam, triggering similar problems of morphological instability (Section 1.2.1).

1.2.4 Navigation

Channel improvements to facilitate navigation include capital and maintenance dredging and the construction of weirs, locks and dams. Dredging is generally a maintenance operation while dams and locks enable the headward navigation limit to be extended or provide increased navigation depth for larger draught vessels. Dredging maintains uniform conditions and destroys benthic and instream habitats, associated aquatic plant and invertebrate communities. It is, therefore, detrimental to fisheries and conservation. Weir pools create artificially uniform conditions within the ponded backwater area.

Vessel operations create additional problems since propellers increase turbulence and near-bed velocities, causing scouring and resuspension of fine sediments which elevate suspended sediment loads. Boat wash can be responsible for accelerated bank erosion, either directly or by destroying vegetation and allowing fluvial and subaerial erosion processes to become more effective (Garrad and Hey, 1987, 1988).

River Stabilisation and Training

Procedures for stabilising aggrading, particularly braided, rivers or degrading channels aim, generally, to control sediment transport rates. By creating a narrower canalised channel to prevent deposition or an artificially ponded river to control erosion, the conservation and amenity value is, at best, seriously impaired.

Bank protection measures for training and maintaining channels which are ostensibly in regime can also be very intrusive and environmentally damaging, particularly when protection measures are over-designed. Heavy engineering solutions have been widely applied, often inappropriately, with limited opportunity to soften their environmental and aesthetic impacts (Hemphill and Bramley, 1989; Hey et al., 1991).

1.3 ENVIRONMENTALLY SENSITIVE RIVER ENGINEERING

River management in England and Wales has, over the last few years, evolved into a truly multifunctional operation in which river engineering, pollution control, water resource development, conservation and fisheries management interrelate in a holistic catchment-based process (Gardiner, 1988; Newson, 1992). This recognises that the potential impact of river and catchment management plans has to be considered not only locally but also on a catchment scale; for example, the impact of afforestation or urbanisation on flood runoff and levels of flood protection further downstream. Equally, consideration needs to be given to longer term effects since any increase in sediment yield resulting from such development will, over time, be transmitted through the river system. In addition, integrated management also recognises the inherently interdisciplinary nature of catchment planning and, in particular, the effect of any environmental change on conservation and fisheries. Clearly, the wider and longer term implications of catchment plans need to be fully appraised, especially in the light of environmental impact legislation, and formal procedures are now being applied and refined to enable this goal to be achieved (Gardiner, 1991).

Traditional river engineering works have created major instability and environmental problems, principally because they impose an unnatural condition on the river by modifying the bankfull dimensions and/or the discharge and sediment transport regimes. In an attempt to maintain an unnatural condition, heavy capital engineering works and/or heavy operational maintenance are often required.

Environmentally sensitive engineering works aim to achieve design objectives by creating a more natural type of channel, with an accompanying range of instream and riparian habitats which preserve both the ecological diversity of the river system and its natural stability. Where these objectives are achieved, river engineering has the potential to either preserve or actually enhance the environmental and conservation value of the river. In addition, an approach based on working with the natural processes and form of the river minimises maintenance requirements. Today, there is a need to restore and rehabilitate heavily engineered reaches by, for example, recreating meandering channels with their associated pool-and-riffle bed morphology and by recreating 'river corridors' within which the channel can be allowed to migrate freely. The basis of this approach is an understanding of river and catchment processes over varying timescales as only by designing with nature will it be possible to achieve this objective (Hey, 1990, 1994). Viewed in this light, sound applied fluvial geomorphology is a cornerstone of environmentally friendly river engineering.

1.4 BOOK CONTENTS

This guidebook is not an engineering design manual. Instead the intention is to provide a basic understanding of fluvial geomorphology and river mechanics to enable the reader to recognise the environmental constraints on engineering designs, and to provide guidance on the application of geomorphologic principles in engineering design. This book represents the collective knowledge of a group of geomorphologists with considerable research experience, who have been directly involved in a range of river engineering projects.

Rivers are inherently unstable, particularly over long timescales (10^3-10^6 years), but can also be unstable over timescales which span the design life of an engineering scheme (Section II). While it is obvious that consideration should be given to catchment-scale processes over time periods of up to 100–200 years, the requirement to consider longer term instability is less obvious. In the Demonstration Erosion Control Project in northern Mississippi, recognition that channel instability cannot be attributed solely to recent human interventions and land-use changes, but has been a feature of the bluff-line streams for at least 15 000 years, has led to a fundamental reappraisal of channel stabilisation methodology. Equally, it is often necessary to identify the cause of channel instability or predict when instability might affect an engineering structure in order to effect a solution or design and undertake preventative measures for new schemes.

Reach-scale river channel and valley processes are covered in Section III. This section reviews the flow processes operating in straight and meandering channels with particular reference to secondary flows and shear stress distributions, flow resistance, sediment transport, erosion, deposition and bank erosion mechanics. Equations are presented which describe these processes together with limits to their range of application. Essentially, this section provides the basis for engineering design because these processes control erosion, deposition and, hence, the three-dimensional geometry of alluvial channels. Emphasis is placed on the role of field measurements and the interpretation of field and survey data to identify both eroding and depositing reaches, and the cause of bank retreat. In addition, Section III provides techniques for determining channel discharge capacity, and how allowance can be made for non-uniform flow, for maintaining stability within engineered reaches, and for establishing maximum scour depths in meander bends. Finally, Section III indicates how knowledge of secondary flows can be used to develop underwater structures, vanes, to control bank erosion and create scour pools or to promote meandering.

River channel morphology and channel adjustments during periods of erosion and deposition are covered in Section IV. A number of different river classification systems are outlined and procedures reviewed for predicting the overall shape and dimensions of natural alluvial channels. Emphasis is placed on mobile-bed meandering channels, as all rivers transport some form of sediment load and are to some degree sinuous. In the past, failure to maintain sediment transport continuity, coupled with the imposition of uniform straight channels, has resulted in instability of the majority of engineered reaches. Many rivers are unstable due to natural or anthropogenic changes and this produces characteristic styles of channel change. Simple mapping procedures can be used to identify eroding and depositing reaches on the basis of the style of change together with an indication of the degree of instability. Geomorphological approaches are also available for predicting channel response to imposed changes to river systems at both catchment and reach scales. Channel instability generates considerable maintenance costs and, even though every effort should be made to minimise the requirement for maintenance by adopting natural design methods, it is inevitable that some operational and maintenance works will be necessary. Working with nature, using geomorphological principles, enables the cause of instability to be treated rather than the symptoms, and proves to be a more environmentally acceptable procedure than traditional approaches based on simplified models of river flow and sediment transport.

Case studies in applied fluvial geomorphology are presented in Section V. These illustrate how the geomorphological procedures and practices, outlined in Sections II–IV, have been applied to a range of river engineering works. For each investigation the river engineering works are described together with the terms of reference for the geomorphological input, the techniques employed, the strategies recommended to the client, and the uptake and/or success of these strategies.

Finally, Chapter 13 outlines how the guidebook can be used by engineers and river managers faced with practical site problems. It identifies a blueprint for geomorphological studies as a component of a feasibility study and indicates which techniques are required for particular investigations and where in Sections II–IV to locate the relevant information and guidance.

1.5 REFERENCES

Brookes, A. 1987a. River channel adjustment down-stream from channelization works in England and Wales. *Earth Surface Processes and Landforms*, **12**, 337–351.

Brookes, A. 1987b. Recovery and adjustments of aquatic vegetation within channelization works in England and Wales. *Journal of Environmental Management*, **24**, 365–382.

Brookes, A. 1988. *Channelized Rivers: Perspectives for Environmental Management.* Wiley, Chichester, 336pp.

Gardiner, G.L. 1988. Environmentally sensitive river engineering: Examples from the Thames catchment. *Regulated Rivers: Research and Management*, 2 (3). Wiley, Chichester, 445–469.

Gardiner, G.L. (Ed.) 1991. *River Projects and Conservation, A Manual for Holistic Appraisal.* Wiley, Chichester, 236pp.

Garrad, P.N. and Hey, R.D. 1987. Boat traffic, sediment resuspension and turbidity in a Broadland river. *Journal of Hydrology*, **95**, 289–297.

Garrad, P.N. and Hey, R.D. 1988. River management to reduce turbidity in navigable Broadland rivers. *Journal of Environmental Management*, **27**, 273–288.

Hemphill, R.W. and Bramley, M.E. 1989. *Protection of River and Canal Banks.* Butterworths/CIRIA, London, 200pp.

Hey, R.D. 1976. River response to inter-basin water transfers: Craig Goch Feasibility Study. *Journal of Hydrology*, **85**, 407–421.

Hey, R.D. 1990. Environmental river engineering. *Journal of the Institution of Water and Environmental Management*, **4**(4), 335–340.

Hey, R.D. 1994. Environmentally sensitive river engineering. In: Calow, P. and Petts, G.E. (Eds), *River Handbook*, Vol. 2. Blackwell, Oxford, 337–362.

Hey, R.D., Heritage, G.L. and Patteson, M. 1990. *Flood Alleviation Schemes: Engineering and the Environment*. MAFF, London, 176pp.

Hey, R.D., Heritage, G.L., Tovey, N.K., Boar, R.R., Grant, A. and Turner, R.K. 1991. *Streambank Protection in England and Wales*. R & D Note 22, NRA, London, 75pp.

Milner, N.J., Scullion, J., Carling, P.A. and Crisp, D.T. 1981. The effects of discharge on sediment dynamics and consequent effect on invertebrates and salmonids in upland rivers. In: Coaker, T.H. (Ed.), *Advances in Applied Biology*, Vol. 6. Academic Press, London, 154–220.

Newson, M.D. 1992. River conservation and catchment management: a UK perspective. In: Boon, P.J., Calow, P. and Petts, G.E. (Eds), *River Conservation and Management*. Wiley, Chichester, 385–396.

Raynov, S., Pechinov, D., Kopaliany, Z. and Hey, R.D. 1986. *River Response to Hydraulic Structures*. UNESCO, Paris, 115pp.

Sear, D.A. and Newson, M.D. 1991. *Sediment and Gravel Transportation in Rivers Including the Use of Gravel Traps*. Project Report 232/1/T, National Rivers Authority, Bristol, 93pp.

Thorne, C.R. and Osman, A.M. 1988. Riverbank stability analysis. II: Applications. *Journal of Hydraulic Engineering, American Society of Civil Engineers*, **114**(2), 151–172.

Williams, G.P. and Wolman, M.G. 1984. *Downstream Effects of Dams on Alluvial Rivers*. US Government Printing Office, Washington DC.

Winkley, B. 1982. Response of the Lower Mississippi to river training and realignment. In: Hey, R.D., Bathurst, J.C. and Thorne, C.R. (Eds), *Gravel-Bed Rivers*. Wiley, Chichester, 659–680.

Section II

Natural Channel Stability and Time Perspectives

SECTION CO-ORDINATOR: M. G. MACKLIN

2 Channel, Floodplain and Drainage Basin Response to Environmental Change

MARK G. MACKLIN[1] and JOHN LEWIN[2]
[1]School of Geography, University of Leeds, UK
[2]Institute of Geography and Earth Sciences, University College of Wales, Aberystwyth, UK

2.1 INTRODUCTION

Arguably the greatest challenge currently facing engineers, scientists and policy-makers concerned with river engineering and catchment management is developing sustainable solutions to river problems at a time of rapid and, in recent geological terms, unprecedented global environment change. The assumption of constancy of climate and environment over 'engineering time' (around a century or less), which was the basis of much engineering and hydrological forward planning until recently, is now widely felt to be unsatisfactory (e.g. Knox, 1984; Lewin et al., 1988). Understanding and managing the impacts of environmental change in a drainage basin, however, requires an appreciation of river processes that operate over longer timescales and larger space scales than those which are generally familiar to the river engineer. For example, the physical landscape has a longer memory than either the climate system or biosphere and it is Late Quaternary (the last 125 000 years of Earth history) environmental changes that provide the 'initial' conditions for present-day river processes, their activity rates and the resulting channel morphology. This is clearly true in river basins that are little affected by human activity. But such environments are increasingly rare, and it is now known that human action can not only result in the transformation of a drainage basin, but may also modify global hydroclimatic systems. This is most graphically illustrated by changes in atmospheric composition associated with the growth this century of world population and industry, which have led to significant increases in the concentration of greenhouse gases and have contributed to recent global warming. Also, in the developing world, deforestation at a hitherto unprecedented scale and rate is widely perceived as being primarily responsible for increased sedimentation and flooding in a number of the world's great rivers (see Haigh, 1994).

In times like these, which are of particular environmental uncertainty, engineers and catchment planners need to consider, and solve, problems of river instability within a global framework. Although human activity can demonstrably exacerbate erosion, sedimentation and flooding, accurate prediction of river and catchment response to human interference is often problematic, primarily because the sensitivity of geomorphic and hydrologic systems to change is highly variable both in time and space. One of the

Applied Fluvial Geomorphology for River Engineering and Management, Edited by C. R. Thorne, R. D. Hey and M. D. Newson.

main tasks of a geomorphologist is identifying those river basins, or reaches, that may be potentially susceptible to future environmental change and those presently subject to dynamic adjustment to altered channel or climatic conditions.

Significant progress, however, has been made in quantifying river response to past, present and future environmental change by more effective integration of process-based and historical approaches to river dynamics. Over the last decade there has been a shift in the focus of evaluating river response to environmental change from solely the extended timescales of the Pleistocene (*c*. 2 000 000 – 10 000 years before the present (BP)), towards the Holocene (the last 10 000 years) where channel transformation in the last few hundred years can be directly related to systematic records of hydroclimate and land-use change (e.g. Rumsby and Macklin, 1994; Warner, 1992). In the USA there is a longer tradition of collaboration between process and historical geomorphologists and a greater wealth of empirical experience and record collation for river managers to draw on when confronted with instability (e.g. Knox, 1972; Lyons and Beschta, 1983). This is, perhaps, best exemplified by the hotly contested and ongoing debate concerning arroyo formation in southwestern USA that occurred at the end of the 19th century (see Box, 2.1). Although the chronology and pattern of channel entrenchment are generally well documented, at present there is no consensus as to whether land mismanagement, hydroclimate change or some combination of both these factors was responsible for river metamorphosis. Identification of the principal causative agent(s) of past and present change, and the disentanglement of 'natural' from human impacts on fluvial processes are fundamental prerequisites for alleviating present river problems. They constitute benchmarks from which sound management frameworks can be developed, that enable the river engineer to be proactive as opposed to reactive when faced with imminent environmental change.

The geologic, tectonic, climatic and cultural histories of a drainage basin are, consequently, of concern in any fluvial hazard evaluation and must be considered in river engineering design. This is a difficult task but we begin in this chapter by examining at a global scale the legacy of Quaternary environmental change and neotectonics, modification of the environment by human action in both prehistoric and historic times, and river response to Holocene climate change. All of these can be considered as extrinsic controls of river instability over a 'geomorphic' timescale (10 to 10 000 years) through their influence on sediment supply and flood magnitude and frequency. Drainage basin and network adjustments that occur over a century or less are also examined, with particular reference to those recorded over the last 200 years or so in North America, Europe and Australasia. In these regions systematic hydrological and meteorological records are commonly available for this period, together with information on past river engineering/regulation, land-use data and topographic surveys from which the causes of river channel change can be assessed. Increasingly, the analyses of valley floor morphology and alluvial sedimentary sequences are being used to supplement historical sources on magnitude and frequency characteristics of floods, as field indicators of channel instability (past, present and future) and for establishing catchment-scale sediment budgets. These budgets define sediment storage, transfer and provenance patterns within a drainage basin and enable changes in sediment fluxes between hillslopes, floodplains and channels to be quantified. Sediment supply variations strongly influence channel morphology and activity rates and, in the case of sediment-associated pollutants, can have severe environmental consequences. By studying the recent geological record and the geomorphology of river valleys we can also gain insights into the mechanisms, patterns, magnitudes and rates of past river instability that may provide an analogue ('lessons from the past') and, in turn, 'empirical forecasts' to future environmental change.

2.2 QUATERNARY ICE AGE INHERITANCE, TECTONICS AND LONGER TERM RIVER DYNAMICS

Over the last two million years or so the Earth has experienced alternately relatively warm (interglacial) and cold (glacial) climate periods that have reflected changes in solar radiation controlled by variation of the Earth's orbit. This last glacial stage ended about 10 000 years ago when the present interglacial, usually termed the Holocene or the Postglacial, formally began. Interglacials such as the present one, however, are periods of comparatively unusual climate that in total constitute less than 10% of the last 750 000 years of Earth history. A high proportion of existing drainage basin landforms and sediments were, therefore, actually formed under very different conditions to those that exist today. Consequently hillslope and river geomorphic systems, particularly in former glaciated catchments, have had to adjust to markedly different catchment and climate boundary conditions during the Holocene. In many drainage basins this process is still continuing and a state of 'passive disequilibrium' exists (Ferguson, 1981).

During the Pleistocene 'glacial' periods, the Earth was colder and drier than at present, with extensive ice sheet and permafrost development in high and middle latitude areas and greater aridity over much of the tropics. Reduced vegetation cover generally increased runoff and sediment supply in the form of clastic detritus to rivers (at least on a seasonal basis) and this caused widespread valley floor aggradation. Today, these deposits are of considerable agricultural and economic importance because they form some of the world's most fertile land, constitute an important source of minerals and act as valuable aquifers. The major glaciers and ice sheets of North and South America, Europe and Asia that melted between 9500–8000 BP left a significant worldwide legacy in terms of crustal deformation, sediment deposits and supply sources which, to varying degrees, continue to influence present river processes. Each of these factors is briefly reviewed and their influence on recent river instability is discussed.

2.2.1 Crustal Effects of Glaciation

During the Late Quaternary, the growth and decay of ice sheets and related changes in global sea level led to the redistribution of mass across the surface of the Earth. Within the margins of the former British, Fennoscandinavian and Laurentide ice sheets, crustal uplift of 40, 700 and 250 m, respectively, has occurred since the early Holocene. In the Gulf of Bothnia area of the northern Baltic Sea, land is still rising at 8–9 mm yr^{-1} and near the centre of the former Laurentide ice sheet in North America, current rates of isostatic recovery are around 11 mm yr^{-1} (Bell and Walker, 1992).

In soft rock terrains (particularly those underlain by Pleistocene or early Holocene marine or lacustrine deposits) within regions of continuing crustal rebound (see Figure 2.1), high rates of channel incision, bank erosion and failure are typical. A recent investigation

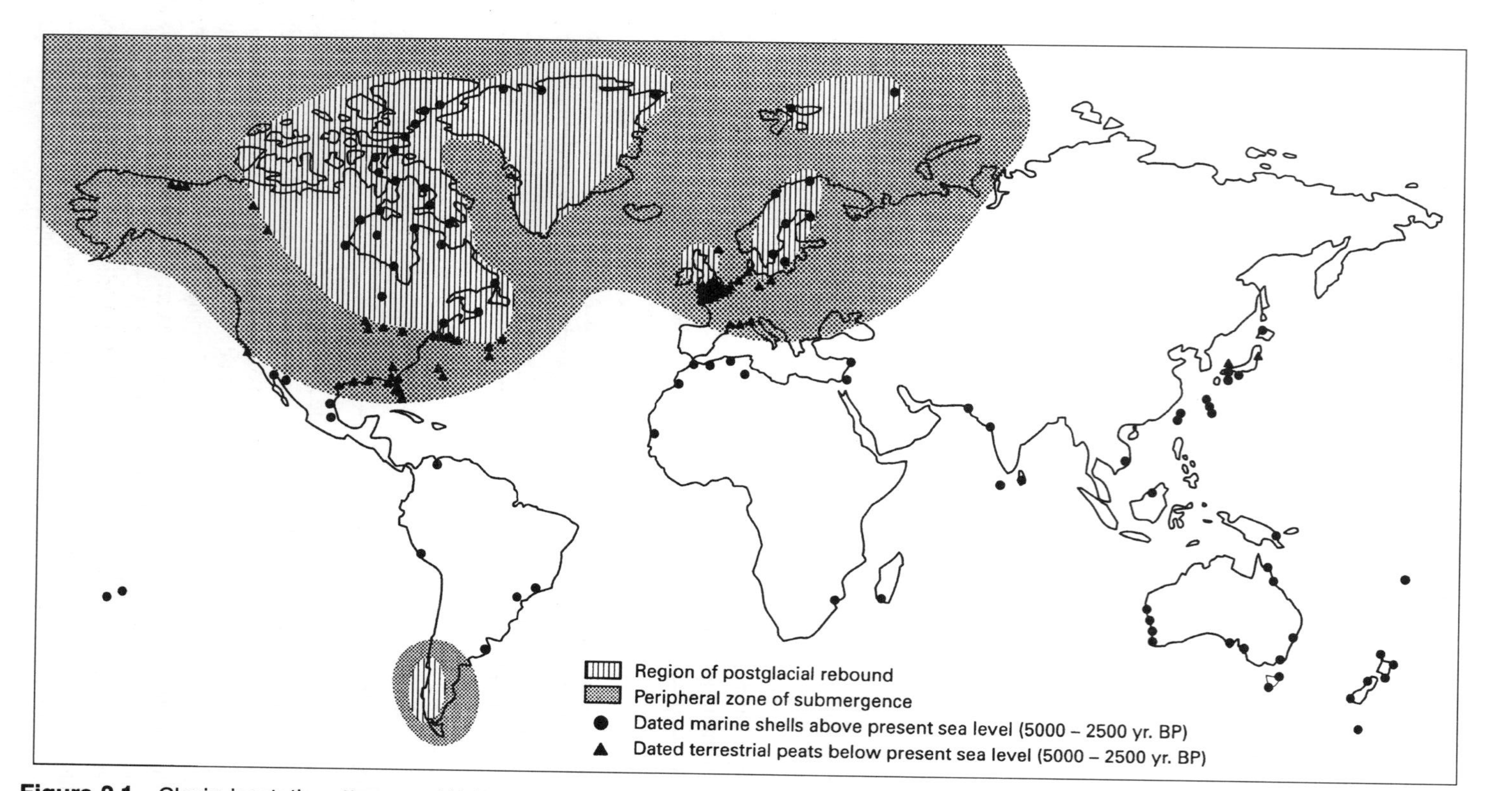

Figure 2.1 Glacio-isostatic patterns and Holocene sea level change (modified from Walcott, 1972)

by Mansikkaniemi (1991) in southern Finland illustrates some of the river engineering problems associated with these environments.

Several hundred kilometres beyond the margin of the major ice sheets, compensatory subsidence has taken place during the Holocene (Figure 2.1). Areas affected include the US eastern seaboard and the southern North Sea in Europe. Subsidence, in conjunction with glacio-eustatic related sea level rise during the Holocene (the result of changes in global ice volume), has increased the risk of inundation and river flooding (by reducing river gradients) in these areas (e.g. lower River Thames valley, UK). Subsidence is also caused by sediment loading. For example, the Mississippi Delta area has subsided at an average rate of 15 mm yr^{-1} since the beginning of the Holocene (Fairbridge, 1983).

2.2.2 Sediment Sources and Supply

Currently and Formerly Glaciated Catchments

In catchments over-run by glaciers and ice sheets and also in river systems that drained former ice margins, Pleistocene glaciation generated large volumes of erodible sediment. Bell and Laine (1985) suggest that continental glaciation in North America led to erosion rates that were an order of magnitude greater than those of today. They point out that much of the sediment produced by this erosion is now deposited offshore, where it amounts to about 15 times the volume of glacial sediment preserved on land. The fact that glacierised basins produce about 10 times as much sediment as those without glaciers is also confirmed by comparative data from rivers in British Columbia (Church et al., 1989). Holocene basin sediment yields and river behaviour in these areas (comprising much of the high and middle latitude areas of the northern hemisphere, as well as the Andes, Patagonia and South Island, New Zealand (see Figure 2.2), prior to human disturbance, have largely reflected the secondary mobilisation of Pleistocene glacial sediment. Climate change since the early Holocene has re-established a vegetation cover and, generally, has led to decreasing sediment supply rates under natural conditions (Church and Ryder, 1972). In presently unglaciated river systems of moderate to high relief, beyond the influence of postglacial sea-level rise, this has resulted in progressive valley floor incision. Rates of incision were, initially, very rapid. This was the case particularly during the late-glacial–Holocene transition (about 13 000–9000 BP) but has slowed, or even reversed, in the latter part of the Holocene, following, for example, glacial advances, volcanic activity, large-scale slope failure or landslips and human disruption. River instability induced by these factors is well documented in the mountain areas of Scandinavia (Grove, 1972), the Northern American Rockies (Jordan and Slaymaker, 1991), the European Alps and the South Island Alps of New Zealand (Griffiths, 1979). One notable hazard in currently glacierised river basins is floods resulting from catastrophic drainage of glacial lakes impounded by ice or morainic material. Maizels and Russell (1992) provide a review of the effect of such floods in Iceland, where they are termed 'jokulhlaups', and how they might best be predicted. Ryder (1991) has documented moraine-dammed glacial lake failures in the coastal mountains of British Columbia, Canada.

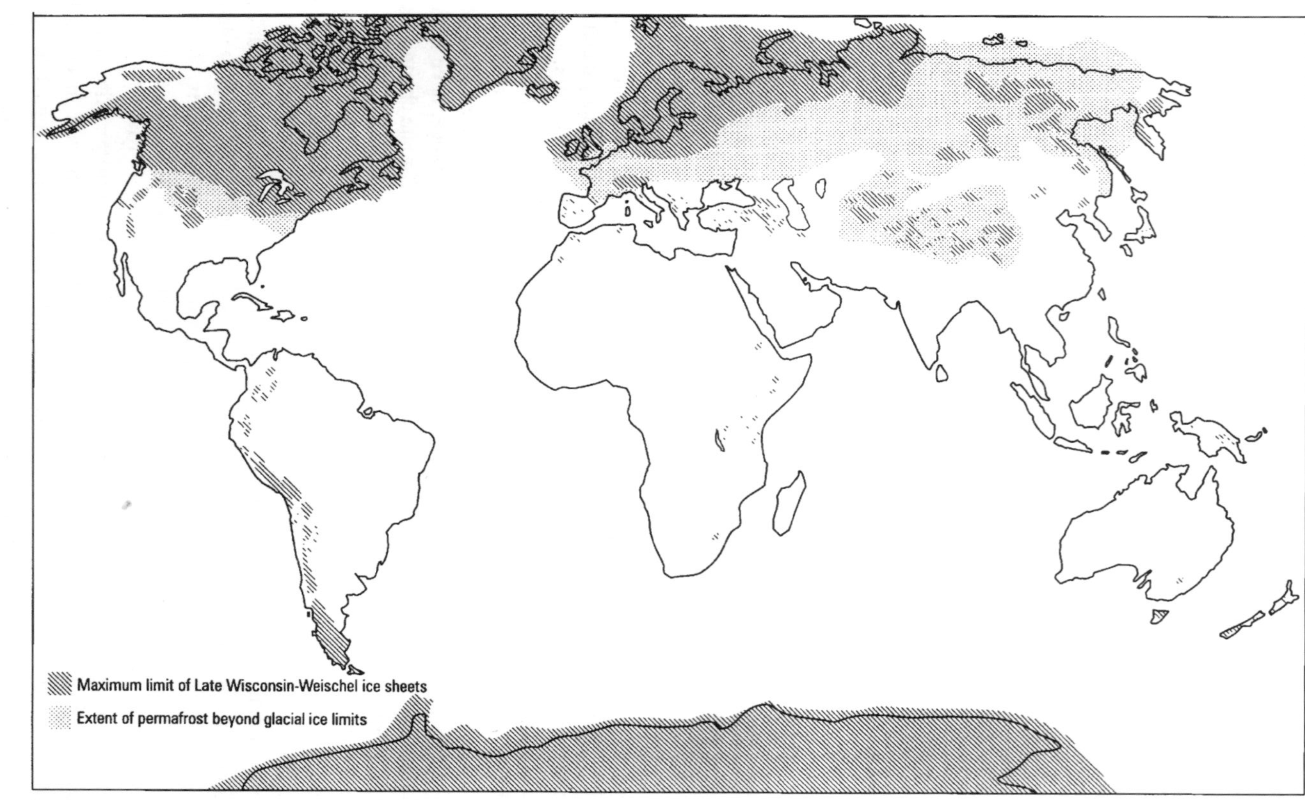

Figure 2.2 Maximum extent of ice sheets and permafrost at the Last Glacial Maximum (18 000 BP) (modified from Williams et al., 1993)

Unglaciated Catchments

Increased aeolian activity during the last glaciation, due to the expansion of arid and semi-arid regions, resulted in the formation of extensive sand dune fields throughout large areas of Africa, India, Australia, and North and South America (Figure 2.3). In other mid-latitude areas, great thicknesses of unstratified silt, known as loess, were deposited over wide areas (Figure 2.4). Most of the windblown loess sediment was derived from glacial outwash plains that bordered the mid-latitude ice sheets. However, in some areas, such as China, the formation of loess may have originated in deserts due to the combined influence of frost and salt weathering.

Both deserts and associated sand dunes were more widespread during the Last Glacial Maximum (about 18 000 BP) than they are today. They characterised almost 50% of the land area between 30°N and 30°S, forming two vast belts. Together, aeolian sand and loess in regions beyond the margin of the last Pleistocene ice sheets constitute the single most important source of sediment for many mid- and low-latitude rivers today. Their high erodibility, particularly when protective vegetation is removed or damaged by human activities, results in extremely high sediment yields. The Yellow River (Huang He) catchment in China, with its tributaries draining thick loess deposits in Shaanxi Province, is probably the best example of this, and stand outs clearly in Figure 2.5 (Lvovich et al., 1991). Loess deposits have also been important sources for floodplain deposition in the southeastern USA (Schumm et al., 1984), and in central and eastern Europe (Figure 2.4).

2.2.3 Tectonically Active Environments

Globally, tectonics exert a strong influence on large-scale drainage basin morphology and river development through time. Figure 2.6 shows the general distribution of earthquakes around the world. This pattern delineates tectonically active areas which are generally located on, or immediately adjacent to, active plate margins. There is a clear association between tectonically active regions and high basin sediment yields (Figure 2.5) in areas such as New Zealand, the northern margin of the Mediterranean Sea, southern and southeast Asia, the Californian and Alaskan coastal ranges, and the Andes. High rates of slope erosion, river instability and rapid siltation are 'natural' characteristics of these upland environments, although in some cases geomorphic processes have been accelerated by recent human activity.

Where rivers flow across an area of rapid uplift or subsidence, drainage networks and long profiles can be disrupted, leading to channel pattern changes and altered vertical and lateral tendencies. Tectonics can also affect river stability through earthquakes and faulting, which can damage river bed and bank structures, cause landslips and form natural dams (Costa and Schuster, 1988). The last two processes often result in a large injection of sediment into a river system, which in turn causes major downstream adjustments in channel morphology.

Slow epeirogenic uplift (the warping of large areas of the Earth's crust without significant deformation), that occurs away from active plate margins, can also affect channel characteristics and river stability. A good example of this is the Monroe dome in the lower Mississippi basin where uplift has resulted in recent channel incision and an increase in sinuosity (Burnett and Schumm, 1983). Repeated levelling surveys during the

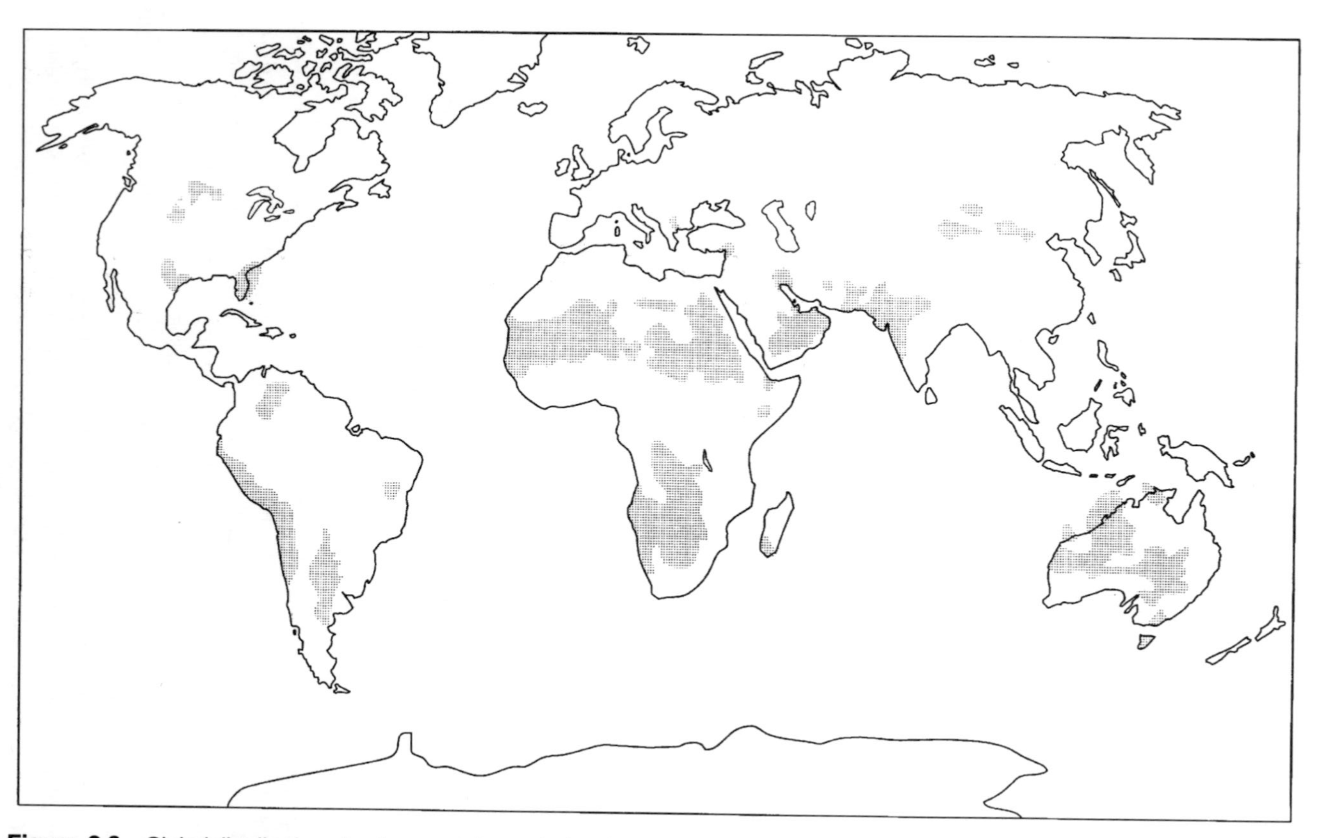

Figure 2.3 Global distribution of active sand dunes during the Last Glacial Maximum (modified from Sarnthein, 1978)

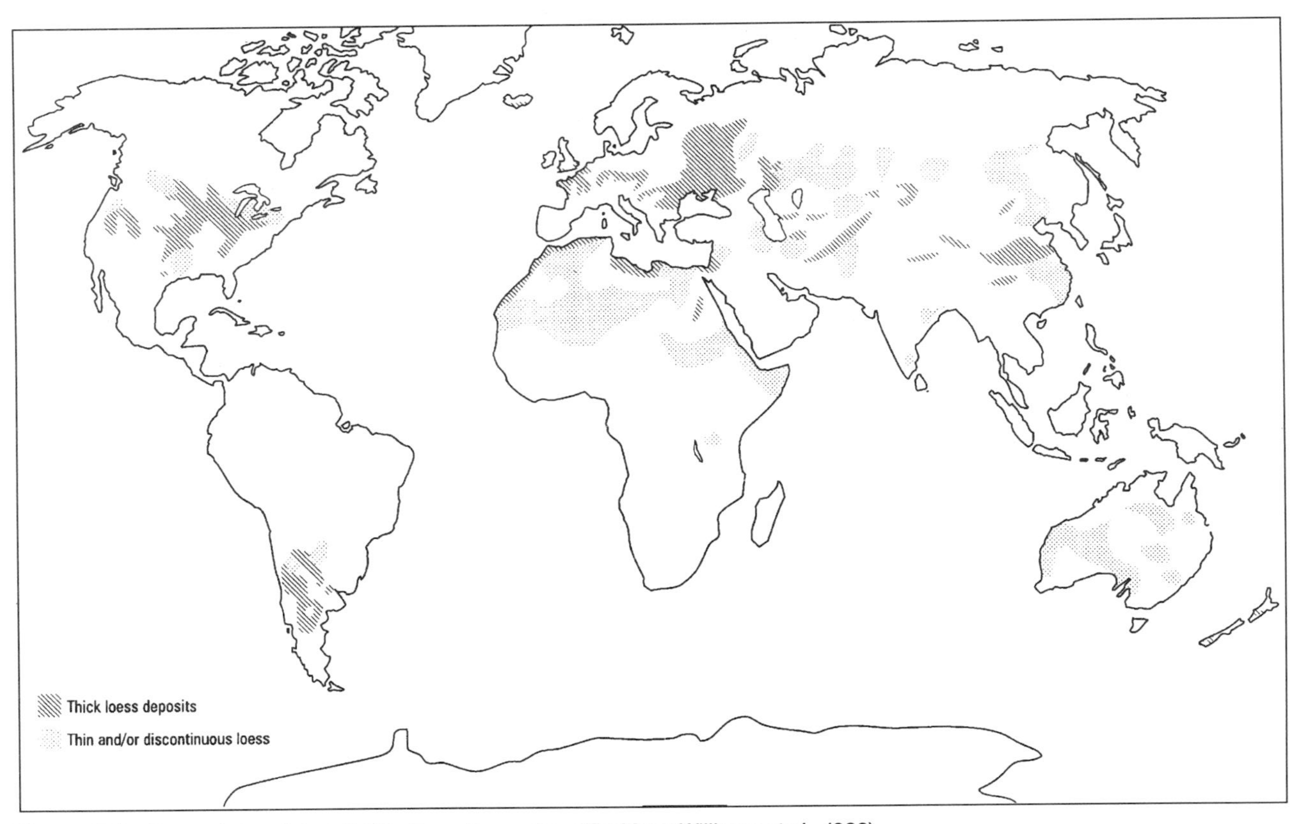

Figure 2.4 Present-day global distribution of loess (modified from Williams et al., 1993)

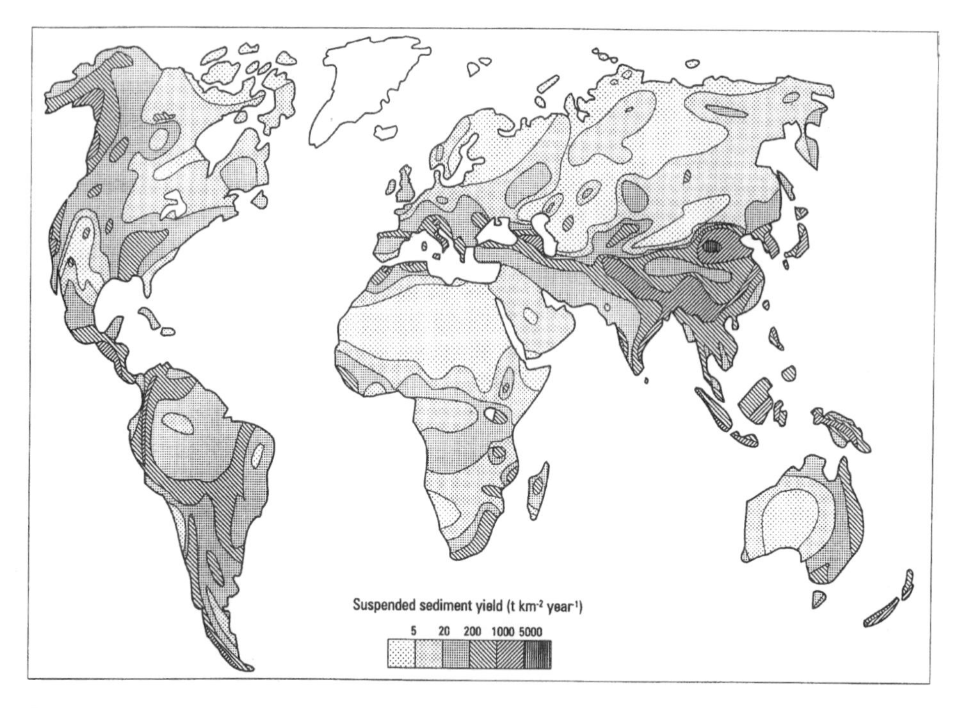

Figure 2.5 Global sediment yields (modified from Lvovich et al., 1991)

period 1934–1966 have shown that surface uplift is still continuing and is sufficiently high to warrant consideration in future river engineering projects.

2.2.4 Lessons for River Engineering and Management

Pleistocene environmental change and neotectonic activity have left a strong imprint on basin physiography, sediment sources and supply which, in turn, control present-day channel and floodplain characteristics and activity rates. These factors should be evaluated as a matter of routine because the implementation of appropriate, sustainable and cost-effective river engineering often hinges on correctly identifying 'natural' versus human causes of current river instability.

2.3 HUMAN DISTURBANCE OF RIVER BASINS

2.3.1 Introduction

Human environmental impact can be considered as inadvertent – when arising, for example, from basin land-use change or the input of mining waste into a river – or direct – when intervention is planned, as in the case of construction of a dam for river

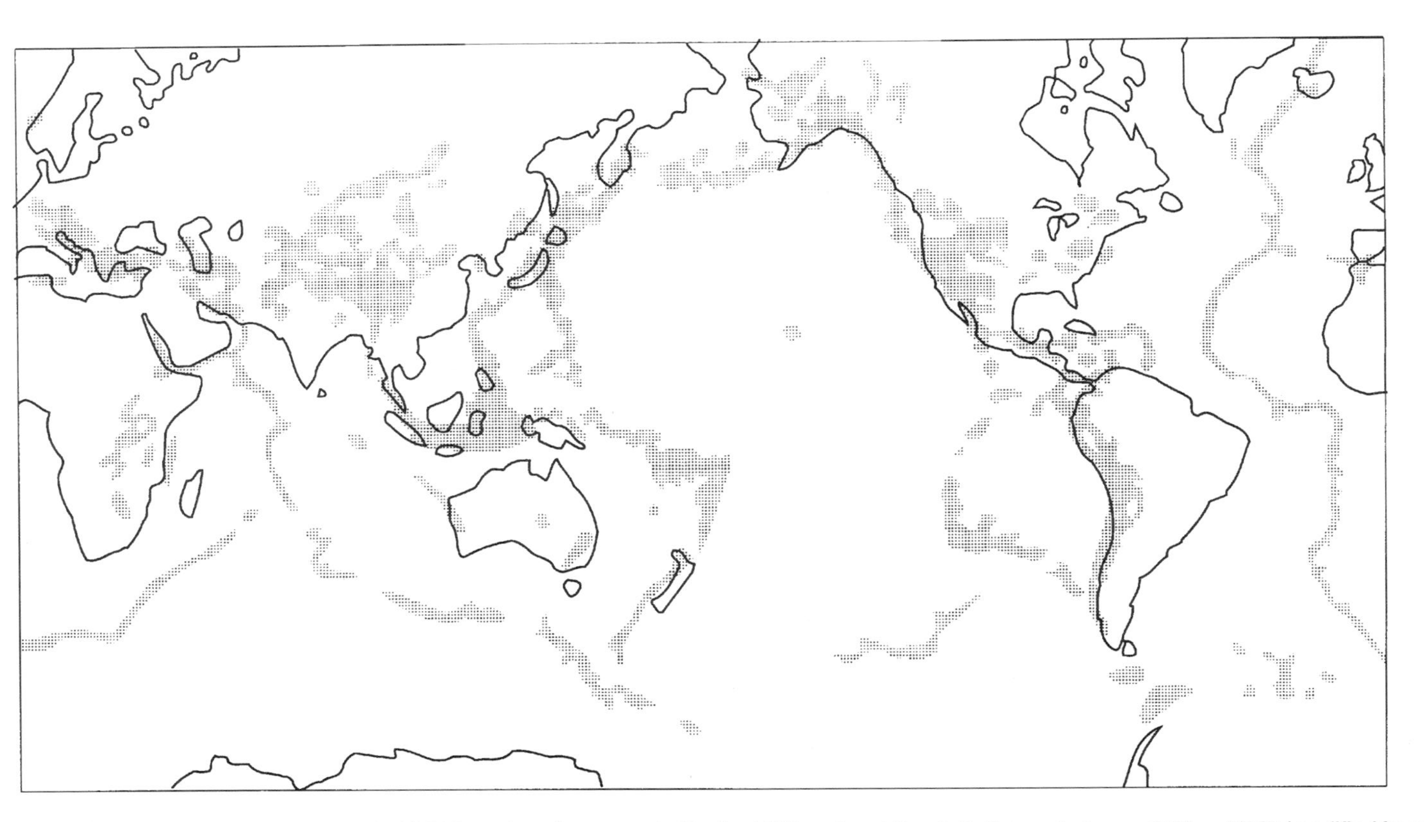

Figure 2.6 Location of approximately 30 000 earthquakes recorded by the US Coast and Geodetic Survey between 1961 and 1967 (modified from Summerfield, 1991)

regulation or embankments for flood protection. The effects of indirect or inadvertent changes on river systems are often delayed until well after the original activity and depend on the sensitivity of the system to change and the existence of geomorphic threshold conditions. The effects of direct changes are usually more rapid and do not depend so strongly on the crossing of thresholds. Environmental change induced by human action is not a new phenomenon, but for the purposes of river basin management it is important to evaluate human intervention within a historical context, and to learn some of the lessons that river histories can teach.

Design modifications of river reaches and basins are considered in Chapter 12 and discussion in this section will focus on inadvertent (indirect) changes arising from agriculture, mineral extraction, forestry and urbanisation, each of which can alter river water and sediment regimes. The effects of these activities on water quality lie outside the remit of this volume, although dispersal mechanisms and storage patterns of sediment-associated contaminants are considered under Metal Mining in Section 2.3.3.

2.3.2 Agriculture

Demonstrable evidence for human transformation of the environment began with the Neolithic agricultural 'revolution' in the Near East around 8000 BC, although fire has apparently been used by hominids as a tool for management of environmental resources and the procurement of game for much longer, probably for the last million years. There have, however, been marked differences in the nature, scale and timing of human environmental impacts over the last 10 000 years. Similarly, the sensitivity and response of river basins to human-induced changes have varied considerably, ranging from accelerated soil degradation and erosion (Dearing, 1991) to river erosion and siltation problems (Trimble and Lund, 1982), and to channel and floodplain metamorphosis (Brown and Keough, 1992).

Farming was practised as early as 6500 BC in Europe and Asia but in other areas its impact was less immediate. In North America, Australia and parts of South America and Africa, hunting and gathering economies survived until the intervention of the European colonial powers in the 16th century. In many areas the farming economy also underwent a process of intensification at that time, when more productive farming methods were accompanied by an increase in population and greater pressure on the environment. In the Near East, Egypt, Europe (excluding Scandinavia) and the Indus Valley this occurred between 3500 and 2500 BC and resulted in locally severe environmental degradation (Renfrew and Bahn, 1991). In Europe (notably around the Mediterranean basin), extensive deforestation in advance of agriculture created an open landscape as early as 3000 BC and this landscape proved highly susceptible to erosion (Bell, 1992; van Andel et al., 1990). Land degradation resulted in impoverishment of upland soil resources (by leaching and by accelerated erosion) and adverse hydrological effects such as channel siltation and flooding in the lowlands. In lower Mesopotamia, over-irrigation appears to have led to salinisation of alluvial soils as early as 2500 BC (Wagstaff, 1985). It is clear from these examples that agriculture-induced degradation was, in some areas, as much of a problem in antiquity as it is today.

Figure 2.7 (after Starkel, 1987) summarises the major impacts that agriculture has on the environment. It highlights the nature of environmental degradation in humid areas

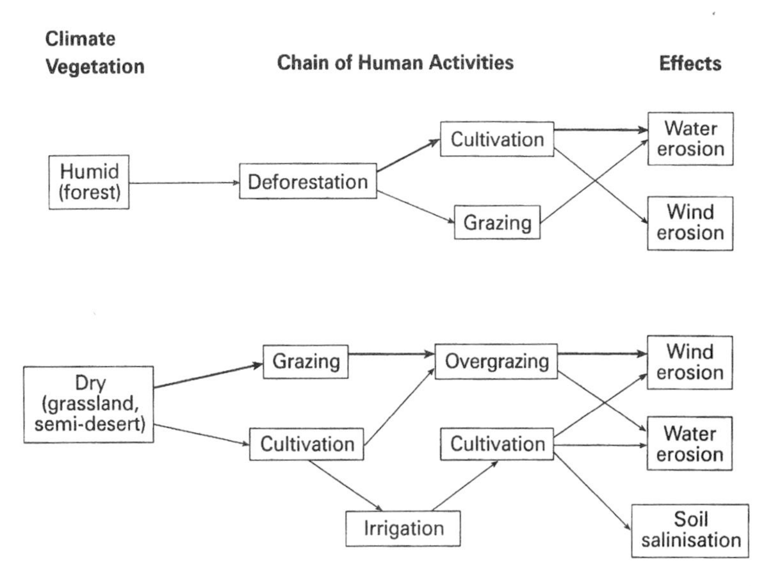

Figure 2.7 Impact of agriculture on erosion and runoff in humid and dry climates (modified from Starkel, 1987)

with a natural forest cover and a surplus of precipitation over evaporation, and semi-arid and arid grassland areas (steppe, savanna) with either a permanent or a seasonal water deficit. In forested catchments, tree clearance was usually necessary before cultivation or intensive grazing could be carried out. This led to a reduction in evapotranspiration and increased runoff that accelerated rates of slope wash, piping, gullying, mass failures and, in some cases, deflation (wind erosion). Flood frequency, duration and the magnitude of flood peaks and river sediment loads and rates of floodplain sedimentation all generally increase following the introduction of agriculture into a drainage basin. Sediment storage on slopes and at the floodplain edge can, however, offset the worse effects of poor farming practices in causing river siltation downstream from the area of disturbance. Indeed, long-term sediment budget studies have shown that usually only a small proportion of soil and sediment mobilised by erosion is actually transported through and out of a river basin (e.g. Trimble, 1983).

Of greater recent concern has been extensive deforestation (for fuel, wood products and in advance of cultivation) in mountain areas such as the Andes and Himalayas. This has resulted in serious water resource management problems in downstream foothill, plain and delta regions. Deforestation can lead both to accelerated flooding from increased surface runoff, and to dry-season drought due to depletion of catchment groundwater storage. Problems are compounded because mountain regions naturally experience high erosion; management strategies have sometimes been ill-advised, and the link between mountain deforestation and lowland flooding is not straightforward (Agarwal and Clark, 1991). In steepland river basins, for example in Oregon (Lyons and Beschta, 1983) and British Columbia (Roberts and Church, 1986), timber production has increased soil erosion rates, frequency of landslides and runoff on hillslopes leading to higher peak discharges, stream sediment loads and unstable rivers. The practice of instream and cross-stream harvesting appears to be particularly damaging, because it destabilises channel banks, adding a significant additional load of organic debris directly to the channel. Large organic debris is an important factor influencing channel morphology and fish habitat in small mountain streams (Lyons and Beschta, 1983). It deflects flows, alters bed roughness and can have a pronounced effect on routing bedload sediment through the channel system. Road construction for access also disturbs the catchment surface, making more sediment available to the stream. Afforestation and reforestation would appear to be logical land-use options for reducing erosion and runoff. However, as experience in the British uplands has shown, where drainage is required, site preparation and planting can themselves increase sediment loads and nutrient loss (e.g. Leeks, 1992; Robinson, 1986).

In drylands, soil tillage and grazing can result in the complete destruction of vegetation cover causing catastrophic wind and water erosion (e.g. creation of the dust bowl in the 1930s in the USA), which ultimately may result in desertification. These changes can also initiate river metamorphosis (see Box 2.1). In catchments with a seasonal or perennial water deficit, irrigation is essential for agricultural intensification, but with prolonged use there is a serious risk of soil salinisation.

'Post-settlement alluvium' in the New World, or 'haugh loams' or 'mada' in the Old World, both of generally a fine-grained character, are a common feature of agricultural landscapes resulting particularly from accelerated soil erosion. Alluvial stratigraphic and archaeological studies in India, China, the Near East, Mediterranean and northern Europe have shown that they were formed after the development of open agricultural landscapes in both prehistoric and early historic times. In North America, Australia and New Zealand, the introduction of alien European agriculture in the 19th century and its effect on catchment hydrology, channel morphology and valley floor sedimentation is particularly well recorded (e.g. Costa, 1975; Knox, 1972, 1977; Trimble, 1983). Knox's studies (1972, 1977) in Wisconsin provide a clear exempler of agricultural impacts in a river drainage basin as channel and floodplain characteristics were documented before and after land-use change. This enabled the rates, nature and directions of river channel and basin response to altered boundary conditions to be traced over a 100 year period. It constitutes a very useful empirical model by which to gauge and forecast the impact of current, or proposed, agricultural development in similar mid-latitude environments. Since World War Two, the replacement of more traditional crop rotation by high technology monocultures, in both the developed and developing world, has seriously depleted organic residues that are important for promoting soil cohesion. Soil erosion, leaching and gullying arising from poor agricultural practices are now recognised as significant problems in Britain and Europe as well as in Australia and North America.

Figure 2.8 (after Starkel, 1987) summarises the global environmental impact of agriculture and highlights significant regional variations which may constrain, or even favour, particular river engineering and basin management options. Eight regions are identified by Starkel and are differentiated on the basis of the length of time agriculture has been part of the human economy, the type of agricultural practice and hydroclimate. This information, although generalised, provides a starting point for developing a historical context and world perspective to current agriculturally accelerated hillslope erosion and associated river sedimentation problems.

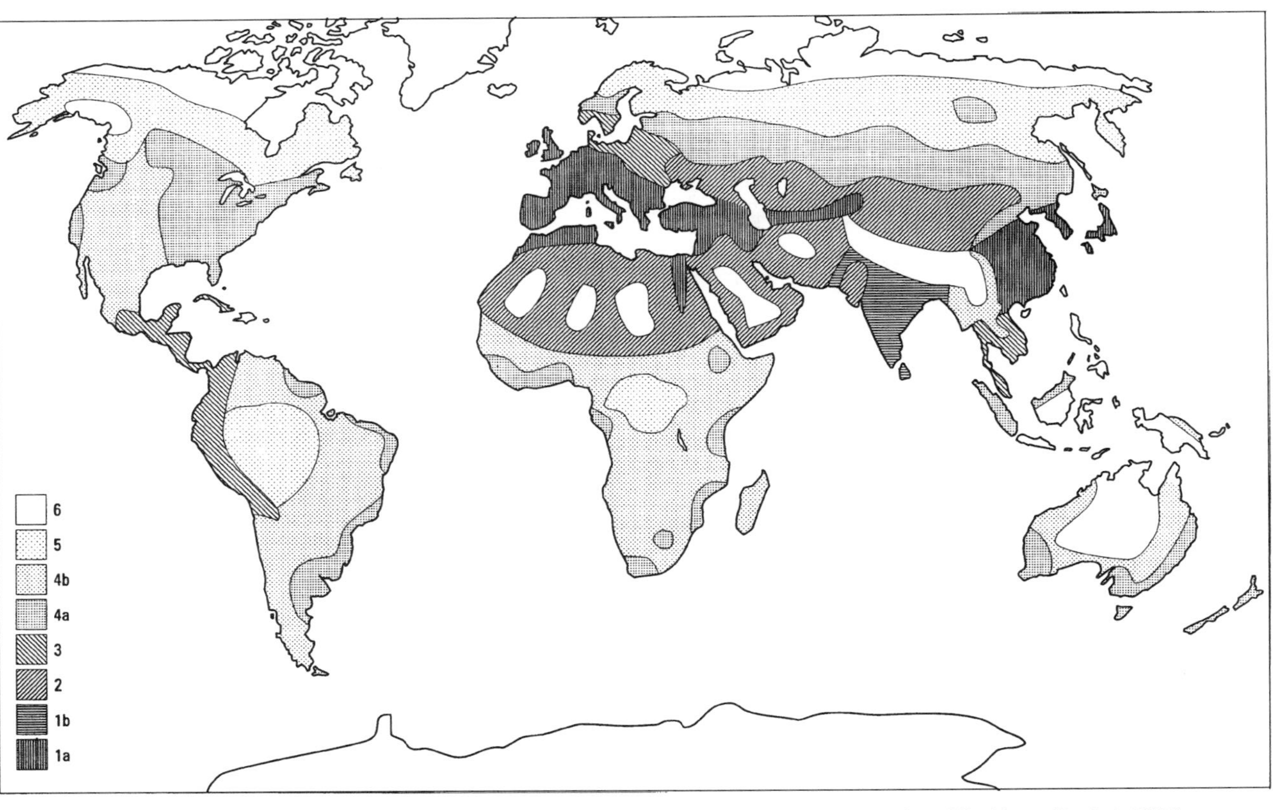

Figure 2.8 Globel environmental impact of prehistoric, historic and present-day agriculture (modified from Starkel, 1987)

Box 2.1 Arroyos

One of the most commonly cited examples of river metamorphosis is the arroyo: 'a trench with a roughly rectangular cross-section excavated in valley-bottom alluvium with a major stream channel on the floor of the trench' (Graf, 1983). Formation of arroyos took place in many of the arid and semi-arid parts of southwestern USA between about 1880 and 1950. Most observers at the time blamed poor land-use practices, especially over-grazing, as the cause of increased runoff and erosion. However, the exposure of palaeoarroyos in the walls of many entrenched streams in the region indicated that previous episodes of erosion had occurred on several occasions during the Holocene, under conditions that did not include livestock grazing and large-scale manipulation of channels and floodplains for agriculture. In attempting to reconcile conflicting evidence, it has been suggested that a variety of causal factors (e.g. over-grazing, agricultural practices) may act to trigger entrenchment in basins that have been rendered unstable by regional mechanisms such as climate change. A similar scenario has been proposed in both northern (Bork, 1988; Harvey et al., 1981; Macklin et al., 1992a) and southern (Macklin et al., 1994) Europe for the inception of gullying and the development of entrenched channels in historic times. Tree ring evidence from Kaneb Creek, southern Utah (a 25 m deep and 80–120 m wide arroyo that developed between 1880–1914), for 1462–1985 (Webb et al., in press) indicates a period of increased flood magnitude between 1866 and 1936 that is unprecedented in the 500 year record. The lack of floods between 1521 and 1866 and after 1916 supports the view that flooding is non-stationary in time. The occurrence of increased precipitation intensity in southern Utah, coupled with increased flood frequency, suggests that short-term fluctuations in climate can be a principal cause of arroyo incision. A similar conclusion has been reached by Balling and Wells (1990) who investigated arroyo cutting and filling in the Zuni river drainage basin, New Mexico, USA.

2.3.3 Mineral Extraction

Mineral extraction for fossil fuels (e.g. coal and peat), non-ferrous metal ores (e.g. lead, zinc, copper) rare (e.g. thorium) and 'noble' (e.g. gold and silver) metals, and aggregates (e.g. alluvial sand and gravel) has a profound effect on river systems in many parts of the world. Mining activities invariably result in the disruption of the hydrological system (through catchment vegetation removal and drainage modification), accelerated slope erosion, increased stream sediment loads, and long-term, large-scale river instability. The extent to which mining adversely affects river stability and the riparian environment, however, depends on two factors: first, whether extraction takes place on the valley sides, the valley floor or within the channel (e.g. placer deposits, alluvial gravel extraction); and second, the degree to which mining waste can be contained and prevented from entering surface or ground waters.

Metal Mining

In Europe and North America, before the introduction of pollution control legislation in the 19th and early part of the 20th century, mining waste was usually discharged in the nearest watercourse, frequently resulting in the transformation of whole fluvial systems (Lewin and Macklin, 1987). Although available mining technology has changed for the better as far as environmental protection is concerned in the developing world, where environmental considerations are not at a premium, mining activity is the cause of some of the most serious environmental and river instability problems. This is perhaps most graphically illustrated by the recent gold rush in Amazonia, notably in the Maderia River (Martinelli et al., 1988), which has resulted in extensive mercury pollution. Major extraction schemes may also involve river systems for waste disposal, as at the Ok Tedi mine in Papua New Guinea (Higgins et al., 1987).

Dramatic, sediment supply-induced channel and floodplain metamorphoses are well documented in many former mining areas including the Sierra Nevada, USA (Gilbert, 1917; James, 1991), Tasmania, Australia (Knighton, 1991), mid-Wales, UK (Lewin et al., 1983) and the northern Pennines, England (Macklin and Lewin, 1989). These metamorphoses generally involved a change from meandering to braided channel patterns, with valley floor aggradation followed by incision and reversion to a single channel in less than a century.

Because mining activity was intermittent and competent discharges for the large-size material occurred episodically, bed material load appears to have moved downstream as an attenuated wave, or a series of sediment 'slugs'. This resulted in the formation of complex-response terraces (Schumm and Parker, 1973) that represent internal adjustments of river systems to changes in bed material load transport rates. The recognition of single or even multiple bed-sediment waves within a drainage basin and information pertaining to their form, wavelength and celerity is of considerable importance to river engineers, particularly those involved in river restoration after mining has ceased. For example, provision could be made to protect a downstream reach in advance of increased bed-sediment supply, while the need for costly intervention may be reduced or even eliminated where instability is judged to be a transient problem and a reach is likely to self-stabilise after the passage of the bed-sediment wave. Bed waves are not restricted to rivers with a history of metal mining, but appear to be large-scale equilibrium bedforms, over engineering timescales, in many gravel- and sand-bed rivers characterised by spatial and temporal variability of sediment transport (Hoey, 1992; Kesel et al., 1992). In coarse, gravel-bed rivers this may reflect a high threshold of sediment motion but more generally appears to be governed by fluctuations in sediment supply caused by major flood events, land-use or hydroclimate change (Passmore et al., 1993; Roberts and Church, 1986). In braided rivers, however, cycles of channel incision followed by a period of bar growth prior to renewed incision produce bedload pulses as a matter of course without any exogenic acceleration of sediment supply (Ashmore, 1991; Hoey and Sutherland, 1991).

One of the most significant environmental and water resource hazards in catchments affected by past and present mining is the dispersal of toxic waste arising from the extraction and processing of metal ores (see Macklin (1996) for a general review of soil and sediment metal pollution). Mine tailings consisting of sand and slurries are particularly susceptible to water erosion and transport and they can be moved tens to hundreds of kilometres downstream of their source, contaminating aquatic biota and floodplain soils. Mining dispersal patterns in alluvial sediments have been widely studied in the UK (Bradley and Cox, 1986; Lewin and Macklin, 1987; Macklin and Dowsett, 1989), in Belgium and The Netherlands (Leenaers, 1989), in Poland (Macklin and Klimek, 1992), in Australia (East et al., 1988) and in North America (Andrews et al., 1987; Axtmann and Luoma, 1991; Graf, 1990; Knox, 1987; Marron, 1989; Miller and Wells, 1986). Especially dangerous are radionuclides (e.g. uranium, thorium) and heavy metals (e.g. cadmium, lead, mercury) which have extended environmental residence times (in some instances approaching thousands of years) and are highly toxic to plants, animals and humans. Indeed, the delayed recovery of a number of unstable reaches in the South Tyne River catchment, northern England, following the cessation of mining, have been shown to be due to high metal levels (notably cadmium, lead and zinc) retarding vegetation development on channel banks and emergent bars (Macklin and Lewin, 1989; Macklin and Smith, 1990).

Alluvial Sand and Gravel Mining

Commercial mining of sand and gravel from rivers often causes considerable localised channel instability and bank erosion, and can also damage (indirectly) riparian vegetation and water resources (Wyzga, 1991). Gravel extraction destroys the bed armour and significantly increases the hydrodynamic roughness of the channel, enhancing bed scouring and transport rates of bed material. Where the rate of bed material removed from a reach exceeds the bed material load supplied from upstream, channel downcutting invariably ensues. Channel incision is self-enhancing and perpetuated by progressive confinement of flood flows and associated increases of in-channel flow velocity and bed-shear stress. This will continue until gravel mining ceases, water and sediment discharges change or the river degrades down to a resistant substrate. Very frequently, the rooting zone of riparian vegetation (an important control of scouring and bank stability) becomes elevated above the low water stage, making banks more susceptible to erosion.

Gravel mining may result in the destruction of fish-spawning gravels in the river, but as a consequence of lowered groundwater levels on the valley floor its effects can also extend well beyond the channel itself. Depressed floodplain and alluvial valley floor groundwater levels can result in the loss of wetlands and riparian forests and can also reduce the root-crop yields where soils are sandy and free-draining.

2.3.4 Urbanisation

While mining activities directly affect rivers by supplying enormous quantities of sediment to the drainage network, the opposite situation occurs following urbanisation, when impervious surface cover decreases sediment availability and increases the amount and rate of runoff (e.g. Leopold, 1990). Such changes may trigger channel erosion or stream incision (Booth, 1990), and cause significant increases in channel capacity (Knight, 1979; Mosley, 1975). In concrete-lined, urban stream channels, large floods may generate values of stream power per unit area in excess of $500\,\mathrm{W m^{-2}}$, leading to removal of concrete slabs from the channel floor and destruction of retaining side walls (Vaughan, 1990). Marked channel changes are typical only of relatively small streams (drainage area $<50\,\mathrm{km^2}$) affected by urban development (Ferguson, 1981). Large rivers, with much smaller proportions of their catchment affected, would be less likely to show such adjustment to urbanisation even if they were not generally canalised where they flow through towns and cities.

2.3.5 Concluding Comments

In both the developed Old and New World, river sedimentation and erosion problems arising from agriculture and deforestation generally preceded the river water and sediment quality deterioration that took place following the Industrial Revolution. Unfortunately, in most developing countries rapid urbanisation, industrialisation and agricultural intensification this century have run together and conspired to produce simultaneous perturbations in river water, sediment and chemical fluxes. Even as we move towards global transformation of the environment by human action, feedback and autovariation within atmospheric, hydrologic, biological and geomorphic systems will ensure the unequal distribution of environmental change over space and through time.

For river engineers and catchment planners existing disequilibrium in fluvial systems presents an obvious problem for predicting and managing continued change in a river and watershed. However, for the geomorphologist it provides an opportunity to make good use of the wealth of empirical research which has documented short- and long-term responses of fluvial systems to anthropogenic changes in the recent past.

2.4 RIVER AND DRAINAGE BASIN RESPONSE TO RECENT CLIMATE CHANGE

2.4.1 Introduction

The possible link between climate change and river instability is one of the most difficult, and certainly the most contentious, issues facing the river engineer at the present time. The debate, fuelled by concern over the impact of anthropogenic enhancement of the 'greenhouse effect' on global climate, has become one of the central issues in water resources management and catchment planning. Although climate change during the Holocene has not been of the same magnitude as that experienced during the Last Glacial–Interglacial Transition (*c.* 18 000–9000 BP), there have been significant fluctuations in precipitation and temperatures that have affected runoff generation and river regime (Knox, 1983, 1988; Starkel, 1991). Because climate variation over the Holocene has been of a scale and duration comparable to social, political and ecological adjustment processes (e.g. industrialisation, urbanisation and demographic change (McDowell et al., 1990)), river instability problems caused by short-term (one month to about 100 years) climatic fluctuations are those which are most likely to be wrongly attributed to human intervention. In addition, while in a drainage basin unmodified by human agencies a climate signal from the recent alluvial record or present river behaviour may be relatively straightforward to decipher, in catchments with a long history of human disturbance it is often very difficult to distinguish between natural and anthropogenic causes of river instability and evolution.

2.4.2 Non-stationarity of Flooding: Controls and Global Patterns

The global warming observed over the past several decades is the culmination of an irregular but progressive increase in global temperature of 0.45°C since the late 19th century (Houghton et al., 1992). Records in central England dating back to the 18th century show that this is part of a longer trend in temperature recovery since the end of the 'Little Ice Age'. The Little Ice Age was the most recent of a series of 'Neoglacials' during the Holocene and was characterised by advance of alpine glaciers, cooler temperatures and expanded areas of sea ice around North America, Europe and in the North Atlantic between 1300 and 1850 AD (Grove, 1988; Wiles and Calkin, 1994). Climatic fluctuations at similar temporal and spatial scales occurred in subtropical and arid regions but are less well documented.

Changes in the frequency of severe floods and droughts during the Holocene would appear to be the clearest manifestations of climate control of river systems (Baker, 1991; Knox, 1983; Starkel, 1983). A consistent pattern in this respect is beginning to emerge from recent studies of historic flooding in Australia (Erskine and Warner, 1988), Britain (Higgs, 1987; Macklin et al., 1992a,b; Walsh et al., 1982), continental Europe (Bravard, 1989), South America (Wells, 1990) and North America (Knox, 1984). They show alternating periods of high and low flood frequency, each of the order of 10–25 years' durations, over the last few hundred years. Analysis of mean annual river discharge records from 50 major rivers around the world has also revealed 'humid' and 'dry' periods this century (Probst and Tardy, 1987) that correlate very closely with established trends of historic climate change.

The clustering of major floods and droughts on this timescale reflects changes in regional climatic conditions controlled by atmospheric circulation, which determines the position of storm tracks and the movement of different air masses, and by the direct effect of insolation changes on heating and evaporation. The non-stationarity this introduces into the mean and variance of flood series makes hydrological forecasting (which commonly assumes that the probability of a flood of a given magnitude will remain the same from year to year), and in turn river engineering, very problematic. Unwittingly, over- or under-designing a channel can have very serious consequences for flood protection and water availability. Although historical and palaeohydrological techniques offer a way of lengthening a short-term data record and reducing the uncertainty in hydrologic analyses (Jarrett, 1991), flood protection also requires an understanding of the meteorological process that generate large floods in a particular region and how these vary over time.

Hayden's (1988) global classification of flooding (Figure 2.9) based on the meteorological causation and atmospheric dynamics, and Hirschboeck's (1991) detailed summary of flood-causing precipitation and runoff for the USA are particularly valuable in this context, for two reasons. First, they delineate regions in which river catchments are internally homogeneous in terms of flood-generating events, moisture availability and other aspects of water resources. Second, and perhaps more importantly for river engineering and management, they identify boundaries between flood climate regions where there may be a greater likelihood of climate-induced river instability as a result of storms entering a region where vegetation, soil and drainage network are equilibrated to different rainfall rates and totals.

The incidence of flood or drought events in high and middle latitude (polewards of latitude 40°N and S) is governed by the latitudinal position, strength, wavelength and amplitude of the circumpolar vortex (or jet stream) that controls the development of, and tracks followed by, surface weather systems. An increased frequency of large floods (particularly those generated by melting snow accompanied by heavy rainfall) in northern Britain during the late 18th and 19th centuries, and in the 1950s and 1960s,

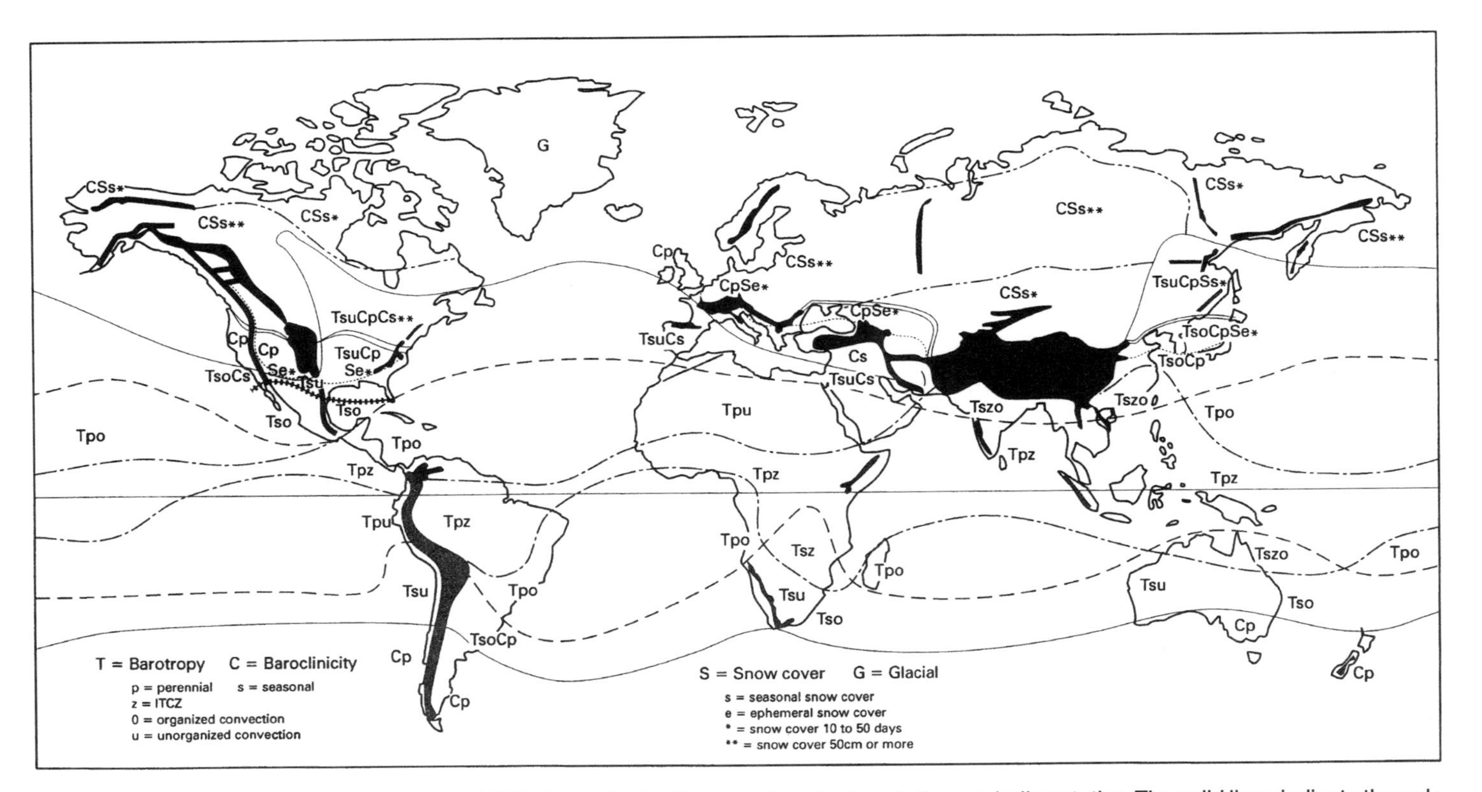

Figure 2.9 Flood climate regions of the world. The legends give the meaning of letters in the symbolic notation. The solid lines indicate the poleward limits of barotropic conditions in summer and dashed lines the same limit in winter. The dash-dot lines are the January and July positions of the ITCZ. The dotted line marks the equatorward limit of winter snow-cover durations of 10 days or more. The dash-dot-dot line indicates regions with more than 50 days of recorded snow cover and more than 50 cm of snow. The cross-hatched solid line marks the equatorward limit of frontal cyclones in the North America sector. Solid areas indicate major mountainous regions (modified from Hayden, 1988)

corresponded with a tendency for westerly winds to decline in vigour and for the circumpolar vortex to become more meridional (Macklin et al., 1992a; Passmore et al., 1993). In the upper Mississippi Valley, relatively high frequencies of large floods in the late 1800s before 1895 and since about 1950 were also associated with a weak westerly circulation in the middle latitudes (Knox, 1984, 1988).

At low latitudes, clustering of flood and/or drought events is dictated by pulses of great atmospheric–ocean systems including the monsoons (Brammer, 1990a), the El Niño–Southern Oscillation (ENSO) (Ely et al., 1993; Wells, 1990) and tropical storms (Spencer and Douglas, 1985). ENSO is the most prominent known source of interannual variability of weather and climate around the world, though not all areas are affected (Houghton et al., 1990). The Southern Oscillation component of ENSO is an atmospheric pattern that extends over most of the tropics. It principally involves a see-saw in atmospheric mass between regions near Indonesia and a tropical and subtropical southern Pacific region centred on Easter Island. It is associated with high sea-surface temperature anomalies that spread eastward across the equatorial Pacific Ocean. The influence of ENSO disruptions of large-scale tropical atmospheric circulation patterns sometimes extends to high latitudes. ENSO events occur every 3–10 years and their major hydrological manifestations are anomalous periods of floods (Ely et al., 1993) and droughts lasting many months. Places especially affected include the tropical and central eastern Pacific Islands, the coast of north Peru, the southwestern United States, eastern Australia, New Zealand, Indonesia, India and part of eastern and southern Africa (Figure 2.10).

ENSO also modulates the frequency of tropical storms in some regions such as Japan, eastern China, and in the south and central Pacific. Cyclones and hurricanes only develop where sea-surface temperatures are greater than 26.5°C and it is well established that cyclones are less frequent at times of lower sea-surface temperature (Spencer and Douglas, 1985). Wendland (1977) has estimated that the annual mean frequency of cyclones in the late 18th and early 19th centuries was about five per year compared with nine per year between 1962 and 1971. Walsh (1977) has shown that cyclone frequencies in the Lesser Antilles, Caribbean, were even lower in the period 1650–1764 which coincided with the Little Ice Age in Europe. Cyclones and hurricanes not only generate coastal and river flooding in the tropics, but can also result in catastrophic flooding in extratropical regions when they interact with mid-latitude frontal systems. Such conditions are common in the Atlantic coastal plain and in southwest and southern parts of the USA (Hirschboeck, 1991) and also in southeast China, where they are responsible for some of the largest recorded peaks in the flood series.

A second planetary-scale, tropical, rain-forming disturbance relates to the annual north–south excursion of the Intertropical Convergence Zone (ITCZ) that produces a pronounced seasonality of rainfall over much of the tropics. The generic term 'monsoon' is often associated with the summer rainy season. In the northern hemisphere, in years when the convergence is weak or fails to penetrate very far north, droughts occur. The great drought and famine of 1981–1985 in Ethiopia marks an episode in the erratic, unreliable history of the ITCZ in this region (Hayden, 1988). Conversely, when the ITCZ extends further north (or south in the case of the southern hemisphere), heavy convectional rain may come to normally arid areas resulting in exceptional flooding (e.g. as occurred in Bangladesh in 1987 and 1988 (Brammer, 1990a,b)).

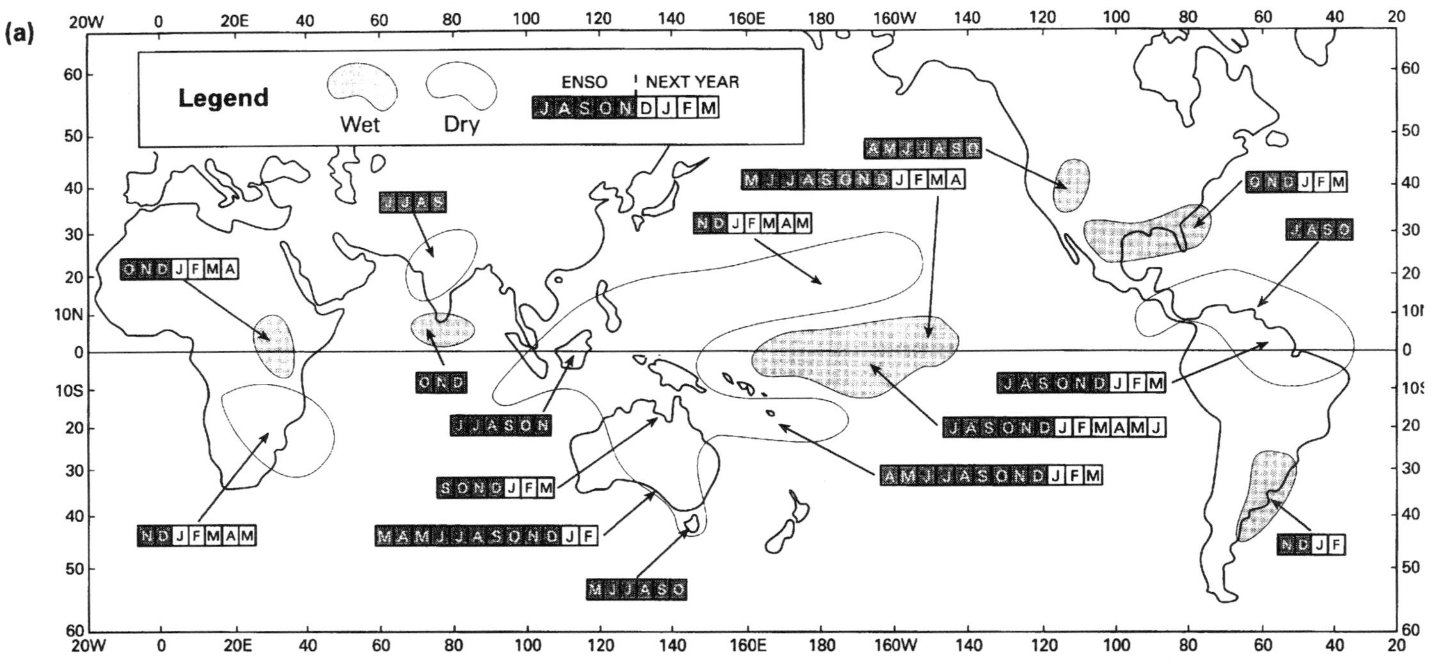

Figure 2.10(a) Schematic diagram of areas and times of the year with a consistent ENSO precipitation signal (modified from Folland et al., 1990)

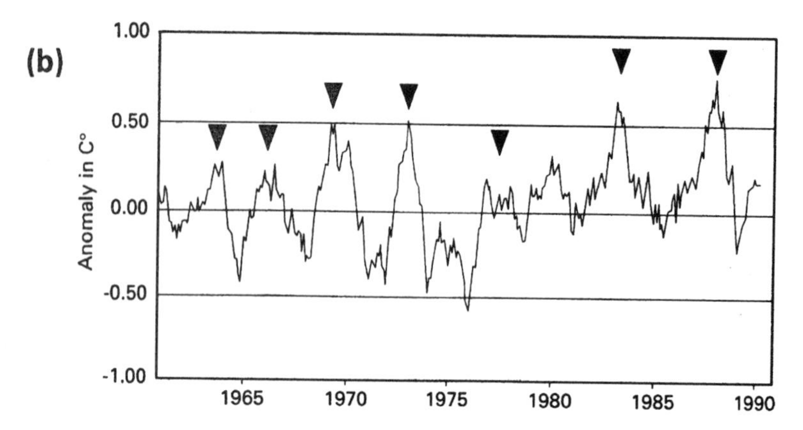

Figure 2.10(b) Monthly tropical sea-surface and land-surface anomalies 1961–1989. Tropics extend from 20° N to 20° S. Arrows mark maximum ENSO warmth in the tropics (modified from Folland et al., 1990)

2.4.3 Climate Change and the River Engineer

Echoing the view of Chorley et al. (1984), it is probably true to say that there is no matter of prime significance to the river engineer (and for that matter the geomorphologist) on which ignorance is so profound as that of climate change and how it affects river form and process. This brief review has shown that non-stationarity in flood frequency and magnitude resulting from short-term climatic fluctuation is probably the rule rather that the exception in most river systems of the world. It is, therefore, important for river engineers to view the climate base as a fluctuating one, and not to regard climate as a fixed backdrop. A global classification of flood climate based on atmospheric dynamics (see Figure 2.9 and Hayden, 1988) provides a very useful conceptual and practical framework for analysing the temporal and spatial variability of floods and droughts, and how these relate to regional or global climate change. Knowledge of the primary causes, geographical patterns and periodicity of fluctuations in hydroclimate is critical if sustainable river engineering is to be achieved.

2.5 GUIDANCE FOR RIVER ENGINEERS

The conclusion to be drawn from this chapter is that it is probably wise for river engineers to assume that a river system is in a state of instability rather than stability, at least until in can be demonstrated otherwise. Rivers may be progressively incising or aggrading in response to series of overlapping environmental changes. The effects of such changes may take time to work their way through catchment systems, so that it is possible to confuse local site-scale instability response to environmental change occurring immediately, with upstream feedback effects that date from some past environmental perturbations, which are only now reaching the site in question.

River engineers and managers are being encouraged to look both upstream and outside the channel for causes of river instability, and downstream to ensure that their intervention does not solve one problem locally, while triggering other problems elsewhere. We would, in addition, urge engineers to consider the antecedence of current 'perceived' river instability and the lessons that may be learnt from examining the effects of past environmental change on river processes. Central to the approach advocated here is the belief that channel instability problems require an appreciation of relict as well as active components of a river system, especially when cause and effect are lagged by at least several decades. There may, as a consequence of complex response of geomorphic systems to environmental change, not always be a readily identifiable 'culprit with a smoking gun', either natural or man-made. This, however, may only become apparent with the benefit of the hindsight provided by historical fluvial geomorphology.

This topic has been explored by Lewin et al. (1988) who concluded that the assumption of an equilibrium channel form over engineering time may need revaluation. They demonstrated, from a review of British river studies, that many channels are sensitive to relatively small changes in sediment supply and runoff. Such rivers adjust their size and shape more frequently, and more rapidly, than is generally appreciated. Indeed, Lewin et al. (1988) went so far as to consider regime theory and environmental change as irreconcilable concepts and to see the replacement of a regime approach by the outcome of greater dialogue and increasing co-operation between engineers and geomorphologists. This is a central and underlying theme not only of this chapter, but also of the whole volume.

The broad climatic fluctuations of the Late Quaternary provide the context in geologic time for river process–response activity; many rivers are now reworking sediments derived from formerly more extreme conditions of physical weathering, water runoff and/or sediment transport. On the whole, the Holocene has been a quieter period in terms of river activity, but it too has seen the effects of climatic fluctuation in sensitive environments, as well as the impact of human activities on fluvial systems in many environments.

Practical considerations, therefore, suggest that environmental investigations are an essential first step for reliable engineering design and management. Such investigations should establish:

- sources for sediment related to environmental changes that have been, or may be, transported through river systems;
- longer term incision or aggradational trends (which could relate, for example, to neotectonics or glacial response) within which contemporary channel dynamics are set;
- past, present and potential future human effects, including the results of accelerated erosion, modified streamflow regime, and mining activity;
- the possibility that past climatic fluctuations may have affected river systems over an engineering timescale and the potential for continued climate change to trigger abrupt process–response through the crossing of geomorphic thresholds.

Fortunately, such investigations are aided by records of past catchment behaviour that are preserved beneath floodplains in the form of alluvial deposits. In many environments these may suggest that foreseeable river behaviour is unlikely to be extreme (assuming that future events which are without precedent do not occur), but this condition does need to be positively established by examination. It would be most unwise to plan on the basis of experience in one environment for development in another. Thus, arroyo changes do not resemble those of fluvial systems in mid-latitude temperate environments, whilst, even on the same river, upstream sites may show river behaviour patterns that have had quite a different history from downstream sites. Local site environmental investigations in addition to catchment-wide studies and assessment are therefore essential.

The design of environmentally sustainable river intervention schemes, especially in the developing world, requires a global understanding of longer term water and sediment regimes, as well as their spatial and temporal patterns and controls. A world perspective provides a first-order guide to likely river instability problems that can be expected in drainage basins where systematic hydrometric records are unavailable and availability of empirical field evidence is limited. Also, and perhaps more importantly, a world perspective helps rebalance the somewhat myopic 'Euro-American' view of erosion and sedimentation problems (and their solution) held by river engineers and fluvial geomorphologists working in the developed world. It therefore encourages the implementation of river engineering, catchment planning and management practices that are environmentally sound in terms both of natural and cultural systems.

2.6 REFERENCES

Agarwal, A. and Clark, A. (Eds) (1991). *Floods, Floodplains and Environmental Myths*. Centre for Science and Environment, New Delhi, 167pp.

Andrews, E.D. 1987. Longitudinal dispersion of trace metals in the Clark Fork River, Montana. In: Averett, R.C. and McKnight, D.M. (Eds), *The Chemical Quality of Water and the Hydrologic Cycle*. Lewis Publishers, 1–13.

Ashmore, P.E. 1991. Channel morphology and bed load pulses in braided, gravel-bed streams. *Geografiska Annaler*, **73A**, 37–52.

Axtmann, E.V. and Luoma, S.N. 1991. Large-scale distribution of metal contamination in the fine-grained sediments of the Clark Fork River, Montana, USA. *Applied Geochemistry*, **6**, 75–88.

Baker, V.R. 1991. A bright future for old flows. In: Starkel, L., Gregory, K.J. and Thornes, J.B. (Eds), *Temperate Palaeohydrology: Fluvial Processes in the Temperate Zone During the Last 15,000 Years*. Wiley, Chichester, 497–520.

Balling, R.C. and Wells, S.G. 1990. Historical rainfall patterns and Arroyo activity within the Zuni River drainage basin, New Mexico. *Annals of the Association of American Geographers*, **80**(4), 603–617.

Bell, M. 1992. The prehistory of soil erosion. In: Bell, M. and Boardman, J. (Eds), *Past and Present Soil Erosion: Archaeological and Geographical Perspectives*. Oxbow Monograph No. 22, Oxbow Press, 21–35.

Bell, M. and Laine, 1985. Erosion of the Laurentide region of North America by glacial and glaciofluvial processes. *Quaternary Research*, **23**, 154–174.

Bell, M. and Walker, M.J.C. 1992. *Late Quaternary Environment Change: Physical and Human Perspectives*. Longman.

Bogen, J., Walling, D.E. and Day, T. 1992. *Erosion and Sediment Transport Monitoring Programmes – River Basins*. IASH Publication 210.

Booth, D.B. 1990. Stream-channel incision following drainage basin urbanisation. *Water Resources Bulletin*, **26**, 407–417.

Bork, H.R. 1988. Mittlelatterliche Relief – sedimentual Bodenentwicklung im Bereich der Wustung Drudewenshusen. *Archalogisches Korrespondenzblatt*, **18**, 89–95.

Bradley, S.B. and Cox, J.J. 1986. Heavy metals in the Hamps and Manifold Valleys, north Staffordshire, UK: distribution in floodplains soils. *Science of the Total Environment*, **50**, 103–128.

Brammer, H. 1990a. Floods in Bangladesh. I Geographical background to the 1987 and 1988 floods. *Geographical Journal*, **156**(1), 12–22.

Brammer, H. 1990b. Floods in Bangladesh. II Flood mitigation and environmental aspects. *Geographical Journal*, **156**(2), 158–165.

Bravard, J.P. 1989. La metamorphose des rivières des Alpes Francaises à la fin du moyenage et à l'époque moderne. *Bulletin de la Société Geographique du Liège*, **25**, 145–157.

Brown, A.G. and Keough, M. 1992. Holocene floodplain metamorphosis in the Midlands, United Kingdom. *Geomorphology*, **4**, 433–445.

Burnett, A.W. and Schumm, S.A. 1983. Active tectonics and river response in Louisiana and Mississippi. *Science*, **222**, 49–50.

Chorley, R.J., Schumm, S.A. and Sugden, D.E. 1984. *Geomorphology*. Methuen, London.

Church, M. and Ryder, J.M. 1972. Paraglacial sedimentation: a consideration of fluvial processes conditioned by glaciation. *Geological Society of America Bulletin*, **83**, 3059–3072.

Church, M., Kellerhals, R. and Day, T.J. 1989. Regional clastic sediment yield in British Columbia. *Canadian Journal of Earth Science*, **26**, 31–45.

Costa, J.E. 1975. Effects of agriculture on erosion and sedimentation in Piedmont province, Maryland. *Bulletin of the Geological Society of America*, **86**, 1281–1286.

Costa, J.E. and Schuster, R.L. 1988. The formation and failure of natural dams. *Geological Society of America Bulletin*, **100**, 1054–1068.

Dearing, J.A. 1991. Erosion and land use. In: Berglund, B.E. (Ed.), *The Cultural Landscape during 6000 Years in Southern Sweden*. Ecological Bulletins, **41**, 283–292.

East, T.J., Cull, R.F., Murray, A.S. and Duggan, K. 1988. Fluvial dispersion of radioactive mill tailings in the seasonally wet tropics, northern Australia. In: Warner, R.F. (Ed.), *Fluvial Geomorphology of Australia*. Academic Press, Sydney, 223–244.

Ely, L.L., Enzel, Y., Baker, V.R. and Cayan, D.R. 1993. A 5000 year record of extreme floods and climate change in the south western United States. *Science*, **262**, 410–414.

Erskine, W.D. and Warner, R.F. 1988. Geomorphic effects of alternating flood-and-drought-dominated regimes on NSW coastal rivers. In: Warner, R.F. (Ed.), *Fluvial Geomorphology of Australia*. Academic Press, Sydney, 303–322.

Fairbridge, R.W. 1983. Isostasy and eustasy. In: Smith, D.E. and Dawson, A.G. (Eds), *Shorelines and Isostasy*. Academic Press, London, 3–28.

Ferguson, R.I. 1981. Channel form and channel changes. In: Lewin, J. (Ed.), *British Rivers*. George Allen & Unwin, London, 90–125.

Folland, C.K., Karl, T.R. and Vinnikov, 1990. Observed climate variations and change. In: Houghton, J.T., Jenkins, G.J. and Ephraums, J. (Eds), *Climate Change: The IPCC Scientific Assessment*. Cambridge University Press, Cambridge, 200–228.

Gilbert, G.K. 1917. *Hydraulic Mining Debris in the Sierra Nevada*. United States Geological Survey Professional Paper, 105.

Graf, W.L. 1983. The Arroyo problem – palaeohydrology and palaeohydraulics in the short term. In: Gregory, K.J. (Ed.), *Background to Palaeohydrology*. Chichester, Wiley, 279–302.

Graf, W.L. 1990. Fluvial dynamics of thorium-230 in the church rock event, Puerco River, New Mexico. *Annals of the Association of American Geographers*, **80**(3), 327–342.

Griffiths, G.A. 1979. Recent sedimentation history of the Waimakariri river, New Zealand. *Journal of Hydrology* (New Zealand), **18**, 6–28.

Grove, J.M. 1972. The incidence of landslides, avalanches and floods in western Norway during the Little Ice Age. *Arctic and Alpine Research*, **4**, 131–138.

Grove, J.M. 1988. *The Little Ice Age*. Methuen, London.

Haigh, M.J. 1994. Deforestation in the Himalayas. In: Roberts, N. (Ed.), *The Changing Global Environment*. Blackwell, Oxford, 440–462.

Harvey, A.M., Oldfield, F., Baron, A.F. and Pearson, G.W. 1981. Dating of post-glacial landforms in the central Howgills. *Earth Surface Processes and Landforms*, **6**, 401–412.

Hayden, B.P. 1988. Flood climates. In: Baker, V.R., Kochel, R.C. and Patton, P.C. (Eds), *Flood Geomorphology*, John Wiley, New York, 13–26.

Higgins, R.J., Pickup, G. and Cloke, P.S. 1987. Estimating the transport and deposition of mine waste at OK Tedi. In: Thorne, C.R., Bathurst, J.C. and Hey, R.D. (Eds), *Sediment Transport in Gravel-Bed Rivers*. Wiley, Chichester, 949–976.

Higgs, G. 1987. Environmental change and hydrological response: flooding in the upper Severn catchment. In: Gregory, K.J., Lewin, J. and Thornes, J.B. (Eds), *Palaeohydrology in Practice*. Wiley, Chichester, 131–159.

Hirschboeck, K.K. 1991. *Climate and Floods*. US Geological Survey Water-Supply Paper 2375, 67–88.

Hoey, T.B. 1992. Temporal variations in bedload transport rates and sediment storage in gravel-bed rivers. *Progress in Physical Geography*, **16**(3), 319–338.

Hoey, T.B. and Sutherland, A.J., 1991. Channel morphology and bedload pulses in braided rivers: a laboratory study. *Earth Surface Processes and Landforms*, **16**, 447–462

Houghton, J.T., Jenkins, G.J. and Ephraums, J.J. 1990. *Climate Change: The IPCC Scientific Assessment*. Cambridge University Press, Cambridge.

Houghton, J.T., Callander, B.A. and Varney, S.K. (Eds), 1992. *Climate Change 1992: The Supplementary Report to the IPCC Scientific Assessment*. Cambridge University Press, Cambridge.

James, L.A. 1991. Incision and morphologic evolution of an alluvial channel recovering from hydraulic mining sediment. *Geological Society of America Bulletin*, **103**, 723–736.

Jarrett, R.D. 1991. *Palaeohydrology and its Value in Analysing Floods and Droughts*. US Geological Survey Water-Supply Paper 2375, 105–116.

Jordan, P. and Slaymaker, O. 1991. Holocene sediment production in Lillooet river basin, British Columbia: a sediment budget approach. *Geographie physique et Quaternaire*, **45**(1), 45–57.

Kesel, R.H., Yodis, E.G. and McCraw, D.J. 1992. An approximation of the sediment budget of the lower Mississippi River prior to major human modification. *Earth Surface Processes and Landforms*, **17**, 711–722.

Knight, C. 1979. Urbanisation and natural stream channel morphology: the case of two English new towns. In: Hollis, G.E. (Ed.), *Man's Impact on the Hydrological Cycle in the United Kingdom*. Geobooks, Norwich, 181–198.

Knighton, A.D. 1991. Channel adjustments along mine-affected rivers of north-east Tasmania. *Geomorphology*, **4**, 205–219.

Knox, J.C. 1972. Valley alluviation in south-western Wisconsin. *Annals of the Association of American Geographers*, **62**(3), 401–410.

Knox, J.C. 1977. Human impact on Winsconsin stream channels. *Annals of the Association of American Geographers*, **67**(3), 323–342.

Knox, J.C. 1983. Responses of river systems to Holocene climates. In: Wright, H.E. (Ed.), *Late Quaternary Environments of the United States*, Vol. 2, *The Holocene*. University of Minnesota Press, Minneapolis, 26–41.

Knox, J.C. 1984. Fluvial responses to small scale climatic changes. In: Costa, J.E. and Fleisher, P.J. (Eds), *Developments and Application of Geomorphology*. Springer-Verlag, New York, 318–342.

Knox, J.C. 1987. Historical valley floor sedimentation in the Upper Mississippi Valley. *Annals of the Association of American Geographers*, **77**, 224–244.

Knox, J.C. 1988. Climatic influence on upper Mississippi valley floods. In: Baker, V.R., Kochel, R.C. and Patton, P.C. (Eds), *Flood Geomorphology*. John Wiley, New York, 279–300.

Leeks, G.J.L. 1992. Impact of plantation forestry on sediment transport processes. In: Billi, P., Hey, R.D., Thorne, C.R. and Tacconi (Eds), *Dynamics of Gravel-bed Rivers*. Wiley, Chichester, 651–670.

Leenaers, H. 1989. *The Disposal of Metal Mining Wastes in the Catchment of the River Guel (Belgium–The Netherlands)*. Geografische Instituut, Rijksumi-versitat utrecht, Amsterdam.

Leopold, L.B. 1990. Lag times for small drainage basins. *Catena*, **18**, 157–171.

Lewin, J. and Macklin, M.G. 1987. Metal mining and floodplain sedimentation in Britain. In: Gardiner, V. (Ed.), *Proceedings First International Conference on Geomorphology*. Wiley, Chichester, 1009–1027.

Lewin, J., Bradley, S.B. and Macklin, M.G., 1983. Historical valley alluviation in mid-Wales. *Geological Journal*, **18**, 331-350.

Lewin, J., Macklin, M.G. and Newson, M.D. 1988. Regime theory and environmental change – irreconcilable concepts? In: White, W.R. (Ed.), *International Conference on River Regime*. Wiley, Chichester, 431–445.

Lvovich, M.I., Karasik, G.Y., Bratseva, N.L., Medvedeva, G.P. and Maleshko, A.V., 1991. *Contemporary intensity of the World land intracontinental erosion*. USSR Academy of Sciences, Moscow.

Lyons, J.K. and Beschta, R.L. 1983. Land use, floods and channel changes: Upper Middle Fork Willamette River, Oregon (1936–1980). *Water Resources Research*, **19**(2), 463–471.

Macklin, M.G., 1996. Fluxes and storage of sediment-associated heavy metals in floodplain systems: assessment and river basin management issues at a time of rapid environmental change. In: Anderson, M.G., Walling, D.E. and Bates, P.D. (Eds), *Floodplain Processes*, Wiley, Chichester, 441–460.

Macklin, M.G. and Dowsett, R.B. 1989. The chemical and physical speciation of trace metals in fine grained overbank flood sediments in the Tyne basin north-east England. *Catena*, **16**, 135–151.

Macklin, M.G. and Klimek, K. 1992. Dispersal, storage and transformation of metal contaminated alluvium in the upper Vistula basin, south-west Poland. *Applied Geography*, **12**, 7–30.

Macklin, M.G. and Lewin, J. 1989. Sediment transfer and transformation of an alluvial valley floor: the River South Tyne, Northumbria, UK. *Earth Surface Processes and Landforms*, **14**, 233–246.

Macklin, M.G. and Smith, R.S. 1990. Historic riparian vegetation development and alluvial metallophyte plant communities in the Tyne basin, north-east England. In: Thornes, J.B. (Ed.), *Vegetation and Erosion*. Wiley, Chichester, 239–256.

Macklin, M.G., Rumsby, B.T. and Heap, T., 1992a. Flood alluviation and entrenchment: Holocene valley floor development and transformation in the British Uplands. *Geological Society of America Bulletin*, **104**, 631–643.

Macklin, M.G., Rumsby, B.T. and Newson, M.D., 1992b. Historical floods and vertical accretion of fine-grained alluvium in the lower Tyne Valley, north-east England. In: Billi, P., Hey, R.D., Thorne, C.R. and Tacconi, P. (Eds), *Dynamics of Gravel-bed Rivers*. Wiley, Chichester, 573–589.

Macklin, M.G., Passmore, D.G., Stevenson, A.C., Davis, B.A. and Benavente, J.A. 1994. Responses of rivers and lakes to Holocene environmental change in the Alcaniz region, Tervel, north-east Spain. In: Millington, A.C. and Pye, K. (Eds), *Effects of Environmental Change in Drylands*. Wiley, Chichester, 113–130.

Maizels, J. and Russell, A. 1992. Quaternary perspectives on Jokulhaup prediction. In: Gray, J.M. (Ed.), *Application of Quaternary Research*. Quaternary Proceedings No. 2, Quaternary Research Association, Cambridge, 133–152.

Mansikkaniemi, H. 1991. Regional case studies in southern Finland with reference to glacial rebound and Baltic regression. In: Starkel, L. Gregory, K.J. and Thornes, J.B. (Eds), *Temperate Palaeohydrology: Fluvial Processes in the Temperate Zone during the last 15,000 years*. Wiley, Chichester, 79–104.

Marron, D.C. 1989. Physical and chemical characteristics of a metal-contaminated overbank deposit, west-central south Dakota, USA. *Earth Surface Processes and Landforms*, **14**, 419–432.

Martinelli, L.A., Ferreira, J.R., Forsberg, B.R. and Victoria, R.L. 1988. Mercury contamination in the Amazon: a gold rush consequence. *Ambio*, **17**, 252–254.

McDowell, P.F., Thompson, W. and Bartlein, P.J. 1990. Long-term environmental change. In: Turner, B.L., Clark, W.C., Kates, R.W., Richards, J.F., Matthews, J.T. and Meyer, W.B. (Eds), *The Earth as Transformed by Human Action*. Cambridge University Press with Clark University, 143–162.

Miller, J.R. and Wells, S.G. 1986. Types and processes of short term sediment and uranium tailings storage in arroyos: an example from the Rio Puerco of the West, New Mexico. In: *Basin Sediment Delivery*. IASH Publication No. 159, Wallingford, UK, 335–353.

Mosley, M.P. 1975. Channel changes on the River Bollin, Cheshire, 1872–1973. *East Midland Geographer*, **6**, 185–199.

Nicholas, A.P., Ashworth, P.J., Kirkby, M.J., Macklin, M.G. and Murray, T. 1995. Sediment slugs: large-scale fluctuations in fluvial sediment transport rates and storage volumes. *Progress in Physical Geography*, **19**(4), 500–519.

Passmore, D.G., Macklin, M.G., Brewer, P.A., Lewin, J., Rumsby, B.T. and Newson, M.D. 1993. Variability of late Holocene Braiding in Britain. In: Best, J. and Bristow, C. (Eds), *Braided Rivers: Form, Process and Economic Applications*. Geological Society, London, 205–229.

Probst, J.L. and Tardy, Y. 1987. Long range streamflow and world continental runoff fluctuations since the beginning of this century. *Journal of Hydrology*, **94**, 289–311.

Renfrew, C. and Bahn, P. 1991. *Archaeology, Theories, Methods and Practice*. Thames and Hudson, London.

Roberts, R.G. and Church, M. 1986. The sediment budget in severely disturbed watersheds, Queen Charlotte Ranges, British Columbia. *Canadian Journal of Forest Research*, **16**, 1092–1106.

Robinson, M., 1986. Changes in catchment runoff following drainage and afforestation. *Journal of Hydrology*, **86**, 71-84.

Rumsby, B.T. and Macklin, M.G. 1994. Channel and floodplain response to recent abrupt climate change: the Tyne basin, northern England. *Earth Surface Processes and Landforms*, **19**, 499–515.

Ryder, J.M. 1991. Geomorphological processes associated with an ice-marginal lake at Bridge Glacier, British Columbia. *Géographie physique et Quaternaire*, **45**(1), 35–44.

Sarnthein, M. 1978. Sand deserts during glacial maximum and climatic optimum. *Nature*, **272**, 43–46.

Schumm, S.A. and Parker, R.S. 1973. Implications of complex response of drainage systems for Quaternary alluvial stratigraphy. *Nature*, **243**, 99–100.

Schumm, S.A., Harvey, M.D. and Watson, C.C. 1984. *Incised Channels*. Water Resources Publications, Littleton, Colorado, 336pp.

Spencer, T. and Douglas, I. 1985. The significance of environmental change: diversity disturbance and tropical ecosystems. In: Douglas, I. and Spencer, T. (Eds), *Environmental Change and Tropical Geomorphology*. George Allen and Unwin, London, 13–38.

Starkel, L. 1983. The reflection of hydrologic changes in the fluvial environment of the temperate zone during the last 15,000 years. In: Gregory, K.J. (Ed.), *Background to Palaeohydrology*. Wiley, Chichester, 213–235.

Starkel, L. 1987. Man as a cause of sedimentologic changes in the Holocene. *Striae*, **26**, 5–12.

Starkel, L. 1991. Long-distance correlation of fluvial events in the Temperate Zone. In: Starkel, L., Gregory, K.J., Thornes, J.B. (Eds), *Temperate Palaeohydrology: Fluvial Processes in the Temperate zone during the last 15,000 years*. Wiley, Chichester, 473–495.

Summerfield, M.A. 1991. *Global Geomorphology*. Longman.

Trimble, S.W. 1983. A sediment budget for Coon Creek basin the driftless area, Wisconsin, 1853–1977. *American Journal of Science*, **283**, 454–474.

Trimble, S.W. and Lund, S.W. 1982. Soil conservation in Coon Creek basin, Wisconsin. *Journal of Soil and Water Conservation*, **37**, 355–356.

van Andel, T.H., Zangger, E. and Demitrack, A. 1990. Land-use and soil erosion in Prehistoric Greece. *Journal of Field Archaeology*, **17**, 379–396.

Vaughan, D.M. 1990. Flood dynamics of a concrete-lined, urban stream in Kansas City, Missouri. *Earth Surface Processes and Landforms*, **15**, 525–537.

Wagstaff, J.M. 1985. *The Evolution of Middle Eastern Landscapes: an Outline to AD 1840*. Croom Helm, London.

Walcott, R.I. 1972. Past sea levels, eustasy and deformation of the earth. *Quaternary Research*, **2**, 1–14.

Walsh, R.P.D. 1977. Changes in the tracks and frequency of tropical cyclones in the Lesser Antilles from 1650 to 1975 and some geomorphological and ecological implications. *Swansea Geographer*, **15**, 4–11.

Walsh, R.P.D., Hudson, R.N. and Howells, K.A. 1982. Changes in the magnitude–frequency of flooding and heavy rainfall on the Swansea Valley since 1875. *Cambria*, **9**, 36–60.

Warner, R.F. 1992. Floodplain evolution in a New South Wales coastal valley, Australia: spatial process variations. *Geomorphology*, **4**, 447–458.

Webb, R.H., Smith, S.S., McCord, V.A.S. In press. *Historic Channel Change of Kanab Creek, Southern Utah and Northern Arizona*. Grand Canyon National History Association Monograph.

Wells, L.E. 1990. Holocene history of the El Nino phenomenon as recorded in flood sediments of northern coastal Peru. *Geology*, **18**, 1134–1137.

Wendland, W.M. 1977. Tropical storm frequencies related to sea surface temperature. *Journal of Applied Meteorology*, **16**, 477–481.

Wiles, G.C. and Calkin, P.E. 1994. Late Holocene, high resolution glacial chronologies and climate, Kenai Mountains, Alaska. *Geological Society of America Bulletin*, **106**, 281–303.

Williams, M.A., Dunkerley, D.L., De Deckker, P., Kershaw, A.P. and Stokes, T. 1993. *Quarternary Environments*. Edward Arnold, London. 329pp.

Wyzga, B., 1991. Present-day downcutting of the Raba River channel (Western Carpathians, Poland) and its environmental effects. *Catena*, **18**, 551–566.

3 Short-term Changes in Channel Stability

ALAN WERRITTY
Department of Geography, University of Dundee, UK

3.1 INTRODUCTION

The timescales involved in the material to be covered in this chapter cover years and decades, rather than the centuries and millennia of the previous chapter. In the terminology of Schumm and Lichty (1965) we are now dealing with present time in which changes in many of the independent variables governing the behaviour of river systems discussed in the preceding chapter (i.e. climate, geology and relief) cease to be directly relevant. The major controls governing the behaviour of the river system at this timescale are sediment supply and flow regime (from immediately upstream), channel and valley morphology (especially gradient), and the nature and volume of sediment supplied to the river from the adjacent slopes and undercut banks.

The key question to be addressed is: 'What determines the stability of the channel over a timespan of years and decades?' In order to answer this question it is necessary to examine the nature of geomorphic thresholds governing the behaviour of rivers and the response of the river to externally imposed changes.

This chapter is structured in the following manner. Initially, the different types of threshold present are examined together with the concept of geomorphic sensitivity. Having identified different types of threshold and distinguished between responsive and robust fluvial landforms, the controls on channel behaviour which are most likely to undergo change (over a timespan of years or decades) are then examined. Generally, these are changes in either the sediment supply or the flow regime and each is considered in turn. The relationship between channel form and flows of varying magnitudes and frequencies is addressed within the concept of dominant and effective discharges. The chapter concludes with an assessment of the impact of rare, large-magnitude floods. In terms of duration such floods are, of course, very short-lived, but their longer term impact may extend well beyond the timespans envisaged in this chapter.

Short-term changes in channel stability can most easily be understood within the context of geomorphic thresholds, robustness, responsiveness and sensitivity (each of these terms is defined in Table 3.1).

Applied Fluvial Geomorphology for River Engineering and Management, Edited by C. R. Thorne, R. D. Hey and M. D. Newson.

Table 3.1 Definition of terms

Term	Definition
Geomorphic thresholds	'a threshold of landform stability that is exceeded either by intrinsic change of the landscape itself, or by a progressive change of an external variable' (Schumm, 1979, p. 488)
Robust landforms	'robust landforms retain a stable identity as they form and reform, under a given process regime, despite being changed as intrinsic thresholds are crossed' (Werritty and Brazier, 1994, p. 103)
Responsive landforms	responsive landforms are those which, in response to externally imposed change, cross extrinsic thresholds to produce a new assemblage of landforms
Sensitivity	'the propensity of a system to respond to a minor external change. If the system is sensitive and near a threshold it will respond to an external influence, but if it is not sensitive it may not respond' (Schumm, 1991, p. 78)

3.2 GEOMORPHIC THRESHOLDS

The concept of geomorphic thresholds was first introduced by Schumm (1973) in a widely cited paper in which a fundamental distinction was made between extrinsic and intrinsic thresholds. Subsequently, these concepts have been revised and redefined by many other researchers. Definitions of these concepts and examples of different types of geomorphic thresholds are provided in Table 3.2.

In order to understand what is meant by the term 'extrinsic threshold', consider what happens as the flow of a river is increased over a bed of potentially mobile particles. As the discharge increases, individual particles on the bed are subject to an increase in applied boundary shear stress. At some point the submerged weight of each particle is no longer sufficient to resist the combined effects of the drag and lift forces exerted upon it. When this occurs the particle begins to move (i.e. it is entrained) and this constitutes the crossing of an extrinsic threshold: the threshold of motion. Other well known examples of extrinsic thresholds are the Froude and Reynolds numbers in open channels which define the conditions at which flow becomes supercritical and turbulent, respectively.

Table 3.2 Definitions and examples of different types of geomorphic thresholds

Definition	Example
Extrinsic threshold: 'one that is exceeded by the application of a force or process external to the system' (Schumm, 1980)	climatic fluctuation; land-use change; base-level change
Intrinsic threshold: 'one in which change occurs without a change in an external variable' (Schumm, 1980)	long-term progressive weathering leading to slope failure; development of a meander cutoff

In these examples an abrupt change in process occurs in response to progressive change in the external variable. This means that a threshold exists, but will not be crossed and hence change will not occur without control being exerted by an external variable (hence *extrinsic* in describing this type of threshold). But intrinsic thresholds also exist in river systems. In this situation change occurs without control being exercised by an external variable (i.e. the capacity to change is *intrinsic* within the system). A well known example of this is the process that leads to the cutoff of a meander bend. Meanders tend over time to become more tortuous. In some cases they become so tortuous that an unstable form develops resulting in a cutoff, an abrupt reduction in sinuosity and the formation of an ox-bow lake.

New terms beyond those originally introduced by Schumm (1973) have also been added to the literature. For example Chappell's (1983) distinction between transitive and intransitive thresholds has extended our understanding of threshold phenomena. This distinction depends upon whether, in response to the change in external conditions, the new state is persistent (transitive) or short-lived (intransitive). Newson (1992), in his case study on the geomorphic impact of the Forest of Bowland floods in 1967, provides a well argued case for Chappell's terminology. This case study also provides a good example of how the concept of geomorphic thresholds can help to identify the responsiveness of river systems.

The use of the threshold concept in fluvial geomorphology now extends back over two decades. During that time the original concept has been so loosely applied by some researchers that its original definition has lost its clarity (see the diversity of definitions used in the 1980 Binghampton Geomorphology Symposium edited by Coates and Vitek (1980)). Given this, the value of threshold concepts has been subject to reappraisal (e.g. Newson, 1992). The question of whether thresholds involve abrupt or gradational change has also been the subject of a particularly vigorous debate. Thus it is now generally agreed that the distinction between braided and meandering channel patterns involves a transition rather than an abrupt threshold change (Carson and Griffiths, 1987; Ferguson, 1987).

Having explored the concept of geomorphic thresholds we now turn to the interrelated concepts of geomorphic robustness, responsiveness and sensitivity.

3.3 GEOMORPHIC ROBUSTNESS, RESPONSIVENESS AND SENSITIVITY

All landforms are subject from time to time to a disturbance in their immediate environment, but this does not necessarily result in the destruction or even a significant modification of the landform. Of crucial importance in predicting the outcome is the balance between the size of the disturbance and the ability of the landform to resist, or accommodate, the impact of the disturbance. The disturbance itself may arise naturally (e.g. in response to a variation in climate, base level or sediment supply) or by human action that may be either deliberate or inadvertent. Irrespective of the type of disturbance involved, the response ultimately depends on the nature of the geomorphic system and its limiting thresholds (Figure 3.1).

Some high energy fluvial systems (such as active, braided channels) are subject to frequent change by processes which occur many times within a few years and involve the

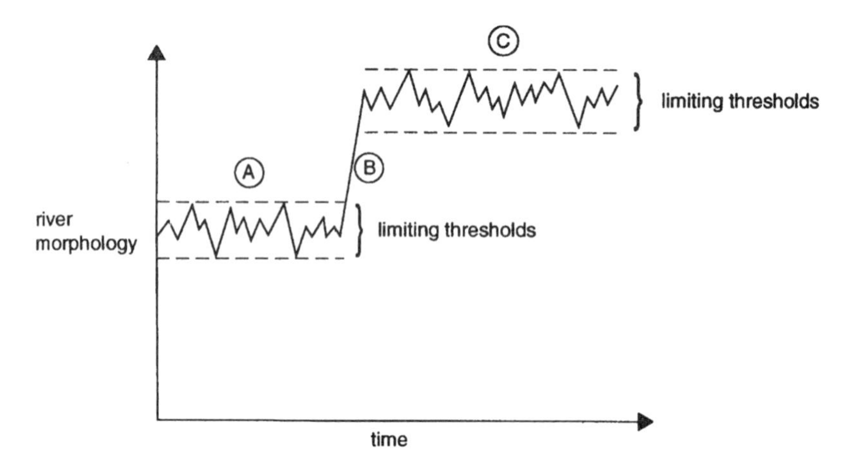

(A), (C) **robust behaviour** - river repeatedly crossing intrinsic thresholds, but overall response stable within limiting thresholds. Negative feedback regulates change. Landforms retain stable identity as they form and reform.

(B) **responsive behaviour** - in response to externally imposed change river moves across extrinsic threshold to new process regime. Landforms in original regime (A) destroyed and replaced by new landforms created in regime (C).

Figure 3.1 The distinctions between robust and responsive behaviour, the two types of geomorphic threshold and stable and unstable behaviour

repeated crossing of intrinsic thresholds. Such processes are an inherent part of the braided system and should be seen as the result of internally imposed change. Thus, the individual landforms (i.e. specific channels and bars) are frequently destroyed but the overall geomorphic system is robust since the new landforms which are created are recognisably similar to the old (see the case study in Box 3.2 concerning the valley floor landforms on the Allt Mor, a boulder-bed stream in the Cairngorms, Scotland (McEwen and Werritty, 1988)). In this situation the morphology is stable but not static, and the landform is said to be geomorphologically robust. If, however, the imposed disturbance causes the system to cross an extrinsic threshold into a new process regime in which a very different assemblage of landforms is likely to develop (e.g. the transition from braided channels to meandering channels across the North European Plain during the Lateglacial and Holocene (Starkel, 1983)), then the initial landform assemblage is deemed to have been geomorphically responsive to the imposed change.

Thus, responsive landforms are those which are destroyed by externally imposed change. This contrasts with robust landforms which, despite being subject to change as intrinsic thresholds are exceeded, retain a stable identity as they form and reform under a given process regime. Sensitive landforms respond to minor external disturbances by crossing an extrinsic threshold. Being already located close to the threshold, they are sensitive because only a minor change in the external environment is needed to produce a major response (Schumm, 1991).

These distinctions between geomorphically robust and geomorphically responsive landforms can also be understood in terms of the magnitude of the imposed change, and their contrasting responses as they cross different types of threshold. Crossing an intrinsic threshold within the normal spectrum of events (low magnitude and high frequency) involves negative feedback (i.e. the impact is self-regulating and stability is retained).

An example of this would be the constant destruction and formation of the braid bars which collectively make up the surface of a proglacial outwash plain created by the meltwater and sediment discharged from an ice sheet. By contrast, crossing an extrinsic threshold (e.g. the change from large-scale meanders to small-scale braiding) results in the river searching for a new equilibrium state in which negative feedback and stability can once more be achieved. The behaviour of a river is termed robust if it only crosses intrinsic thresholds as it adjusts to normal fluctuations in water and sediment supply. However, the same river is termed responsive if, in response to substantial changes in water and sediment supply, it becomes necessary to cross an extrinsic threshold in order to attain a new equilibrium state.

3.4 CHANGES IN SEDIMENT SUPPLY AND FLOW REGIME

The two most important controls determining whether or not a channel is stable over a period of years or decades are sediment supply and flow regime. If either of these undergoes progressive or sudden change, the channel may cross an extrinsic threshold and undergo change. At its most extreme this process has attracted the term 'river metamorphosis' since the morphology of the channel is completely changed. A good example of such behaviour is the Cimarron River in SW Kansas which was transformed from a relatively narrow, sinuous, deep river prior to 1914 to a wide, shallow, straight channel by 1931 in response to a series of major floods (Schumm, 1977). Within a smaller area, and on a much reduced timescale, Harvey (1986) has identified similar behaviour in localised shifts from meandering to braided channels in small streams in the Howgill Fells of NW England. This change occurs in response to sediment being released into the channel from hillslope gullies as they undergo erosion in response to major floods.

The supply of sediment to a channel can vary in response to changes in the source areas: hillslopes, small tributary channels, or the margins of the channel itself. Except for the impact of major floods (see below), most changes in sediment supply relate to changes in land use within the catchment of which the three most important are afforestation, urbanisation and mineral extraction.

In humid, temperate environments, such as the British uplands, two parts of the forestry cycle are associated with major changes in sediment supply: planting (because of upslope/downslope ploughing as part of the land preparation) and felling (when heavy machinery is introduced to extract the timber). As Leeks (1992) demonstrated in his review of the impacts of forestry in the UK, both suspended and bedload transport are elevated (Figure 3.2), but their specific impact crucially depends on whether the sediment sources

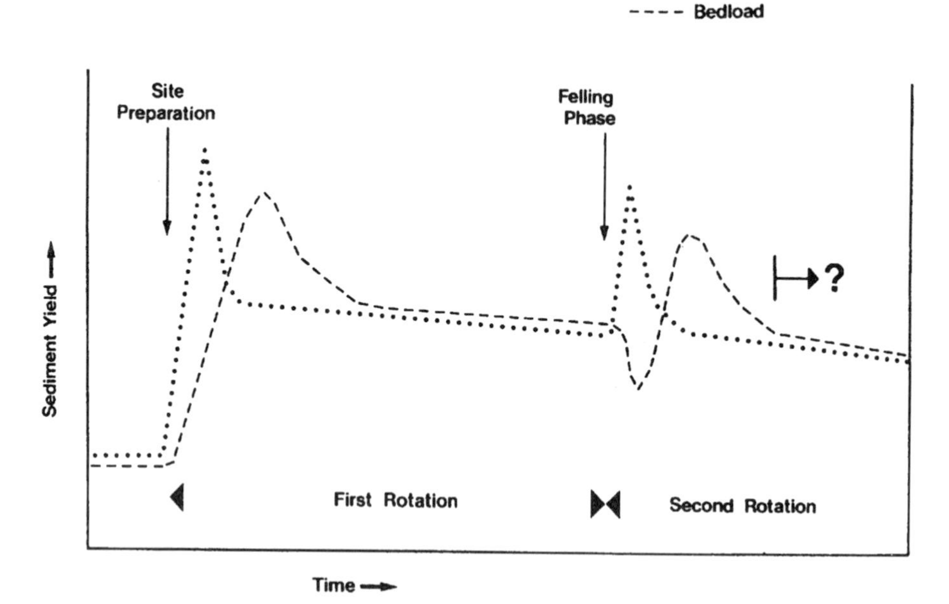

Figure 3.2 Summary diagram of upland stream sediment yields over the forest rotation (Leeks, 1992)

on the slopes (i.e. the eroding ditches) are directly linked by catchwater drains into the permanent stream system (see also Stott et al., 1987).

Another major land-use change with significant implications for sediment supply to the channel is urbanisation and, specifically, the construction phase when local sediment yields can increase by two or three orders of magnitude (Wolman and Schick, 1967). In theory, the impact of urbanisation is two-fold (Wolman, 1967; Roberts, 1989). Initially, there is a reduction in channel capacity due to local aggradation caused by increased sediment supply to the channel during the construction phase. This is followed by erosion and an increase in channel capacity as the local sediment supply is reduced (in response to the newly concreted, asphalted and grassed surfaces) and the frequency of higher flows is increased (in response to overland flow directed into the river via stormwater sewers).

The third major land-use change that can have a profound impact on short-term channel stability is mineral extraction (Lewin and Macklin, 1987). As the previous chapter has demonstrated, mining waste can very easily become incorporated into the downstream movement of sediment. Often, this can be tracked as a sediment slug moving as a wave through the fluvial system. At the scale of a drainage basin the complete passage of such a sediment slug typically involves several centuries, but for individual reaches the resulting instability may last decades (Macklin and Lewin, 1989).

Changes in flow regime are less easily summarised than changes in sediment supply as their potential origins are much more varied. Nevertheless, a useful distinction can be made between those which are inadvertent or are mainly natural in origin and those which are planned. The most significant inadvertent changes in river flow regime arise from climate change or, more accurately for a timescale of years and decades, increased climatic variability. The vexed question as to how far the current increase in climatic variability is due to natural or human causes has been addressed in the previous chapter. Here, it is appropriate to note that changes in flow regime arising from increased climatic variability can register profound impacts on channel stability (for a useful case study from Australia, see Warner (1987)).

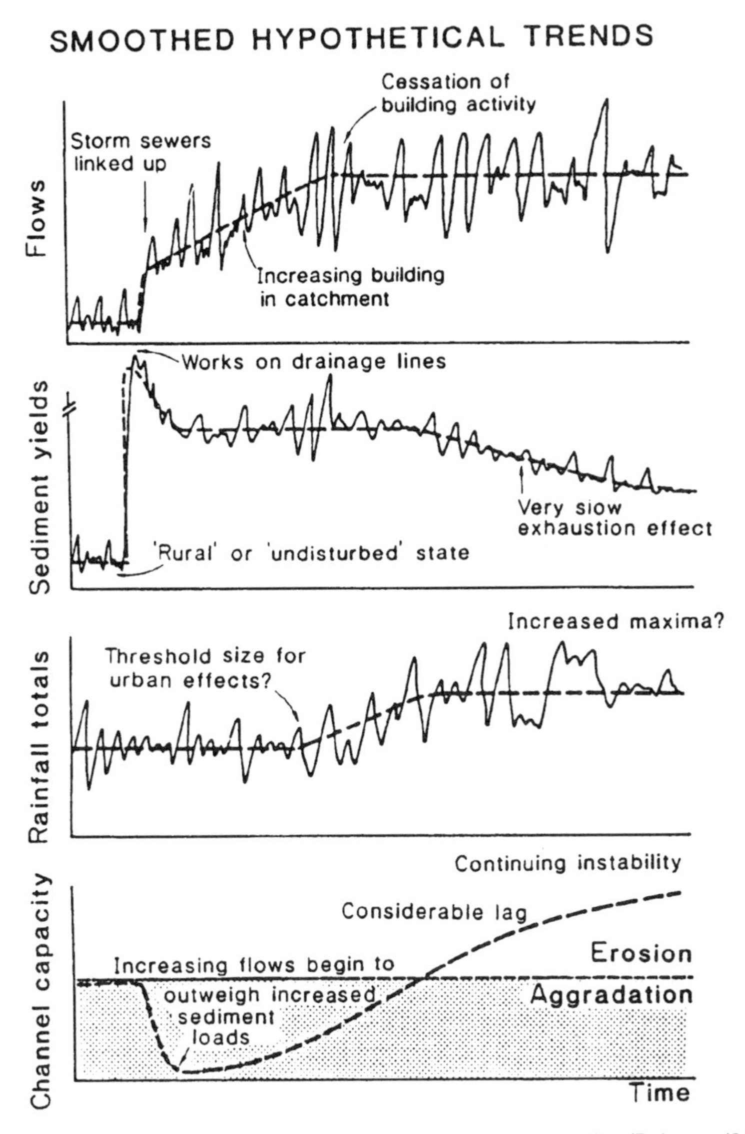

Figure 3.3 Hypothetical trends in the river system following urbanisation (Roberts, 1989)

Inadvertent changes in flow regime can also arise directly from land-use changes. Both urbanisation and, to a lesser extent, afforestation can locally register significant increases in high flows with potential impacts on channel stability. Since such land-use changes typically occur over areas covering hectares rather than square kilometres and across years rather than decades, their geomorphic impact can be both rapid and highly significant. Roberts (1989) provides a succinct summary of the hydrological impact of many of the New Towns built in Britain over the last 50 years (Figure 3.3), whereas Hollis (1974) and Walling (1979) summarise earlier findings. Unit hydrograph analyses suggest that where the proportion of urbanisation within a catchment exceeds 20%, peak flows can be increased by a factor of two or more. The impact of afforestation on flow regimes can be even more dramatic than that of urban sites but, from a geomorphic perspective, it is less significant because it is relatively short-lived.

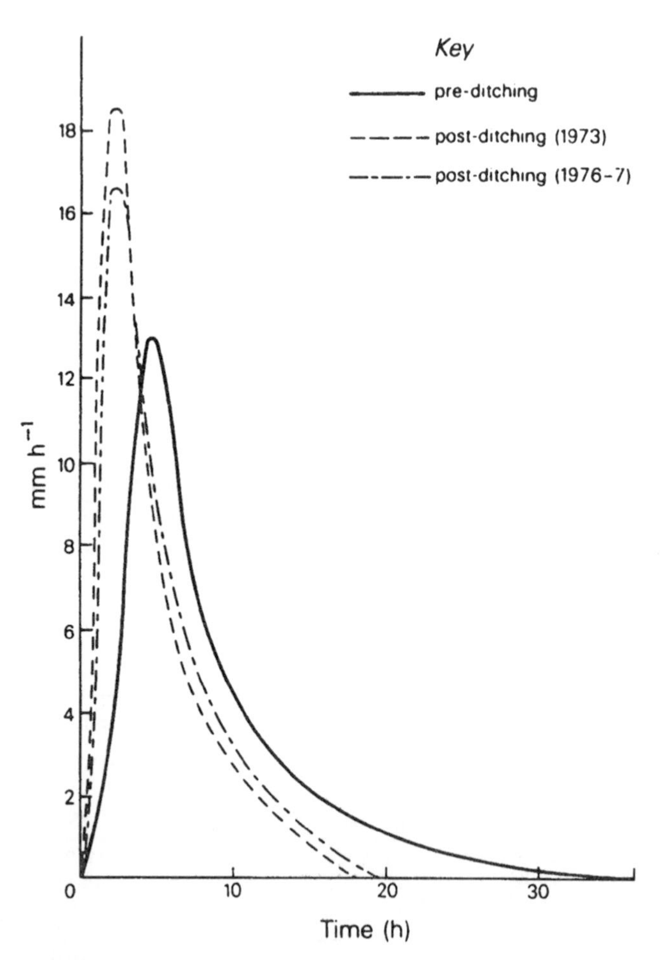

Figure 3.4 Pre- and post-ditching unit hydrographs for a forestry plantation at Coalburn, NW England (Robinson, 1980)

Table 3.3 Geomorphic impacts on channels of changes in water and sediment leading to river metamorphosis (after Schumm, 1977)

Change	River bed morphology	Change	River bed morphology
Qs + Qw =	aggradation, channel instability, wider and shallower channel	Qs + Qw −	aggradation
Qs − Qw =	incision, channel instability, narrower and deeper channel	Qs + Qw +	processes increased in intensity
Qw + Qs =	incision, channel instability, wider and deeper channel	Qs − Qw −	processes decreased in intensity
Qw − Qs =	aggradation, channel instability, narrower and shallower channel	Qs − Qw +	incision, channel instability, deeper, wider? channel

Key: Qs, sediment discharge; Qw, water discharge; +, increase; −, decrease; =, remains constant; ?, uncertain response.

Using a unit hydrograph model, Robinson (1980) reports a 40% increase in peak flows and a halving of the time to peak for the 1.52 km^2 Coalburn site in NW England, results confirmed by Werritty et al. (1993) for the Loch Dee basins in SW Scotland. But in both studies the effects were substantially reduced within a decade of planting the forest (Figure 3.4), whereas the impact of urbanisation on the flow regime appears to be more permanent (Roberts, 1989).

Having identified and documented the circumstances in which changes in sediment supply and flow regime can occur, what are their potential impacts on short-term channel stability? The present state of numerical modelling for fluvial systems cannot yet provide reliable answers to this question. The best that can be achieved is qualitative reasoning along the lines pioneered by Schumm (1977) in his conceptual treatment of river metamorphosis. There are eight possible combinations of changes in water and sediment discharge and their impacts on channel morphology are itemised in Table 3.3. In some cases the change is accommodated by the river as part of its inherent variability. In this case negative feedback will occur and the change will not result in sustained channel instability and irreversible channel change. In other cases, the change will exceed the natural inherent variability of the river, positive feedback will occur and the disturbance will result in the channel becoming unstable, thus potentially crossing an extrinsic threshold. Under such conditions channel metamorphosis can occur. Examples within Britain of local channel metamorphosis which has occurred within a timespan of years or decades include upland streams in the Howgill Fells of NW England subject to localised convective storms (Harvey, 1986); headwater streams of the River Tyne which underwent trenching in response to a severe storm (Newson and Macklin, 1990); and the middle Rheidol Valley in central Wales responding to a sediment wave made up of mining waste (Lewin and Macklin, 1987).

3.5 DOMINANT DISCHARGE

It has long been known that empirical relationships exist between the flow in a channel and its geometry. These ideas were implicit in regime theory (see Blench, 1969) and found a more explicit development in Leopold and Maddock's (1953) theory of 'hydraulic geometry'. A key question in such work was identifying the size of the channel-forming or dominant discharge. This attracted a number of definitions of which the most persuasive is that the dominant discharge is the flow which yields the maximum sediment transport (Wolman and Miller, 1960). It is further claimed that since river cross-sections are adjusted, on average, to a flow that is just contained within the banks, the dominant discharge is the bankfull flow. This has been confirmed in a number of empirical studies which have demonstrated that the frequency of bankfull discharge (approximately one to two years) equates well with the frequency of the flow which cumulatively transports most sediment and thus constitutes the effective discharge (e.g. the Snake River, Wyoming (Andrews, 1980)). 'A link is thus established between dominant discharge, most effective discharge and bankfull discharge' (Knighton, 1984, p. 94).

However, this formulation has not gone unchallenged. As Williams (1978) and others have noted, the unambiguous morphological definition of bankfull discharge is fraught with difficulties, and the frequency with which it occurs can vary between one and 32 years. Hey (1975) observed that whilst the return period for bankfull discharge for gravel-bed rivers was around one year, it was much less than this for sand-bedded streams. Other workers have noted that incision reduces the frequency of bankfull discharge. Thus, the widely reported assertion that bankfull discharge occurs on average once every one to two years is now seen to be an over-simplification. The accompanying claim that the flow which cumulatively transports most sediment (the effective discharge) also occurs on average once every one to two years also requires qualification. The reasons for this are made clear in the next section.

3.6 THE IMPACT OF RARE, LARGE FLOODS

Anyone who has stood beside a large river in flood must have been struck by the awesome nature of such an event. The lay person's reaction to such an experience is to conclude that floods must be the major geomorphic agent responsible for transporting fluvially derived sediment to the sea, and also the major agent creating suites of new landforms within and adjacent to the channel. The purpose of this section is to put that lay person's opinion to the test and evaluate the geomorphic role of floods on a scientific basis.

Wolman and Miller (1960) first posed this question in geomorphology using the following homely metaphor. Imagine a forest inhabited by a single giant and by a dwarf. Each has the task of felling the forest. Every day the dwarf sets to with his small and largely ineffective hand-axe. Bit by bit, but very slowly, the smaller and medium-sized trees are felled, but the larger ones remain beyond his grasp. By contrast, the giant only wakes up very rarely and goes on a rampage pulling trees both large and small up by their roots. But which is the more effective in felling the forest: the person who attacks the task every day on a piecemeal basis, or the one who makes a violent but very short-lived attack on the forest? The metaphor in the context of the geomorphic role of floods, is between small- to medium-sized flows which occur many times a year (dwarfs) and a large, catastrophic flood (the giant) which is immensely powerful, but very rare.

Before attempting to answer the question as to whether medium-sized flows occurring many times a year or very rare catastrophic floods are the more effective as geomorphic agents, we need to place the question within a broader geologic context. Within geology and geomorphology there is a long-standing debate as to whether geologic change is slow, steady and gradual (i.e. 'uniformitarian') or whether it is sudden, rapid and highly episodic (i.e. 'catastrophist'). Whewell, in his review of Lyell's (1830) *Principles of Geology* put it very succinctly (Cunningham, 1977, p. 106):

> 'Have the changes which lead us from one geological state to another been ... uniform in their intensity, or have they consisted of epochs of paroxysmal and catastrophic action interposed between periods of tranquillity?'

As is well known, the answer for the rest of the 19th century and much of the 20th century was that geological change was slow, steady and gradual (Gould, 1977). The 'uniformitarians' won the argument and this view rapidly came to be the accepted paradigm by all Earth scientists (Werritty, 1993). But is the rate of geologic change really uniform? This is now increasingly debated within the Earth sciences and specifically in the geomorphic literature on the impact of rare floods (Baker, 1977; Wolman and Gerson, 1978; Newson, 1989). Nineteenth century 'catastrophism' has re-emerged under the title of 'neo-catastrophism' (Dury, 1975) and the role of large, rare events in forming fluvial landscapes is once again a significant area of geomorphic research.

One of the most significant contributors to this modern debate was J. Harlen Bretz who in the 1920s questioned the validity of the 'uniformitarianism' paradigm in seeking to explain the origins of the landscape of the Channelled Scablands in Eastern Washington, in the NW USA (Bretz, 1923). He proposed a catastrophic flood as the formative agent for these landforms, but was ridiculed by the US geological establishment for whom the ruling paradigm of 'uniformity of rate' was incontestable (Gould, 1977). There is a delightful irony in the fact that Bretz's hypotheses have now largely been vindicated, and the Channelled Scablands stand as the finest example of a landscape sculpted by the largest catastrophic flood yet to be identified from the evidence that it left behind on the Earth's surface.

3.6.1 Magnitude and Frequency of Geomorphic Processes

In order to answer the question posed by Wolman and Miller (1960), the joint roles of the magnitude of a flood and its frequency must be examined. This is most easily undertaken via a simple diagram (Figure 3.5) in which curves A and B are characterised separately, and then jointly (curve C). Two processes must be treated individually:

(i) the sediment transport rate once the threshold of motion has been exceeded (typically a power function): magnitude;
(ii) the frequency of occurrence of flows of varying magnitude: frequency.

If the threshold of motion is relatively low, the peak in the curve defining the product of magnitude and frequency (curve C) typically occurs in the middle range of flows. This led

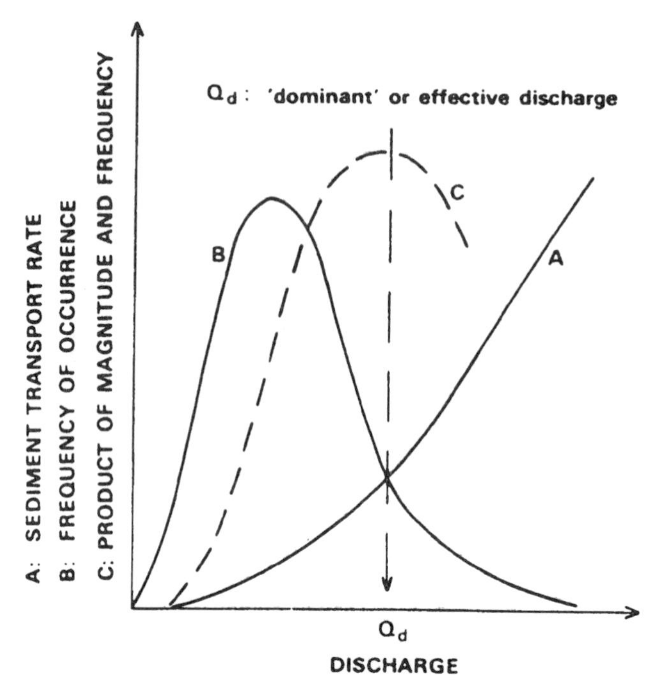

Figure 3.5 Hypothetical magnitude and frequency distributions showing the dominant role of middle-range flows (Knighton, 1984)

Wolman and Miller (1960) to conclude that, contrary to the lay person's view derived in the previous section, low-magnitude, relatively high-frequency events are more important than rare floods in terms of their cumulative sediment transport. This conclusion was supported by a substantial body of empirical evidence mainly from rivers on the east coast, the High Plains or the Rocky Mountains of the USA. For example, in the Bighorn River at Thermopolis in Wyoming, flows above 283 $m^3 s^{-1}$ occur only 6% of the time and transport 9% of the sediment load, whereas flows below 156 $m^3 s^{-1}$ (which occur 90% of the time) transport 57% of the load.

But Wolman and Miller (1960) also attached three important qualifications to their conclusion:

(i) the conclusion is only valid where the entrainment threshold is low, and many of the rivers studied were, in fact, sand-bedded;
(ii) the more variable the flow, the greater the percentage of the load which is transported by more infrequent flows, i.e. floods play a proportionately greater role in semi-arid areas;
(iii) the smaller the drainage basin, the greater the amount of sediment transported by less frequent flows. This is because rare high-magnitude storms can 'blanket' a small basin (e.g. Wells and Harvey, 1987).

It is important to note that the dominant role attributed by Wolman and Miller to low-magnitude, high-frequency flows was based upon data generally derived in humid/

subhumid, temperate, well-vegetated drainage basins with fine-grained alluvial channels. As Baker (1977) observed in central Texas, a change in climate, lithology, vegetation and the timing of floods significantly alters the magnitude–frequency distribution. Wolman and Gerson (1978), on re-examining this balance across a range of environments, came to broadly comparable conclusions to Baker (1977).

How then can contrasts in climate, lithology, vegetation and the timing of floods modify the respective roles of low-magnitude, high-frequency flows and high-magnitude, low-frequency flows? The case study in Box 3.1 provides some of the answers.

The timing of floods is also highly significant in understanding the geomorphic role of rare floods. But this requires the context to be broadened beyond that of sediment transport occurring during individual floods to embrace such concepts as geomorphic effectiveness, and relaxation or recovery times. Only then can the full long-term impacts of rare, large floods be adequately assessed.

3.6.2 Long-term Effectiveness of Rare Floods

Wolman and Gerson (1978, p. 190) further added to our understanding of the magnitude and frequency of geomorphic processes by introducing the concept of geomorphic effectiveness which they defined as 'the ability of an event or combination of events to affect the shape or form of the landscape'. This is significant because it reminds us that floods not only transport sediment, but in so doing also sculpt the Earth's surface. Their true geomorphic impact may, thus, be in long-term erosion or deposition rather than in terms of sediment transport. But the erosive or depositional impact of a flood will only be registered in the long term if its impact is not erased by the restorative effects of lesser flows (e.g. the very short-term impact of Hurricane Agnes in 1972 on the eastern seaboard of the USA (Costa, 1974)). Thus, the effectiveness of a flood as 'a destructive event depends on the force exerted, the return period of the event, and upon the magnitude of the constructive or restorative processes which occur in the intervening period' (Wolman and Gerson, 1978, p. 190).

The return period of an event provides a measure of its comparative rarity and is defined as the inverse of its probability of occurrence. Hence the likelihood of an event with a return period of 100 years occurring in any particular year is one in a hundred. In Figure 3.6 such events are recorded as vertical 'steps' on the 'staircase'. Following a highly disruptive event a period of form adjustment is likely to occur. This healing of the landscape is generally referred to as the recovery time or relaxation time. Its duration will vary between environments and is typically brief in channels with fine-grained bed material in humid temperate climates (see Costa, 1974), but is much longer in semi-arid climates (Wolman and Gerson, 1978). In some environments, such as central Texas (see Baker, 1977), where the entrainment threshold is very high, minimal relaxation occurs and the landscape merely awaits the arrival of the next catastrophic flood. But this is atypical and in most environments relaxation eventually gives way to a temporary equilibrium period, during which the form of the landscape may be constant (static equilibrium) or may evolve under the action of low-magnitude, high-frequency events. This situation lasts until the next catastrophic event. Selby (1974) uses this model to contrast the overall rates of landscape change across contrasting climatic and land-use types. The steepness of the overall 'staircase' (Figure 3.6) is at its greatest in the Himalayas (see Starkel (1976) for

Box 3.1 Geomorphic impact of catastrophic floods in Central Texas (Baker, 1977)

Central Texas is subject to the occasional passage of severe tropical storms tracking inland from the Gulf of Mexico and capable of generating daily rainfalls reaching 1000 mm. Much of the region is underlain by a massive limestone which tends to produce bare, bedrock channels into which blocks of dense, hard limestone are introduced from undercut banks on meander bends. Other parts of central Texas are underlain by granitic bedrock which weathers into sandy granule-sized material often aggraded across the whole of the channel floor. In terms of vegetation, the alluvial bottom lands are characterised by a dense growth of oak, sycamore and pecan, whereas the bare limestone outcrops on the slopes support a scrubby xerophytic cover. In terms of runoff generation, Hortonian (infiltration excess) overland flow is readily generated off these rocky hillslopes. The specific controls exerted by climate, lithology and vegetation on the geomorphic response of this area to catastrophic floods are itemised below.

(a) *Climate.* Rainfall intensities from tropical storms in this area can exceed 300 mm per hour. Given the pattern of runoff generation from impermeable limestone slopes supporting only a scrubby vegetation, peak flows exceeding 1000 $m^3 s^{-1}$ and velocities around 6 ms^{-1} have been reconstructed for sites such as Elm Creek, a catchment of 200 km^2. Flows of this magnitude locally fill the whole valley bottom, generate shear stresses sufficient to entrain boulders >2.5 m in size and locally produce hydraulic plucking of *in situ* bedrock.

(b) *Lithology.* Whereas the limestones generate bed material in which cobbles and boulders dominate the grain-size curve, the granitic terrain only produces granule-sized material. The impact on entrainment thresholds is striking, in that bed material in the limestone areas can only be mobilised by very rare floods, whereas bed material in the granitic areas is transported whenever there is a moderate flow.

(c) *Vegetation.* The dense woody vegetation on the valley floors results in a rapid increase in flow resistance as the flood stage rises to submerge the vegetation cover. This continues until the flood flows are capable of locally uprooting the vegetation at which point the flow resistance is decreased. The flow and shear velocities within the channel at different stages within the flood are thus strongly influenced by the nature and extent of the valley floor vegetation. By contrast, the scrubby xerophytic vegetation on the surrounding hillslopes promotes the generation of overland flow and a very flashy runoff response, especially within the limestone areas.

Central Texas displays two contrasting types of stream response to severe but rare tropical storms.

(i) In limestone areas, with bare bedrock channels mantled with large boulder-sized blocks and dense woody vegetation on the valley floor, the channel can only be shaped and reworked by catastrophic floods.

(ii) In the granitic areas, where the runoff response is less flashy (reduced overland flow) and the thresholds for entrainment are very low, the channel is shaped and reworked by discharges of much smaller and more frequent events.

Thus the geomorphic impact of a severe storm in this area is largely determined by the nature of the underlying rock type and, to a lesser degree, by the local vegetation cover.

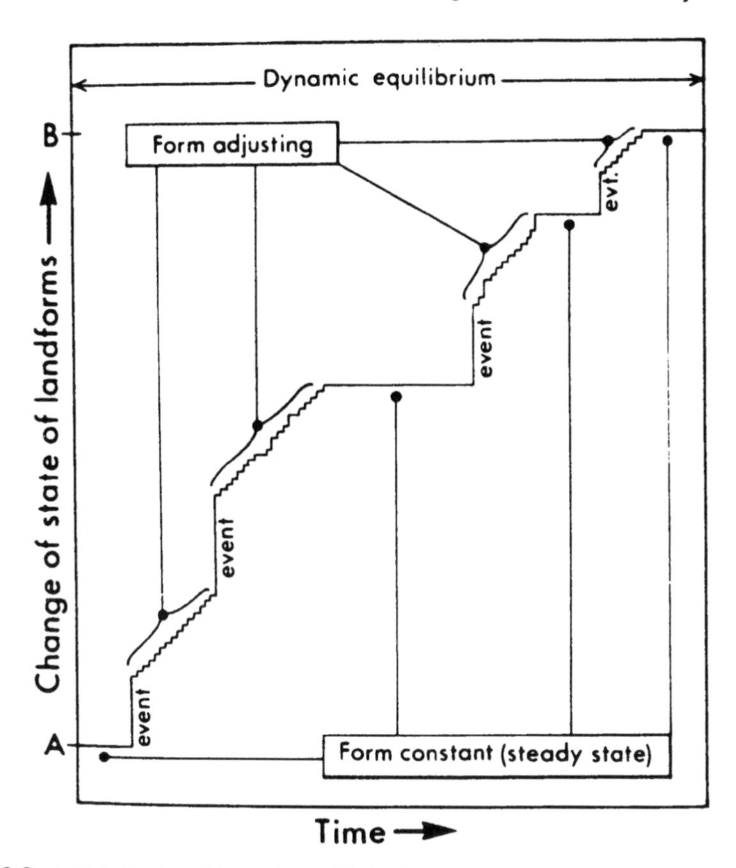

Figure 3.6 Within the term 'dynamic equilibrium' are contained three states: (1) a landforming event; (2) the adjustment of form that follows that event; and (3) a period of steady state in which there is virtually no adjustment of form. The curve which represents the change of landforms with time may, therefore, rise very steeply, gradually or slightly, depending upon the magnitude and frequency of the dominant process (Selby, 1974)

details) and at its lowest in western Europe. Intermediate values are recorded for North Island, New Zealand, with areas under pasture undergoing more rapid transformation than areas under woodland.

This important idea of formative events punctuated by periods of evolution, recovery or even temporary periods of steady state can be illustrated by a case study from the Scottish Highlands outlined in Box 3.2.

On leaving the mountains and reaching the plains, the impact of large, rare floods changes, with deposition replacing erosion as the dominant channel and system response. In a lowland environment the passage of a catastrophic flood is more likely to be recorded by vertical accretion on the nearby floodplain (see Macklin et al. (1992) for a detailed example). An unusual variant on this is Nanson's (1986) example of floodplain stripping which occurs in selected sites in Australia where valleys are confined by bedrock. However, this is the exception rather than the rule.

Box 3.2 Impact of large, rare floods on mountain torrents (McEwen and Werritty, 1988)

In the mountain torrents that make up many of the headwater streams of upland Britain, the entrainment threshold for bed material is very high. Thus, formative events capable of mobilising sediment within the channel are relatively rare (only several a century). The meteorological conditions capable of generating the shear stresses necessary to exceed the entrainment threshold are limited to intense, localised, convective storms (e.g. the Noon Hill flash floods (Carling, 1986)) or frontal storms of unusual rainfall intensity (e.g. Hurricane Charley (Newson and Macklin, 1990)).

The Allt Mor, which drains the northern slopes of the Cairngorms, is one such upland stream. Over the past half century it has been subject to formative events in 1956 and in 1978. During both floods the peak discharges exceeded $60\,m^3\,s^{-1}$, well above the entrainment threshold, and much of the valley floor was reworked. Nevertheless, the respective geomorphic impacts of the two events were strikingly different. In 1956 many slope failures occurred immediately adjacent to the channel, with the result that the sediment supplied to the stream during the flood exceeded its capacity to remove it. Following Newson's (1980) terminology this could be regarded as both a 'slope' and a 'channel' flood. After the passage of the flood wave most of this sediment was deposited to create a new valley floor and incidentally produced a protective 'apron' at the foot of slope failures. By 1978, when another severe but lesser flood occurred, the sites of the former slope failures had become partially stabilised and no new sediment was supplied to the valley floor. The slopes and the channel were no longer 'coupled'. This event was, thus, only a 'channel' flood for only the immediate channel and part of the valley floor was reworked.

In assessing the geomorphic impact of the two catastrophic floods at this site it is important to note the sequence of the floods (the larger being first) and the time interval between the floods, during which some sediment sources ceased to be accessible to fluvial reworking.

3.7 CONCLUSION

This chapter has sought to demonstrate the nature of the controls which determine channel stability over a timespan of years and decades. Crucial to this discussion has been the concept of geomorphic thresholds and the realisation that some channels are robust in responding to changes in their controlling variables. However, other channels are much more sensitive and respond dramatically to changes in their external controls, even to the extent of undergoing channel metamorphosis.

Changes in external controls can arise from natural causes (e.g. climatic change) or man-made causes (e.g. a land-use change). In terms of the latter, urbanisation (which has a dramatic and long-term impact), afforestation (contrasting impacts at different stages in the afforestation process) and mineral extraction (downstream impacts which can last for centuries) are of particular importance in assessing short-term channel instability. The impact of climatic change on channel stability is more difficult to characterise, but there is increasingly evidence that changes in the frequency of floods can be very significant.

Floods themselves exercise an inevitable control on channel stability. But the conventional wisdom that it is rare, catastrophic floods which determine how fluvial landforms develop, is an over-simplification. The geomorphic significance of such floods is now known to depend on many other factors including climate, lithology, vegetation and the actual timing of catastrophic flows. The impact of a rare flood in terms of sediment transport and channel stability can thus vary on a regional basis in response to the local setting. While in some parts of the world river channels are largely shaped by catastrophic floods which occur regularly, in other parts of the world such events are rare and of limited significance in terms of channel geometry and stability. Careful scrutiny of the local environmental setting is the key to correctly assessing the role of such floods.

3.8 REFERENCES

Andrews, E.D. 1980. Effective and bankfull discharges of streams in the Yampa river basin, Colorado and Wyoming. *Journal of Hydrology*, **46**, 311–330.

Baker, V.R. 1977. Stream channel response to floods, with examples from Central Texas. *Bulletin of the Geological Society of America*, **88**, 1057–1071.

Blench, T. 1969. *Mobile-bed Fluviology*. University of Alberta Press, Edmonton.

Bretz, J.H. 1923. The channelled scabland of the Columbia Plateau. *Journal of Geology*, **31**, 617–649.

Carling, P.A. 1986. The Noon Hill flash floods; July 17th 1983. Hydrological and geomorphological aspects of a major formative event in an upland landscape. *Transactions of the Institute of British Geographers*, NS, **11**, 105–118.

Carson, M.A. and Griffiths, G.A. 1987. Bedload transport in gravel channels. *Journal of Hydrology* (NZ), **26**, 1–151.

Chappell, J. 1983. Thresholds and lags in geomorphic changes. *Australian Geographer*, **15**, 357–366.

Coates, D.R. and Vitek, J.D. 1980. Perspectives on geomorphic thresholds. In: Coates, D.R. and Vitek J.D. (Eds), *Thresholds in Geomorphology*. George Allen & Unwin, London, 3–23.

Costa, J.E. 1974. Response and recovery of a piedmont watershed from tropical storm Agnes. *Water Resources Research*, **10**, 106–112.

Cunningham, F.F. 1977. *The Revolution in Landscape Science*. Tantalus Research, Vancouver.

Dury, G.H. 1975. Neocatastrophism? *Anals da Academia de Ciencias dos Brasil*, **47**, 135–151.

Ferguson, R.I. 1987. Hydraulic and sedimentary controls of channel pattern. In: Richards, K.S. (Ed.), *River Channels: Environment and Process*. Blackwells, Oxford, 129–158.

Gould, S.J. 1977. The great Scablands debate. In: *The Panda's Thumb*. Penguin, Harmondsworth, 194–203.

Harvey, A.M. 1986. Geomorphic effects of a 100-year storm in the Howgill Fells, Northwest England. *Zeitschrift für Geomorphologie*, **30**, 71–91.

Hey, R.D. 1975. Design discharges for natural channels. In: Hey, R.D. and Davies, J.D. (Eds), *Science and Technology in Environmental Management*. Saxon House, Farnborough, 73–88.

Hollis, G.E. 1974. The effects of urbanisation on floods in the Canon's Brook, Harlow, Essex. In: *Fluvial Processes in Instrumented Catchments*. Institute of British Geographers Special Publication, 6, 123–139.

Knighton, D. 1984. *Fluvial Forms and Processes*. Arnold, London.

Leeks, G.J.L. 1992. Impact of plantation forestry on sediment transport processes. In: Billi, P., Hey, R.D., Thorne, C.R. and Tacconi, P. (Eds), *Dynamics of Gravel-bed Rivers*. Wiley, Chichester, 651–668.

Leopold, L.B. and Maddock, T. 1953. *The hydraulic geometry of stream channels and some physiographic implications*. United States Geological Survey, Professional Paper, 252.

Lewin, J. and Macklin, M.G. 1987. Metal mining and floodplain sedimentation in Britain. In: Gardiner, V. (Ed.), *International Geomorphology*. Wiley, Chichester, 1009–1027.

Macklin, M.G. and Lewin, J. 1989. Sediment transfer and transformation of an alluvial valley floor, the River South Tyne, Northumbria, UK. *Earth Surface Processes and Landforms*, **14**, 233–246.

Macklin, M.G., Rumsby, B.T. and Newson, M.D. 1992. Historical floods and vertical accretion of fine-grained alluvium in the lower Tyne valley, Northeast England. In: Billi, P., Hey, R.D., Thorne, C.R. and Tacconi, P. (Eds), *Dynamics of Gravel-bed Rivers*. Wiley, Chichester, 573–588.

McEwen, L.J. and Werritty, A. 1988. The hydrology and long-term geomorphic significance of a flash flood in the Cairngorm Mountains, Scotland. *Catena*, **15**, 361–377.

Nanson, G.C. 1986. Episodes of vertical accretion and catastrophic stripping: a model of disequilibrium floodplain development. *Bulletin of the Geological Society of America*, **97**, 1467–1475.

Newson, M.D. 1980. The geomorphological effectiveness of floods–a contribution stimulated by two recent events in Wales. *Earth Surface Processes*, 5, 1–16.

Newson, M.D. 1989. Flood effectiveness in river basins: progress in Britain during a decade of drought. In: Beven, K. and Carling, P. (Eds), *Floods: Hydrological, Sedimentological and Geomorphological Implications*. Wiley, Chichester, 151–169.

Newson, M.D. 1992. Geomorphic thresholds in gravel-bed rivers: refinement for an era of environmental change. In: Billi, P., Hey, R.D., Thorne, C.R. and Tacconi, P. (Eds), *Dynamics of Gravel-bed Rivers*. Wiley, Chichester, 3–15.

Newson, M.D. and Macklin, M.G. 1990. The geomorphologically-effective flood and vertical instability in river channels. In: White, W.R. (Ed.), *Flood Hydraulics*.Wiley, Chichester, 123–140.

Roberts, C.R. 1989. Flood frequency and urban-induced channel change: some British examples. In: Beven, K. and Carling, P. (Eds), *Floods: Hydrological, Sedimentological and Geomorphological Implications*. Wiley, Chichester, 57–82.

Robinson, M. 1980. *The Effect of Pre-afforestation Drainage on the Stream Flow and Water Quality of a Small Upland Catchment*. Institute of Hydrology Report 73, Wallingford.

Schumm, S.A. 1973. Geomorphic thresholds and the complex response of drainage systems. In: Morisawa, M. (Ed.), *Fluvial Geomorphology*. State University of New York, Binghampton, 299–310.

Schumm, S.A. 1977. *The Fluvial System*. Wiley, New York.

Schumm, S.A. 1979. Geomorphic thresholds: the concept and its applications. *Transactions of the Institute of British Geographers* (NS), **4**, 485–515.

Schumm, S.A. 1980. Some applications of the concept of geomorphic thresholds. In: Coates , D.R. and Vitek, J. (Eds). *Thresholds in Geomorphology*. George Allen & Unwin, London, 473–485.

Schumm, S.A. 1991. *To Interpret the Earth: Ten Ways to be Wrong*. Cambridge University Press, Cambridge.

Schumm, S.A. and Lichty, R.W. 1965. Time, space and causality in geomorphology. *American Journal of Science*, **263**, 110–119.

Selby, M.J. 1974. Dominant geomorphic events in landform evolution. *Bulletin of the International Association of Engineering Geology*. **9**, 85–89.

Starkel, L. 1976. The role of extreme (catastrophic) meteorological events in the contemporary evolution of slopes. In: Derbyshire, E. (Ed.), *Geomorphology and Climate*. Wiley, Chichester, 203–224.

Starkel, L. 1983. The reflection of hydrological changes in the fluvial environment of the temperate zone during the last 15,000 years. In: Gregory, K.J. (Ed.), *Background to Palaeohydrology*. Wiley, Chichester, 213–255.

Stott, T. A., Ferguson R. I., Johnson, R.C. and Newson, M.D. 1987. Sediment budgets in forested and non-forested basins in upland Scotland. In: Hadley, R.F. (Ed.), *Drainage Basin Sediment Delivery*. International Association of Scientific Hydrology, Publication 159, 57–68.

Walling, D.E. 1979. The hydrological impact of building activity: a case study near Exeter. In: Hollis G.E. (Ed.), *Man's Impact on the Hydrological Cycle in the U.K.* Geobooks, Norwich, 135–151.

Warner, R. 1987. Spatial adjustments to temporal variations in flood regime in some Australian rivers. In: Richards, K.S. (Ed.), *River Channels: Environment and Process*. Blackwells, Oxford, 14–40.

Wells, S.G. and Harvey, A.M. 1987. Sedimentologic and geomorphic variations in storm-generated alluvial fans, Howgill Fells, northwest England. *Bulletin of the Geological Society of America*, **98**, 182–198.

Werritty, A. 1993. Geomorphology in the UK. In: Walker, H.J. and Grabau, W.E. (Eds), *The Evolution of Geomorphology.* Wiley, Chichester, 457–468.

Werritty, A. and Brazier, V. 1994. Geomorphic sensitivity and the conservation of fluvial geomorphology SSSIs. In: Stevens, C., Gordon, J.E., Green, C.P. and Macklin, M.G. (Eds.) *Conserving Our Landscape,* English Nature, Peterborough, 100–109.

Werritty, A., Harper, F. and Burns, J.C. 1993. Rainfall and runoff response in the Loch Dee sub-catchments. In: Lees, F.M. (Ed.), *Acidification, Forestry and Fisheries Management in Upland Galloway, Proceedings of the Loch Dee Symposium.* Foundation for Water Research, Marlow, 13–21.

Williams, G.P. 1978. Bankfull discharge of rivers. *Water Resources Research*, **14**, 1141–1158.

Wolman, M.G. 1967. A cycle of sedimentation and erosion in urban river channels. *Geografiska Annaler*, **49A**, 385–395.

Wolman, M.G. and Gerson, R. 1978. Relative scales of time and effectiveness in watershed geomorphology. *Earth Surface Processes*, **3**, 189–208.

Wolman, M.G. and Miller, J.P. 1960. Magnitude and frequency of geomorphic processes. *Journal of Geology*, **68**, 54–74.

Wolman, M.G. and Schick, A.P. 1967. Effects of construction on fluvial sediment, urban and suburban areas of Maryland. *Water Resources Research*, **3**, 117–125.

Section III

River Channel and Valley Processes

SECTION CO-ORDINATOR: J. C. BATHURST

4 Environmental River Flow Hydraulics

JAMES C. BATHURST
Department of Civil Engineering, University of Newcastle upon Tyne, UK

4.1 INTRODUCTION

A thorough understanding of flow characteristics and their interaction with channel geometry and planform is essential for almost any engineering, ecological, economic or management study involving rivers. Flood hazard, bank erosion, sediment transport, contaminant dispersion, aquatic habitat and aesthetic considerations are all linked to flow hydraulics. Flow characteristics such as average depth, maximum depth, mean velocity and secondary circulation are all determined by channel properties such as cross-sectional shape, channel slope, bed and bank material size distributions and riparian vegetation, which channel properties are themselves in turn modified by the flow characteristics. Of particular interest is the problem of flow resistance, concerned with the prediction of the mean velocity of flow in terms of those channel properties and flow characteristics which act as a resistance, or an energy loss, to the flow. Through its effect on velocity, and thence depth, flow resistance determines the quantity of water a channel can carry: that is, its conveyance. An accurate prediction of flow resistance is, therefore, essential in such problems as flood routing, slope-area discharge gauging, stable channel design and the calculation of channel flood capacity. More generally, when building structures such as bridges and water offtakes, developing bank areas, training river channels or otherwise modifying the channel environment, it is important to understand the flow characteristics and their likely response to imposed change if undesirable and expensive consequences are to be avoided.

This chapter examines the development of secondary circulation, its effect on boundary shear stress distribution, and the calculation of flow resistance. As flow characteristics change with the type of channel, the chapter opens with a brief review of the different types of channel which may be found in the river system.

Because of space limitations this chapter can provide no more than a brief review of environmental flow hydraulics. It is hoped that the reader will be alerted to the range of flow characteristics which are found in natural channels and which should be considered in river engineering schemes, and the methods which may be deployed in measuring or otherwise accounting for those characteristics. However, the reader is strongly urged to examine the original references which are quoted, before applying any of the methods in practice.

Applied Fluvial Geomorphology for River Engineering and Management, Edited by C. R. Thorne, R. D. Hey and M. D. Newson.

4.2 CHANNEL TYPES

An idealised river system with upland headwaters, piedmont zone and alluvial plain contains several channel types. These are distinguished principally by bed material size and channel slope, both of which usually decrease in the downstream direction (e.g. Leopold et al., 1964, pp. 189–190). The varying interaction of the flow with these two properties means that each channel type has its own characteristic bedforms, velocity profiles, flow resistance and sediment transport.

4.2.1 Sand-bed Channels

Sand-bed channels are particularly prevalent in the middle and lower parts of a river system, in areas characterised by alluvial plains. Channel slopes tend to be relatively low (less than 0.1%), bed material diameter is less than 2 mm and relative submergence (the ratio of mean depth to median sediment size) can, therefore, be large, often exceeding 1000. Both bedload and suspended load sediment transport take place at practically all flows and the bed is therefore continuously forming and reforming into a range of characteristic bedform features – ripples, dunes, plane bed and antidunes, depending on hydraulic conditions.

4.2.2 Gravel-bed Channels

In the piedmont zone the dominant bed material ranges from pea gravel (about 2–10 mm in diameter) in the downstream reaches to cobbles (up to 250 mm) in the upstream reaches. Relative submergence usually lies in the range of about 5 to 100. Slopes lie in the approximate range 0.05–0.5%. Bedload transport occurs irregularly, only at high flows, and small-scale bedforms such as ripples and dunes are not generally present. Instead, the characteristic small-scale bedform consists of assemblages of particles called pebble clusters. At a larger (channel) scale the dominant bed feature is the pool/riffle sequence, consisting of a succession of pools and intervening rapids.

4.2.3 Boulder-bed Channels

In the upland zone, channel slopes steepen to the range 0.5–5% and bed material coarsens into the boulder range (sizes greater than 250 mm), although all sizes down to sand may be present. Relative submergence may fall below unity, creating rough flow conditions. Significant bedload transport involving the larger bed material sizes takes place only at the more extreme flows and no distinctive bedform is therefore associated with boulder-bed channels.

4.2.4 Steep Pool/Fall Channels

In the small channels at the upstream end of the river system, slopes are steep (exceeding 5%) and the flow cascades from one obstruction (such as a boulder cluster or step, fallen log or bedrock outcrop) to another, via intervening pools. At low flows this gives the appearance of a series of steps: hence the description as step-pool or steep pool/fall series.

Table 4.1 Characteristics of different channel types and characteristic values of their flow resistance coefficients. From Bathurst (1993)

Type of channel	Approximate range of			
	Channel slope (%)	Bed material D_{50}[a] (mm)	Darcy–Weisbach f	Manning n
Sand-bed	≤ 0.1	≤ 2	0.01–0.25	0.01–0.04
Gravel/cobble-bed	0.05–0.5	10–100	0.01–1	0.02–0.07
Boulder-bed	0.5–5	≥ 100	0.05–5	0.03–0.2
Steep pool/fall	≥ 5	variable	0.1–100	0.1–5

[a] D_{50} = bed material particle size for which 50% of the material is finer.

Considerable energy may be dissipated in hydraulic jumps and ponding in the pools, although this effect decreases as the pool/fall structure is drowned out at high flows.

Table 4.1 summarises the characteristics of the various channel types and indicates typical values of the resistance coefficients discussed later (Section 4.5). However, the given values are only approximate and there may also be an overlap of values between the different channel types.

4.3 VERTICAL VELOCITY PROFILE

In sand-bed and gravel-bed channels, where the flow depth typically exceeds the bed material size by a factor of 10 or more, the vertical velocity profile often exhibits an approximate dependency on the logarithm of distance from the bed as:

$$\frac{u}{u_*} = \frac{2.303}{\kappa} \log\left[\frac{y}{k_s}\right] + B \tag{4.1}$$

where u = velocity at distance y above the bed; u_* = shear velocity; κ = the von Kármán constant (0.41); B = a constant which depends on bed characteristics; and k_s = a measure of the bed roughness, often called the equivalent uniform sand roughness height. From turbulent boundary layer theory this dependency strictly applies only to the bottom 10–20% of the flow depth. However, the divergence of the profile from this law in the upper 80–90% of the flow is often relatively small in shallow open channels and can be ignored as a first approximation. The depth-averaged velocity can then be shown to be the same as the point velocity at a distance above the bed of 0.37 times the depth (usually approximated as 0.4 times the depth), advantage of which is taken in gauging discharge by current meter.

Several flume studies have shown that the addition of suspended load to a flow alters the shape of the velocity profile. Velocities near the bed are slightly decreased relative to the clear water case while velocities away from the bed are increased. Aspects of this effect are discussed by Coleman (1981, 1986), Itakura and Kishi (1980) and Lau (1983).

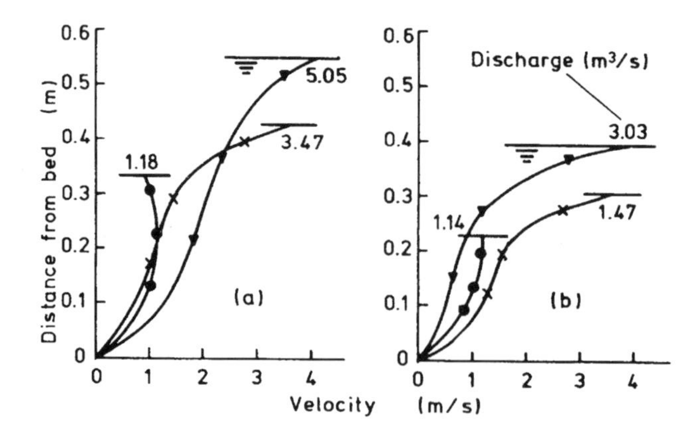

Figure 4.1 Development of the S-shaped velocity profile. Velocity profiles are shown for three discharges at approximately the same vertical at two sites on the Roaring River, Colorado, USA. Symbols indicate the measured points for each profile and the profile curves are drawn by eye (from Bathurst, 1988a)

In boulder-bed channels, where depths are comparable with the larger bed material sizes, the semi-logarithmic velocity profile breaks down. Instead, an S-shaped profile appears in which that part of the flow below the tops of the major boulders has low velocities, resulting from drag and other resistance effects, while that part of the flow above the boulders, which is relatively unimpeded, develops high velocities, especially at steep slopes (Figure 4.1) (e.g. Marchand et al., 1984; Bathurst, 1988a). Attempts to represent this pattern with a semi-logarithmic velocity law can lead to a significant underestimate of mean flow velocity. Discharge gauging by the 0.4-depth method may then underestimate the true discharge by 10–15% (Bathurst, 1988a).

Appearance of the S-shaped profile is likely to be limited to flows with d/D_{84} in the range 1 to 4, depending on channel slope (d = flow depth; D_{84} = the size of bed material intermediate axis for which 84% of the material is finer). This range may seem narrow but it encompasses many of the flood flows which occur in mountain rivers with slopes above about 1%. At present there is no generally accepted mathematical formula with which to describe the S-shaped velocity profile.

4.4 SECONDARY CURRENTS AND THE BOUNDARY SHEAR STRESS DISTRIBUTION

Secondary circulation occurs in the plane normal to the local axis of the primary flow. It combines with the primary flow to produce a spiral flow about the primary axis. The circulation arises because of the effects of the channel geometry and planform on the flow. Equally, the circulation, by altering isovel and boundary shear stress patterns, can affect the geometry and planform and has consequences for the dispersion of contaminants and the sorting of mobile sediments.

4.4.1 Straight Channels

A weak form of secondary circulation may develop in long, straight, non-circular channels, in which the spirals are scaled on the flow depth. The most common examples are circulation acting to direct water into channel corners (Figure 4.2a) and a series of streamwise cells, each cell rotating in a sense opposite to that of its neighbour (Figure 4.2b). Where the spirals direct water towards the bed or banks, the near-boundary isovels are compressed and the boundary shear stress is increased relative to the value that would be measured in the absence of secondary circulation. The impact on channel geometry is relatively minor since shear stresses are only marginally increased by this process. However, the circulation can alter the local pattern of bedload transport by concentrating the sediment into streamers at zones of flow convergence.

Secondary circulation of this nature is apparent in straight canals and artificially straightened river channels with rectangular and trapezoidal cross-sections. Natural river channels, however, rarely feature the necessary straight, uniform reaches and are unlikely, therefore, to exhibit this type of circulation.

4.4.2 Channel Bends

Secondary circulation in single-thread rivers is most evident at bends, where it appears as a cell directing surface water towards the outside of the bend and near-bed water towards the inside. In alluvial channels the cross-section at the bend apex is characteristically

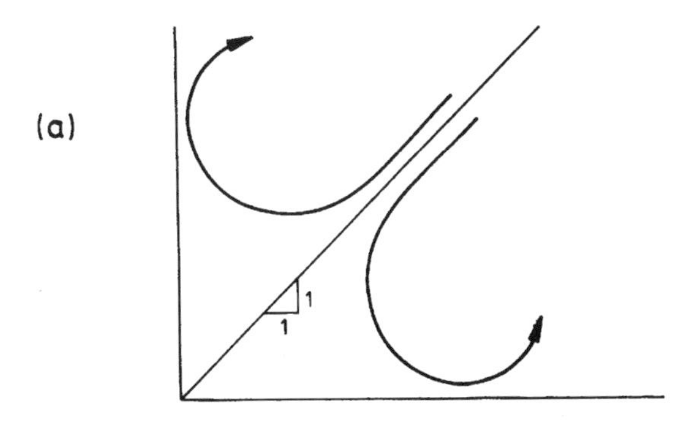

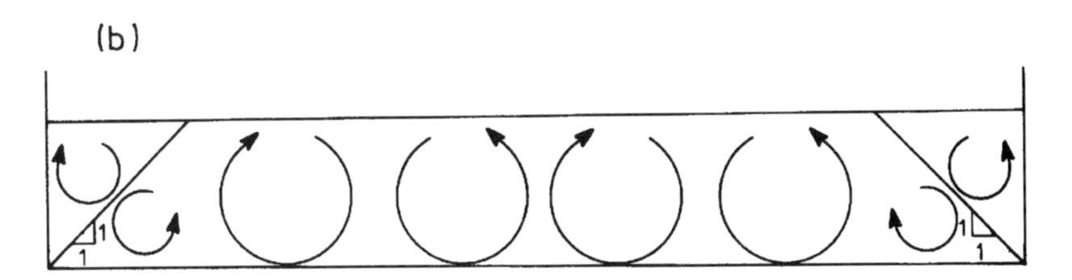

Figure 4.2 Patterns of weak secondary circulation in straight channels (primary flow normal to the page): (a) channel corner cells; (b) multicell formation

triangular, with a point bar on the inside of the bend and a pool lying against a steep outside bank. The circulation pattern may then also include a small cell of reverse circulation at the steep outside bank (Bathurst et al., 1979) and a dominance of outward flow near the inside bank caused by a progressive longstream decrease in depth along the point bar (Dietrich and Smith, 1983; Pitlick and Thorne, 1987; Thorne et al., 1989; Markham and Thorne, 1992) (Figure 4.3). In these circumstances the main secondary flow cell occupies only part of the channel width, where the depth is greatest.

At the entrance to a bend the core of maximum velocity tends to move towards the inside bank as free vortex flow develops. However, as secondary circulation develops and a net outward cross-stream discharge occurs over the point bar, the core of maximum velocity is pushed towards the outside bank and remains in the outer half of the bend for the remainder of the bend. Downstream of the bend the circulation decays and, in a sequence of meander bends, the secondary circulation from one bend is replaced by the new circulation for the next bend downstream. Flume experiments show that, for high width/depth ratios, the new cell originates near the bed and grows upwards. The old cell is thus displaced upwards and sideways (Prus-Chacinski, 1954; Tamai et al., 1983). Consequently flow at the inflection point between bends can consist of two cells, one above the other (e.g. Thorne and Hey, 1979). However, the pattern is very sensitive to any pre-existing circulation (Prus-Chacinski, 1954) and also to overbank flow (Toebes and Sooky, 1967). Consequently a variety of responses can be observed. Bed topography and bend morphology are also important and formation of the new cell is probably influenced by the switch of the thalweg from one bank to the other. In some cases, the bed topography and bend morphology may be such as to break up the circulation and leave a more complex pattern (e.g. Thorne et al., 1989).

Boundary shear stress peaks in the channel cross-section are associated with the core of maximum velocity and with downwelling near the outside bank, especially when the downwelling is reinforced by a cell of reverse circulation next to the bank (Figure 4.4). The longstream variation of shear stress is closely linked with the movement of the core of maximum velocity. At the bend entrance the highest boundary shear stress is, therefore, near the inside bank but within the bend it is transferred outwards. Consequently, there is a downstream increase in shear stress along the outside bank and a decrease along the inside bank (e.g. Hooke, 1975; Dietrich et al., 1979; Nouh and Townsend, 1979; Bridge and Jarvis, 1982) (Figure 4.5). Thus, the pool (and the bank) is scoured near the outside bank and the point bar is deposited along the inside bank (Hooke, 1975; Dietrich and Smith,

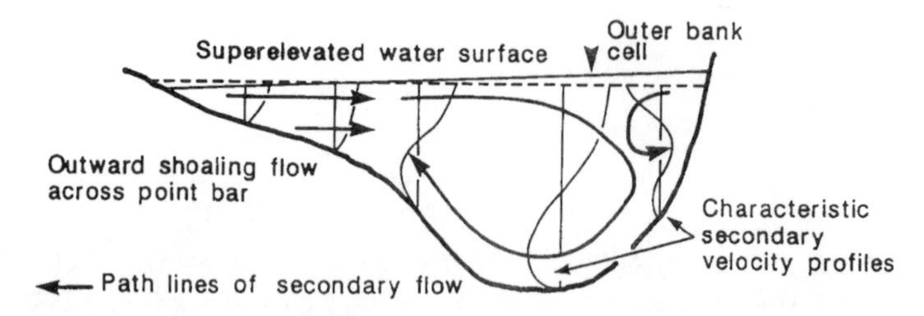

Figure 4.3 Secondary circulation pattern at a river bend cross-section, showing the main circulation cell restricted to the deepest part of the section, the cell of reverse circulation at the outside bank and shoaling-induced outward flow over the point bar at the inside bank (from Markham and Thorne, 1992)

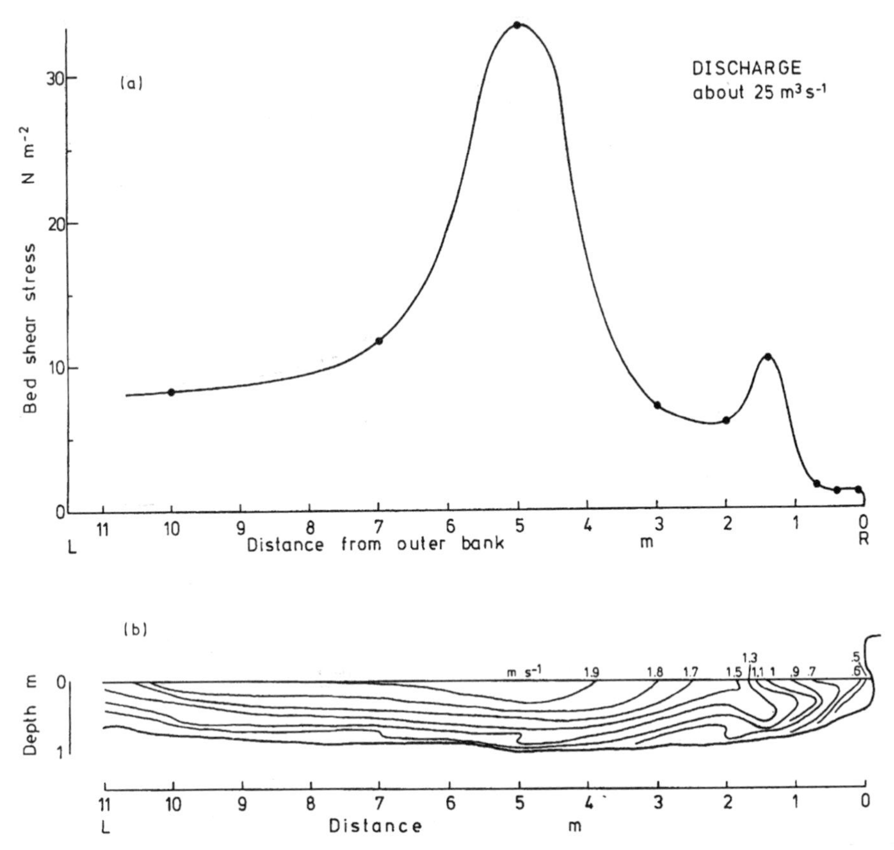

Figure 4.4 Example of (a) measured bed shear stress and (b) isovel pattern at a discharge of about 25 $m^3 s^{-1}$ at Llandinam Bend, River Severn, Wales. Shear stress peaks are associated with the core of maximum velocity and with downwelling at the outside bank (indicated by the isovel pattern) (from Bathurst, 1977)

1983). The region of maximum sediment transport follows the line of maximum shear stress and the distribution of shear stress also affects the size sorting of the bed material (Hooke, 1975; Dietrich and Smith, 1984).

4.4.3 Changing Discharge at Bends

Measurements generally show that the core of maximum velocity follows the thalweg at low flows but shortens its path by cutting across the point bar at high flows (Simons and Şentürk, 1977, p. 60). Consequently, the point at which the region of high shear stress is transferred towards the outside bank shifts downstream as discharge increases towards bankfull. Maximum bed scour and bank erosion are, therefore, likely to be observed in the downstream part of the bend, or even downstream of the bend, at high discharges (Bathurst, 1979) (e.g. Figure 4.5). This pattern has a direct impact on bend migration and bend planform evolution.

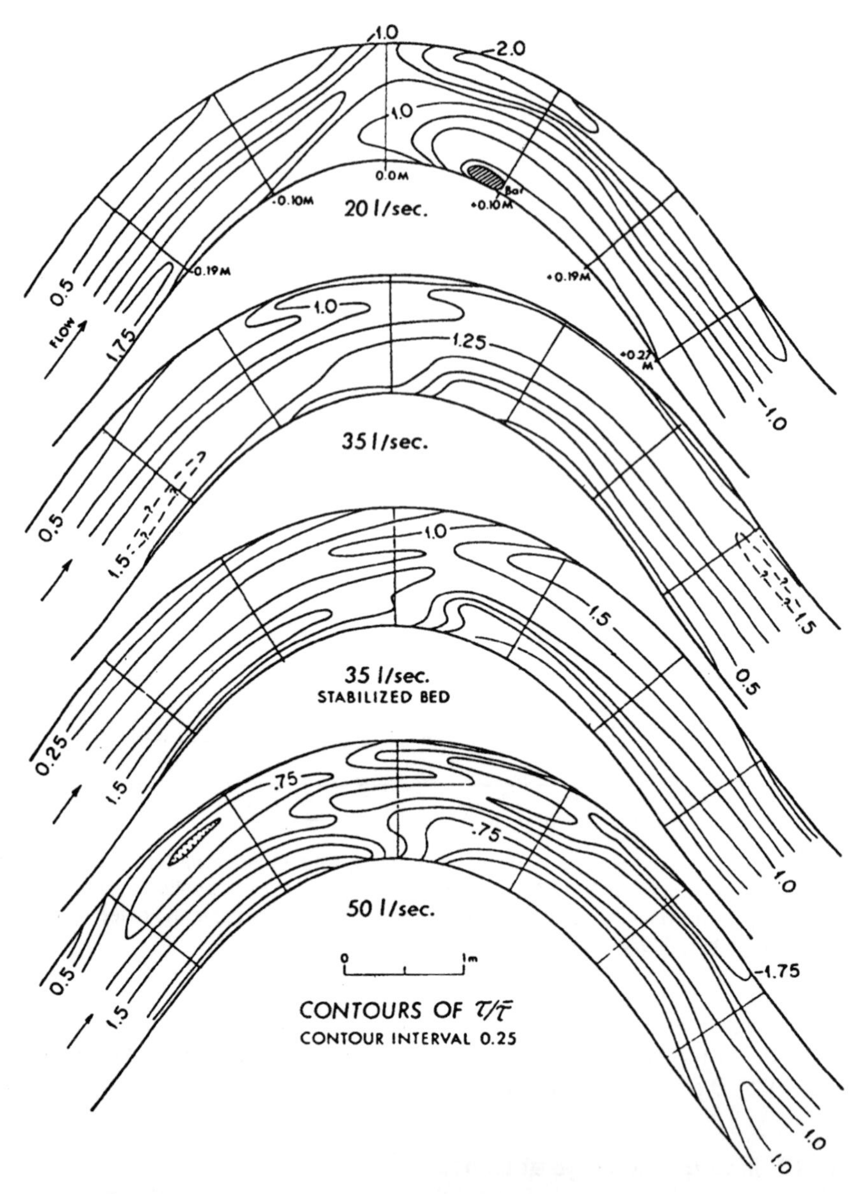

Figure 4.5 Distribution of bed shear stress around a bend in a 1-m flume with mobile sand bed at various discharges. Shear stress is shown normalised by mean bed shear stress round the bend. The pattern shows the transfer of maximum shear stress from the inside bank to the outside bank through the bend; it also shows the downstream drift of the location of the maximum shear stress at the outside bank as discharge increases (from Hooke, 1975, in *J. Geology.* Reproduced with permission of The University of Chicago Press)

4.4.4 Confluences

The convergence of two inflowing streams leads to the formation of two counter-rotating secondary flow cells, which act to drive water downwards at the centre. A strong scouring action develops at the confluence centre, creating a scour hole with steep avalanche faces (e.g. Mosley, 1976; Ashmore and Parker, 1983; Best, 1986, 1987) (Figure 4.6). The two streams remain generally separate along the scour zone, with limited mixing at the centre. Most of the bedload similarly passes in two separate streams round the sides of the scour hole. Downstream of the scour zone, the two flow and transport streams converge and the bed elevation rises. Mid-channel and lateral bars may then form, fed by the material eroded from the scour zone. As a result, bifurcation of the flow can occur, the two streams subsequently rejoining downstream of the mid-channel bar. In braided channels the repeated pattern of flow convergence, scour hole development, bar formation, bifurcation and again flow convergence is responsible for the characteristic morphology of such channels (e.g. Ashworth et al., 1992).

Orientation and depth of the scour hole are observed to vary with the confluence angle between the two inflowing streams, the relative discharges and, less importantly, sediment input.

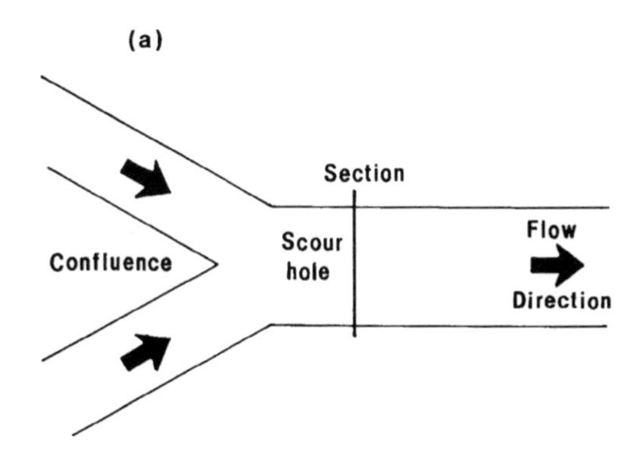

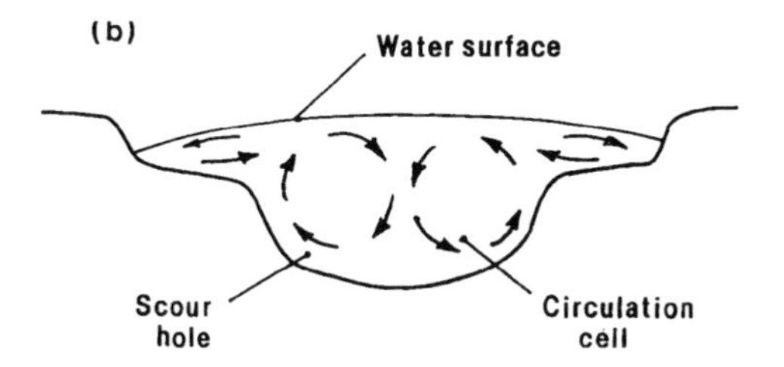

Figure 4.6 Secondary circulation pattern at a channel confluence: (a) plan view of a symmetrical confluence, showing location of section for: (b) circulation pattern in the scour zone downstream of the confluence (from Bathurst (1988b) based on Mosley (1976))

4.4.5 Overbank Flow

A number of recent studies have shown that there can be a significant apparent shear stress at the interface between overbank flow and the main channel flow. This is caused by an intensive momentum exchange from the main channel flow to the adjacent overbank flow. Failure to account for this phenomenon in discharge calculations results in serious overestimation of the channel-carrying capacity for any given depth of flow (e.g. Myers, 1978; Rajaratnam and Ahmadi, 1981; Wormleaton et al., 1982; Knight and Demetriou, 1983; Knight and Shiono, 1990; Myers and Brennan, 1990; Wormleaton and Merrett, 1990). Means of accounting for the momentum exchange in discharge calculations are proposed in several of these references.

Where the main channel is meandering, it usually follows a longer path down the valley than that taken by floodplain flow. The interaction of the floodplain and channel flows then produces secondary circulation cells strikingly different from and more pronounced than those occurring at bends with inbank flow (e.g. Toebes and Sooky, 1967; Elliott and Sellin, 1990) (Figure 4.7). These accentuate the exchange of momentum between the floodplain and channel flows (e.g. Ghosh and Kar, 1975; Smith, 1978).

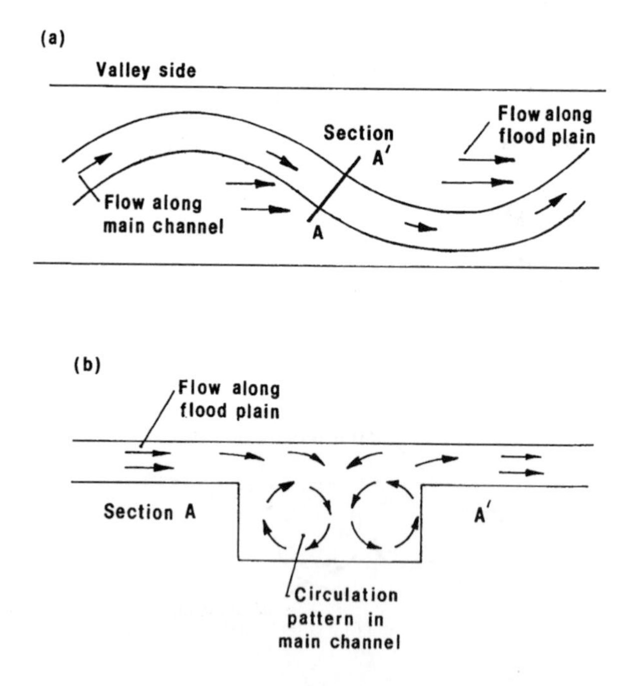

Figure 4.7 Secondary circulation pattern in a meandering channel with floodplain flow: (a) plan view of meandering channel within a floodplain, showing location of section AA′ for: (b) circulation pattern in the main channel (from Bathurst (1988b) based on Toebes and Sooky (1967))

4.4.6 Field Techniques

Very accurate measurements are necessary to distinguish between the primary and secondary velocities, since the latter are generally one or two orders of magnitude lower than the former. Under field conditions such accuracy is generally beyond the capability of conventional (propeller or cup) current meters but electromagnetic meters which make simultaneous measurements in two dimensions, with a potential accuracy of $\pm 0.01\,\mathrm{m\,s^{-1}}$, have been found to be suitable (Bathurst et al., 1979, 1981). A means of orienting the instrument precisely is required. Recently, acoustic-doppler meters have become available. These measure the flow velocity in three dimensions simultaneously, do not intrude physically into the flow and are accurate to millimetres per second. They may therefore potentially support detailed and accurate measurements of secondary circulation.

Point values of bed shear stress can be calculated from near-bed velocity profile measurements using Eqn 4.1. A plot of velocity against the logarithm of distance from the bed yields shear velocity (and thence boundary shear stress) from the plot gradient. However, particular care is required in locating the zero velocity datum (e.g. Jackson, 1981). As the equation applies strictly only to the bottom 10 to 20% of the flow, the velocity measurements should be made in that region. In the presence of secondary circulation the use of the resultant velocity of the primary and secondary flows gives the resultant shear stress vector. Because of unsteady and non-uniform flow effects and the inaccuracies associated with field measurements, the method is unlikely to achieve an accuracy of better than 10% (e.g. Wilkinson, 1984).

4.5 BED AND BAR FLOW RESISTANCE

Relationships linking velocity and flow resistance account for the resistance with a single coefficient. The most commonly used relationships are the Darcy–Weisbach equation:

$$U = \left[\frac{8gRS_{\mathrm{f}}}{f}\right]^{1/2} \tag{4.2}$$

the Manning equation (in SI units):

$$U = \frac{R^{2/3}S_{\mathrm{f}}^{1/2}}{n} \tag{4.3}$$

and the Chézy equation:

$$U = C(RS_{\mathrm{f}})^{1/2} \tag{4.4}$$

where U = mean velocity; R = hydraulic radius; S_{f} = friction slope; g = acceleration due to gravity; and f, n and C are, respectively, the Darcy–Weisbach, Manning and Chézy resistance coefficients, related by:

$$\frac{U}{u_*} = \left[\frac{8}{f}\right]^{1/2} = \frac{R^{1/6}}{ng^{1/2}} = \frac{C}{g^{1/2}} \tag{4.5}$$

where $u_* = (gRS_{\mathrm{f}})^{1/2}$ = the mean shear velocity.

Evaluation of the resistance coefficient is the principal source of error in the application of Eqns 4.2–4.4 to natural channels. Quantitative understanding of the processes involved is relatively limited and all available equations for calculating the resistance coefficient have an empirical element.

Several components of the total flow resistance of a channel have been identified (Rouse, 1965; Bathurst, 1982):

1. boundary resistance, resulting from the skin friction or form drag of the bed and bank materials, bedforms and vegetation;
2. channel resistance, incorporating the effects of channel cross-sectional shape and longitudinal non-uniformities in bed slope, water surface slope and channel planform;
3. free surface resistance, accounting for energy losses in the distortion of the free surface by waves and hydraulics jumps and for the effect of the free surface on the flow turbulence structure and velocity profile.

Suspended sediment load is also thought to influence flow resistance by altering the turbulence characteristics of the flow. Most studies have indicated that the presence of suspended load reduces resistance compared with the equivalent conditions in clear water (e.g. Vanoni and Nomicos, 1959; Itakura and Kishi, 1980; Lau, 1983). However, the effect may be small compared with the increased drag caused by the bedforms generated by the sediment transport.

Most available resistance formulae consider only boundary resistance. The effects of the other components are known qualitatively but have proved less susceptible to quantification. Also, the relative importance of the various sources of resistance varies through the river network according to channel characteristics such as cross-sectional shape, bed material size distribution and slope. It is important, therefore, to use a resistance equation that has been designed and validated for application to the type of channel under consideration. A wider consideration of the subject than is possible here is given in Bathurst (1993).

4.5.1 Sand-bed Channels

Important features are the dependency of resistance on the bedform (ripples, dunes, plane bed, antidunes) and the variation of the shape of the velocity profile with varying suspended sediment load concentration. Available resistance equations are empirical, derived to a greater or lesser extent from field and laboratory data. No single equation has been found to be generally acceptable for the full range of conditions met in sand-bed rivers and a multiplicity of equations has, therefore, appeared. Even the more reliable ones, though, are unlikely to predict velocity with an accuracy better than $\pm 30\%$. It is often recommended, therefore, that if field measurements relevant to a particular problem are available, these should be used to test and, if necessary, modify a chosen method prior to its application.

The available equations fall into two main groups: (1) those that give the overall bed resistance (without distinguishing between bed material grain roughness and bedform drag) as a logarithmic or power law; and (2) those that separate the grain roughness and the form drag. A further subdivision can be made according to whether the effect of each type of bedform is considered separately. Engelund's (1966, 1967) method (involving separation of grain roughness and form drag) is one of the more popular means of determining flow resistance. However, the range of equations which have been developed prevents a detailed review here and the reader is referred to the literature. Reviews of the available equations are given by Vanoni (1975), Simons and Şentürk (1977) and Garde and Ranga Raju (1985). Tests of certain equations are reported by Simons and Şentürk (1977, pp. 348–352), White et al. (1979), Willis (1983) and Klaassen et al. (1986). Because of their empirical content the equations should be applied only within the range of conditions for which they were originally derived.

4.5.2 Gravel-bed Channels

Flow resistance depends mainly on the relative submergence of the bed (the ratio of depth to bed material size) and on the ponding or backwater effects of pool/riffle sequences and bars.

A popular form for the resistance relationship, based on the semi-logarithmic velocity profile (Eqn 4.1), is a dependency on the logarithm of relative submergence. A variety of such relationships have been derived empirically, differing mainly in the values of their coefficients (e.g. Limerinos, 1970; Bray, 1979; Griffiths, 1981). For uniform flows (for example at riffles but not in pools) Hey's (1979) equation is recommended as the most satisfactory:

$$\left[\frac{8}{f}\right]^{1/2} = 5.75 \log\left[\frac{aR}{3.5D_{84}}\right] \tag{4.6}$$

where f = the Darcy–Weisbach resistance coefficient; R = hydraulic radius; D_{84} = bed material size for which 84% of the material is smaller; and a = a function of channel cross-sectional shape varying between 11.1 and 13.46 and calculated from (R/d_m) either graphically (Hey, 1979) or approximately as (Bathurst, 1982):

$$a = 11.1\left[\frac{R}{d_m}\right]^{-0.314} \tag{4.7}$$

where d_m = maximum depth at the section.

Using UK river data, Hey found that the resistance coefficient could be calculated using Eqn 4.6 with a standard error of estimate of $\pm 12.7\%$ for riffle flows, compared with $\pm 153.7\%$ for pool flows. This pattern is reflected in Figure 4.8 where values of $(8/f)^{1/2}$ calculated by Eqn 4.6 for pools and riffles are compared with measured values. Equation 4.6 may not always apply where depth is comparable with bed material size ($R/D_{84} < 4$) and where three-dimensional effects are present ($w/d < 15$, where w = width, and at bends).

In the non-uniform flow characteristic of a natural river with a pool/riffle sequence, the riffles or bars keep the flow in the pools deeper and slower than would be the case for uniform flow conditions. Figure 4.9 compares the measured variation of resistance for a pool/riffle sequence on the River Swale, UK (Bathurst, 1977) with the variation calculated from Eqn 4.6 for the same flow conditions but as if the flow were uniform and only bed material roughness were important. The comparison indicates the considerable degree to which pool/riffle sequences increase flow resistance at low flows and the way in which the effect diminishes as the sequence is drowned out at high flows. Resistance equations such

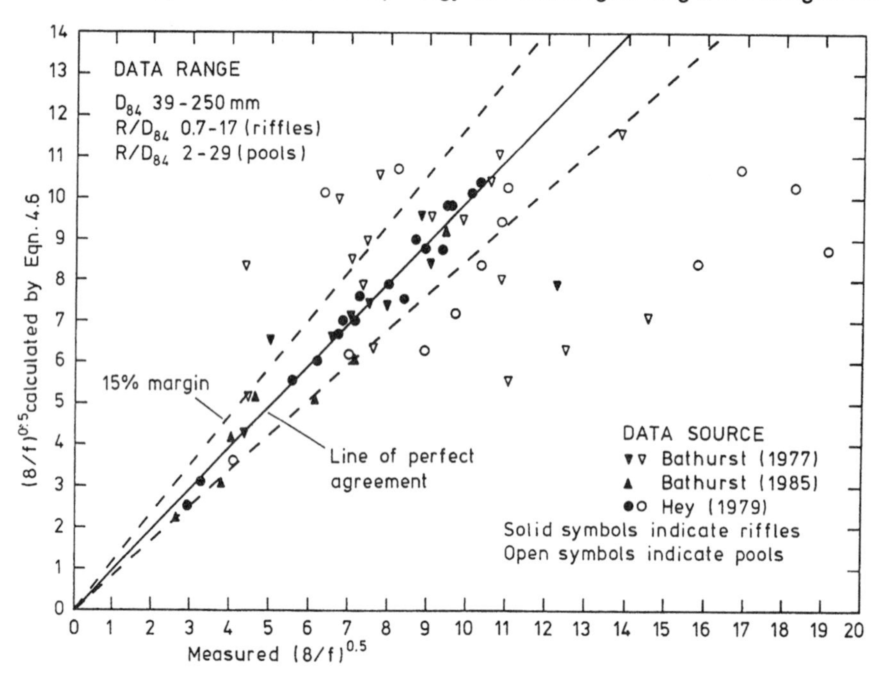

Figure 4.8 Comparison of measured values of $(8/f)^{1/2}$ with values calculated by Eqn 4.6 for pool and riffle flows in mainly gravel-bed rivers. The 15% margin is provided for visual comparison and has no statistical significance (from Bathurst, 1993)

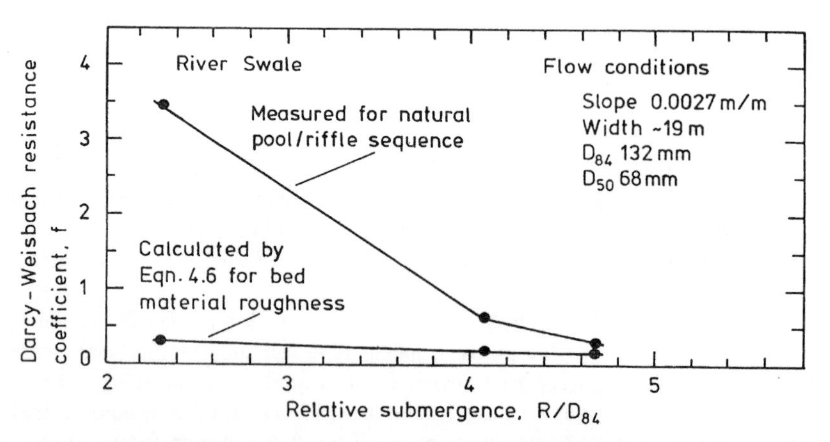

Figure 4.9 Comparison of measured variation in the Darcy–Weisbach resistance coefficient for a pool/riffle sequence on the River Swale, UK, with the variation calculated by Eqn 4.6 as if the flow were uniform and only bed material roughness were important. From left to right on each curve, points represent discharges of 0.7, 4.36 and 7.84 $m^3 s^{-1}$ (from Bathurst, 1993)

as Eqn 4.6 and other similar equations (e.g. Bray, 1979; Griffiths, 1981) which do not allow for bar resistance may therefore be approximately applicable to pool/riffle sequences at flows approaching bankfull (e.g. Beven and Carling, 1992). For wider applicability, though, it is necessary to account directly for bar resistance. Hey (1988) modified Eqn 4.6 to give:

$$\left[\frac{8}{f}\right]^{1/2} = 5.75 \log\left[\frac{ad}{D_t}\right] \tag{4.8}$$

where f = the average Darcy–Weisbach resistance coefficient for the reach; d = average flow depth in the reach; a is as defined for Eqn 4.6 but for the reach; and D_t = total roughness height of the bed material and bars, given by:

$$D_t = ad\left[\frac{D_g}{a_r d_r}\right]^F \tag{4.9}$$

where a_r is the shape coefficient as defined for Eqn 4.6 but for the riffles; D_g = roughness height of the bed material, given by 3.5 D_{84}; d_r = average flow depth for the riffles; and

$$F = \left[\frac{f_r}{f}\right]^{1/2} = \left[\frac{d_r^3 W_r^2 S_r}{d^3 W^2 S}\right]^{1/2} \tag{4.10}$$

where f_r = the average Darcy–Weisbach resistance coefficient for the riffles; W and W_r = average reach and average riffle widths respectively; and S and S_r = average reach and average riffle water surface slopes respectively.

Further data on bar resistance are provided by Parker and Peterson (1980), Prestegaard (1983) and Jaeggi (1984).

4.5.3 Boulder-bed Channels

Where depth is of the same order of magnitude as the bed material size ($R/D_{84} < 4$), flow resistance varies primarily as a function of the drag forces of the boulders, including wave drag arising from the distortion of the water surface by protruding boulders. The determining factors include channel slope, bed material size distribution and relative submergence. Figure 4.10 shows examples of observed variations in flow resistance but as yet the available data are insufficient to disentangle the complex relationships. Various resistance equations have been derived empirically to account for the effects of the boulders but they tend to be more complex than is warranted by their accuracy. Semi-logarithmic equations such as Eqn 4.6 are not theoretically applicable because of the breakdown of the semi-logarithmic velocity profile (Eqn 4.1) but may apply empirically at relative submergences less than about 1.5 (Figure 4.10) (Thorne and Zevenbergen, 1985). At higher relative submergences, though, the semi-logarithmic resistance relationship significantly overestimates the flow resistance coefficient (e.g. Figure 4.10).

Noting the difficulty of developing a practical method for calculating flow resistance, Bathurst (1985) suggested an empirical approach in which the likely limits to the resistance variation are delineated and the flow resistance is calculated from the available

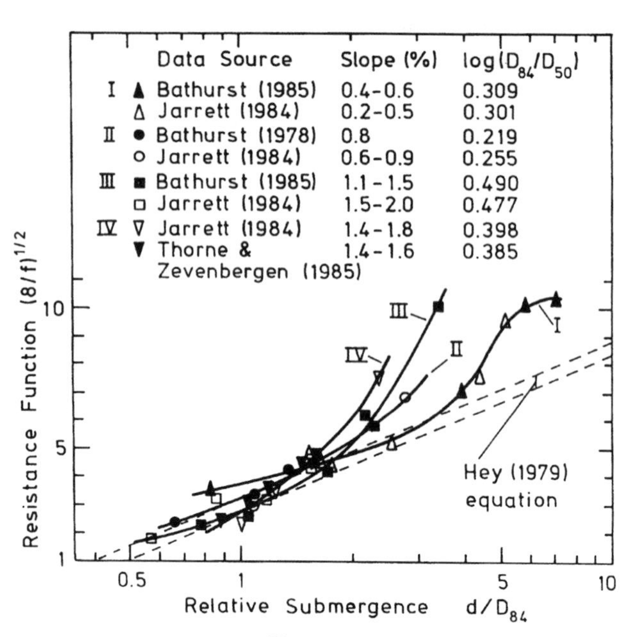

Figure 4.10 At-a-site variation of $(8/f)^{1/2}$ with relative submergence d/D_{84} for four combinations of channel slope and standard deviation of bed material size distribution for boulder-bed rivers (from Bathurst, 1994. Reproduced by permission of the American Society of Civil Engineers)

data. He presented the following equation, which lay along the centre of his envelope of data:

$$\left[\frac{8}{f}\right]^{1/2} = 5.62 \log\left[\frac{d}{D_{84}}\right] + 4 \tag{4.11}$$

The equation was derived for channel slopes in the range 0.4–4% and $d/D_{84} < 10$. Because of its semi-logarithmic nature the equation represents the overall trend of the data envelope (for a range of sites) but is less representative of the at-a-site variation for individual sites. Thus the equation tends to underestimate the resistance coefficient f at low flows and overestimate it at high flows (Figure 4.11).

Equation 4.11 assumes that depth d is determined from the water surface width, w, and the cross-sectional area of the water only, not including boulders projecting into the flow. However, surveys of cross-sections in boulder-bed rivers are usually made relative to the channel bed on which the boulders lie and therefore yield a total cross-sectional area A_T composed of the flow cross-sectional area A and the cross-sectional area of the projecting boulders. An approximate means of obtaining A is given by Bathurst's (1985) empirical formula:

$$\frac{A_T - A}{A_T} = 0.275 - 0.375 \log\left[\frac{d}{D_{84}}\right] \tag{4.12}$$

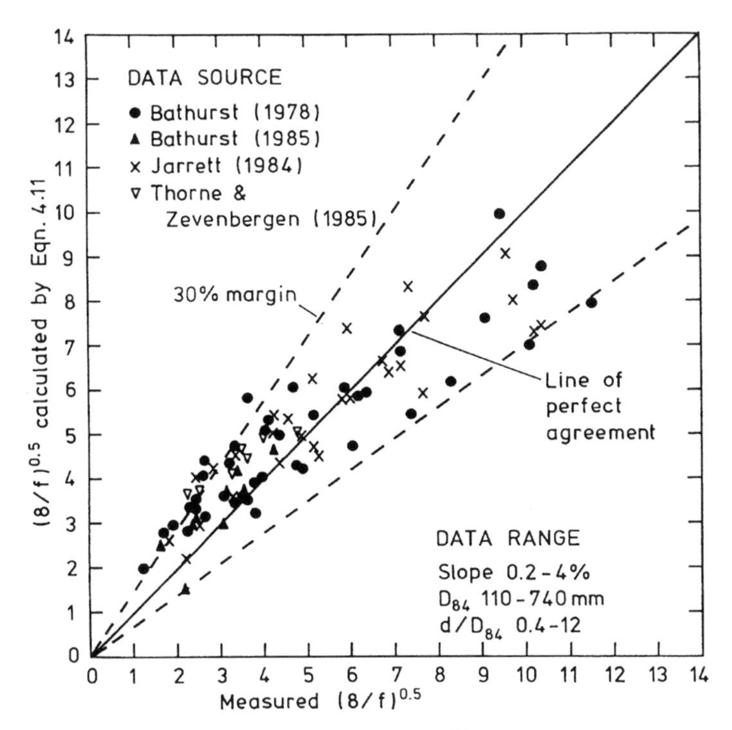

Figure 4.11 Comparison of measured values of $(8/f)^{1/2}$ with the values calculated by Eqn 4.11 for high-gradient, mainly boulder-bed rivers. The 30% margin is provided for visual comparison and has no statistical significance (from Bathurst, 1993)

This applies for d/D_{84} less than 4; at higher values the projection of the boulders into the flow can be neglected. The equation is solved iteratively with $d = A/w$.

A second empirical resistance equation which has been field tested for boulder-bed channels is given by Jarrett (1984) as:

$$n = 0.32 S_f^{0.38} R^{-0.16} \tag{4.13}$$

for channel slopes in the range 0.2–4%. The units are SI. As with Eqn 4.11, this may represent the overall variation of flow resistance for an envelope of data from a range of sites, rather than the at-a-site variation at individual sites.

Equations 4.11 and 4.13 both exhibit standard errors of estimate of approximately 30%. They can also give significantly different values of discharge for the same flow data.

4.5.4 Steep Pool/Fall Channels

Flow resistance is determined chiefly by the interaction of the pool/fall structure with the flow, rather than by bed material roughness (e.g. Hayward, 1980; Whittaker and Jaeggi, 1982; Lisle, 1986; Whittaker, 1987). At low flows, resistance is extremely high as a result of ponding in the pools and energy losses in the hydraulic jumps below each fall. At high flows the pool/fall structure is drowned out and, if sediment transport occurs, the pool may even be temporarily partly filled. Flow resistance is then greatly reduced. Data from Beven et al. (1979) show variations in the value of the Darcy–Weisbach coefficient from over 100 at low flow to 1 at high flow. Similarly, the measurements of Newson and Harrison (1978) show a hyperbolic relationship between flow discharge and time of travel for tracer in the flow. Travel times are long at low discharge (high resistance) and short at high discharge (low resistance) (Figure 4.12).

At present (1996) there is no generally accepted method of calculating flow resistance for steep pool/fall channels.

4.5.5 Flow Resistance in Vegetated Channels

Two particular cases can be identified where vegetation has a significant effect on flow resistance: these are channels where the bed is lined with vegetation and channels where the bed is free of vegetation but the banks are vegetated.

(a) *Vegetation-lined channels.* Examples include grass-lined channels and channels with high weed growth. In a manner similar to that used in determining the flow resistance

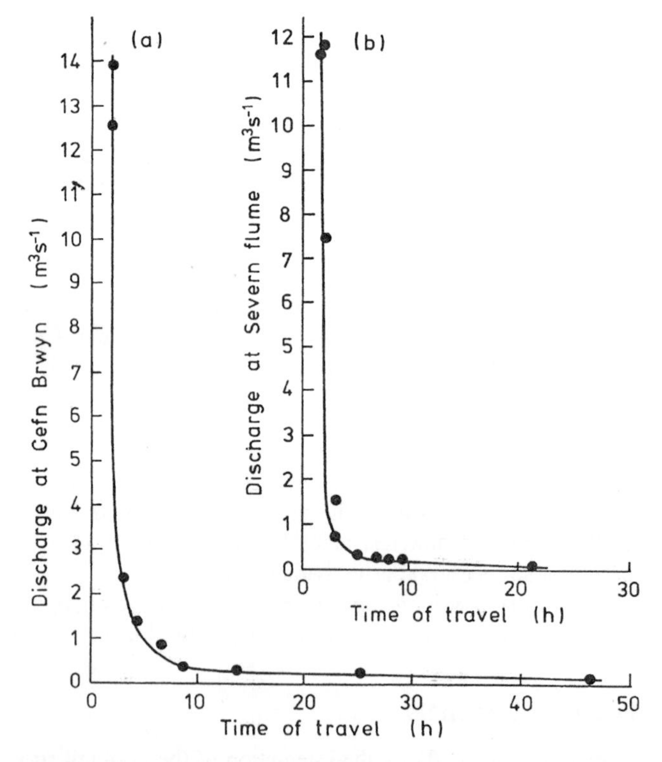

Figure 4.12 Variation of time of travel with discharge in steep pool/fall streams. (a) The upper River Wye; (b) the upper River Severn, near Plynlimon, Wales (from Newson and Harrison, 1978. Reproduced by permission of the Institute of Hydrology)

of bed material roughness, the flow resistance of vegetation-lined channels can be related to the vegetation density and height. In addition, though, the ability of the vegetation to bend in response to the drag force imposed by the flow, reducing its height and decreasing its resistance, must be accounted for. The degree to which a vegetation resists bending depends on its flexural rigidity and density, so that the flow resistance depends on the interaction of the flow and the vegetation flexibility.

Kouwen and others (most recently summarized in Kouwen (1992)) have related the flow resistance to the ratio of depth, d, to deflected vegetation height, k, as:

$$\left[\frac{8}{f}\right]^{1/2} = A \log\left[\frac{d}{k}\right] + B \tag{4.14}$$

where A and B = coefficients dependent on the extent to which the vegetation is bent. Hydraulically there are two regimes: (1) the vegetation is erect, either stationary or displaying a waving motion; and (2) the vegetation is bent prone. Means of determining the regime, the coefficients A and B, and the deflected height k are summarised in Kouwen (1992).

(b) *Channels with vegetated banks*. This is a common case where trees, shrubs and grass growing along the bank can affect the flow resistance, especially at high flows. Masterman and Thorne (1992) have proposed a means of accounting for the effect by dividing the flow cross-section into subareas in which the flow resistance is considered to be affected separately by the bank vegetation and the bed material. In the central subarea (where resistance depends primarily on the bed material) the flow resistance, and thence the subarea discharge, is calculated using an equation such as Eqn 4.6. In the two bank subareas (where resistance depends primarily on the vegetation) the flow resistance, and thence the subarea discharge, is calculated using Eqn 4.14. The total discharge is determined by adding together the subarea discharges. Criteria for defining the subareas are discussed further in Masterman and Thorne (1992) and the effect of bank vegetation on overall flow resistance is shown to be small in flows with width/depth ratios greater than about 15.

The vegetation resistance equations provide a means of examining the effects of changes in vegetation cover, either through seasonal growth or through maintenance programmes, on channel discharge capacity. They also provide a basis for comparing the beneficial effects of bank vegetation (in stabilising banks) with the adverse effects which result from the potential reduction in discharge capacity.

The above method based on Eqn 4.14 does not currently include the effects of trees or other non-flexible vegetation taller than the water depth.

4.5.6 Field Techniques

Application of the flow resistance equations to obtain flow velocity and discharge requires the following information:

1. flow mean depth or hydraulic radius;
2. channel slope;
3. bed material size distribution;
4. vegetation characteristics (if vegetation effects are significant).

The first two parameters can be obtained by standard surveying techniques. The basic requirement is that the channel reach should be approximately straight and uniform and that the flow depth and velocity should be determined by channel controls (boundary resistance and channel cross-sectional shape) and not be affected by downstream section controls such as bars, rock outcrops or hydraulic structures. In a gravel-bed river the resistance equation might therefore be applied at a riffle but not at a pool. Other surveying criteria are discussed in, for example, Barnes (1967), Barnes and Davidian (1978) and Bathurst (1986).

The size distribution for sand-sized sediments is obtained by collecting a bulk sample, sieving it and deriving a size distribution from the weights held within each sieve size range. For coarse sediments, though, the number of particles required to give acceptable accuracy by this method usually involves a sample too big to be manageable (e.g. Church et al., 1987). Instead, a more limited sample of particles is collected on a grid basis from the bed surface and its size distribution is analysed by number rather than weight (Wolman, 1954; ISO, 1992). The D_{84}, or other desired sizes, are taken from the resulting cumulative size distribution.

The sample size should be sufficient for the sample to represent the equivalent population of size parameters with desired accuracy. Hey and Thorne (1983) present the following relationship for calculating the accuracy of a sample. At the 95% confidence level:

$$\delta = \frac{ts}{M^{0.5}} \tag{4.15}$$

where δ = the difference between the sample mean ($\log x$) and the population mean ($\log X$) for the sample reach [$\delta = \pm(\log X - \log x)$]; s = the standard deviation of the sample of bed material composed of M particles, in log units ($s = \log D_{84} - \log D_{50}$); and t = the value of Student's t for $M - 1$ degrees of freedom. (Log units are used since surface gravel samples are generally log normally distributed.) A convenient sample size is $M = 100$ since this automatically provides the percentage of the sample in each size class. For a typical sample standard deviation of $s = 0.3$, and with $t = 1.985$, the accuracy is about 15% at the 95% confidence level ($\delta = \pm \log 1.15$).

Zevenbergen (1984) noted that Eqn 4.15 requires that samples be split into subsets to estimate confidence intervals for particular percentile values. To avoid this he proposed the equation:

$$\delta_n = t\left[\frac{s^2}{M} + \frac{0.5Z^2s^4}{M-1}\right]^{1/2} \tag{4.16}$$

where δ_n = confidence interval for size D_n (as defined for Eqn 4.12); and Z = the number of standard deviations that D_n is removed from the mean (i.e. $Z = 1$ for D_{84}). For D_{50} (i.e. $Z = 0$), Eqn 4.16 reduces to Eqn 4.15. Equation 4.16 is valid only for log-normal bed material size distributions.

4.6 SUMMARY

Flow characteristics and their interaction with channel geometry and planform are central to all problems of river engineering. However, the characteristics vary with channel type (sand-bed, gravel-bed, boulder-bed, steep pool/fall and vegetated). It is important, therefore, to understand the differences between channel types and to ensure that calculation techniques are appropriate to the type under consideration.

The vertical velocity profile characteristically has an approximately semi-logarithmic dependency on distance above the bed for sand- and gravel-bed rivers (Eqn 4.1). In boulder-bed rivers, though, where depths are comparable with the larger bed material sizes, it adopts an S-shaped dependency. For such rivers the use of a semi-logarithmic velocity law can lead to a significant underestimate of the mean flow velocity.

Secondary circulation is of most significance at channel bends, where it affects the flow velocity distribution, bed shear stress pattern and sediment scour and deposition. As a result of changes in circulation strength and pattern between low and high flows, maximum bed scour and outside bank erosion are likely to occur in the downstream part of the bend, or even downstream of the bend, at high discharges. Successful management of channel planform therefore requires a good understanding of the discharge dependency of secondary circulation effects. Very accurate current-meters, preferably able to measure in two or three dimensions simultaneously, are needed to measure secondary flow velocities.

Flow resistance is determined: as a function of bed material and bedform roughness for sand-bed rivers (with a variety of equations in the literature); as a function of bed material roughness and for uniform flow conditions for gravel-bed rivers (e.g. Eqn 4.6); and by empirically derived formulae for boulder-bed rivers (e.g. Eqns 4.11 and 4.13). No equation is currently available for steep pool/fall channels. Vegetation flow resistance is determined as a function of vegetation density, height and flexural rigidity. The basic means of obtaining the data to apply these equations are discussed in Section 4.5.6 but readers should examine the original references for full details before applying the equations in practice. In particular, if equations are applied outside the range of conditions for which they were determined, the results should be interpreted with great care.

Finally, it will be apparent that much of our understanding of flow resistance, and indeed of flow characteristics in general, is incomplete. Research is continually adding to our knowledge and the interested reader should therefore keep an eye on the literature for new methodologies and approaches to problems.

4.7 REFERENCES

Ashmore, P. and Parker, G. 1983. Confluence scour in coarse braided streams. *Water Resources Research*, **19**(2), 392–402.

Ashworth, P.J., Ferguson, R.I. and Powell, M.D. 1992. Bedload transport and sorting in braided channels. In: Billi, P., Hey, R.D., Thorne, C.R. and Tacconi, P. (Eds), *Dynamics of Gravel-bed Rivers*. Wiley, Chichester, 497–513.

Barnes, H.H. 1967. *Roughness Characteristics of Natural Channels*. Water Supply Paper 1849, US Geological Survey, Washington DC, 213pp.

Barnes, H.H. and Davidian, J. 1978. Indirect methods. In: Herschy, R.W. (Ed.), *Hydrometry.* Wiley, Chichester, 149–204.

Bathurst, J.C. 1977. *Resistance to Flow in Rivers with Stony Beds*. PhD thesis, University of East Anglia, Norwich, 402pp.

Bathurst, J.C. 1978. Flow resistance of large-scale roughness. *Proceedings of the American Society of Civil Engineers, Journal of the Hydraulics Division*, **104**(HY12), 1587–1603.

Bathurst, J.C. 1979. Distribution of boundary shear stress in rivers. In: Rhodes, D.D. and Williams, G.P. (Eds), *Adjustments of the Fluvial System*. Kendall/Hunt, Dubuque, Iowa, 95–116.

Bathurst, J.C. 1982. Theoretical aspects of flow resistance. In: Hey, R.D., Bathurst, J.C. and Thorne, C.R. (Eds), *Gravel-bed Rivers*. Wiley, Chichester, 83–105.

Bathurst, J.C. 1985. Flow resistance estimation in mountain rivers. *Proceedings of the American Society of Civil Engineers, Journal of Hydraulic Engineering*, **111**(4), 625–643.

Bathurst, J.C. 1986. Slope-area discharge gaging in mountain rivers. *Proceedings of the American Society of Civil Engineers, Journal of Hydraulic Engineering*, **112**(5), 376–391.

Bathurst, J.C. 1988a. Velocity profile in high-gradient, boulder-bed channels. *Proceedings of the International Association of Hydraulic Research International Conference on Fluvial Hydraulics '88*, Budapest, Hungary, 29–34.

Bathurst, J.C. 1988b. Flow processes and data provision for channel flow models. In: Anderson, M.G. (Ed.), *Modelling Geomorphological Systems*. Wiley, Chichester, 127–152.

Bathurst, J.C. 1993. Flow resistance through the channel network. In: Beven, K. and Kirkby, M.J. (Eds), *Channel Network Hydrology.* Wiley, Chichester, 69–98.

Bathurst, J.C. 1994. At-a-site mountain river flow resistance variation. *Proceedings of the American Society of Civil Engineers 1994 National Conference on Hydraulic Engineering*, Buffalo, New York, Vol. 1, 682–686.

Bathurst, J.C., Thorne, C.R. and Hey, R.D. 1979. Secondary flow and shear stress at river bends. *Proceedings of the American Society of Civil Engineers, Journal of the Hydraulics Division*, **105**(HY10), 1277–1295.

Bathurst, J.C., Thorne, C.R. and Hey, R.D. 1981. Closure to 'Secondary flow and shear stress at river bends'. *Proceedings of the American Society of Civil Engineers, Journal of the Hydraulics Division*, **107**(HY5), 644–647.

Best, J.L. 1986. The morphology of river channel confluences. *Progress in Physical Geography,* **10**(2), 157–174.

Best, J.L. 1987. *Flow Dynamics at River Channel Confluences: Implications for Sediment Transport and Bed Morphology.* Society of Economic Paleontologists and Mineralogists.

Beven, K. and Carling, P.A. 1992. Velocities, roughness and dispersion in the lowland River Severn. In: Carling, P.A. and Petts, G.E. (Eds), *Lowland Floodplain Rivers: Geomorphological Perspectives*. Wiley, Chichester, 71–93.

Beven, K., Gilman, K. and Newson, M. 1979. Flow and flow routing in upland channel networks. *Hydrological Sciences Bulletin*, **24**(3), 303–325.

Bray, D.I. 1979. Estimating average velocity in gravel-bed rivers. *Proceedings of the American Society of Civil Engineers, Journal of the Hydraulics Division*, **105**(HY9), 1103–1122.

Bridge, J.S. and Jarvis, J. 1982. The dynamics of a river bend: a study in flow and sedimentary processes. *Sedimentology,* **29**(4), 499–541.

Church, M.A., McLean, D.G. and Wolcott, J.F. 1987. River bed gravels: sampling and analysis. In: Thorne, C.R., Bathurst, J.C. and Hey, R.D. (Eds), *Sediment Transport in Gravel-bed Rivers*. Wiley, Chichester, 43–79.

Coleman, N.L. 1981. Velocity profiles with suspended sediment. *Journal of Hydraulic Research*, **19**(3), 211–229.

Coleman, N.L. 1986. Effects of suspended sediment on the open-channel velocity distribution. *Water Resources Research*, **22**(10), 1377–1384.

Dietrich, W.E. and Smith, J.D. 1983. Influence of the point bar on flow through curved channels. *Water Resources Research*, **19**(5), 1173–1192.

Dietrich, W.E. and Smith, J.D. 1984. Bed load transport in a river meander. *Water Resources Research*, **20**(10), 1355–1380.

Dietrich, W.E., Smith, J.D. and Dunne, T. 1979. Flow and sediment transport in a sand bedded meander. *Journal of Geology*, **87**, 305–315.

Elliott, S.C.A. and Sellin, R.H.J. 1990. SERC flood channel facility: skewed flow experiments. *Journal of Hydraulic Research*, **28**(2), 197–214.

Engelund, F. 1966. Hydraulic resistance of alluvial streams. *Proceedings of the American Society of Civil Engineers, Journal of the Hydraulics Division*, **92**(HY2), 315–326.

Engelund, F. 1967. Closure to 'Hydraulic resistance of alluvial streams'. *Proceedings of the American Society of Civil Engineers, Journal of the Hydraulics Division*, **93**(HY4), 287–296.

Garde, R.J. and Ranga Raju, K.G. 1985. *Mechanics of Sediment Transportation and Alluvial Stream Problems*, 2nd edn. Wiley Eastern, New Delhi, 618pp.

Ghosh, S.N. and Kar, S.K. 1975. River flood plain interaction and distribution of boundary shear stress in a meander channel with flood plain. *Proceedings of the Institution of Civil Engineers*, Part 2, **59**, 805–811.

Griffiths, G.A. 1981. Flow resistance in coarse gravel bed rivers. *Proceedings of the American Society of Civil Engineers, Journal of the Hydraulics Division*, **107**(HY7), 899–918.

Hayward, J.A. 1980. *Hydrology and Stream Sediments in a Mountain Catchment*. Special Publication 17, Tussock Grasslands and Mountain Lands Institute, Lincoln College, Canterbury, New Zealand, 236pp.

Hey, R.D. 1979. Flow resistance in gravel-bed rivers. *Proceedings of the American Society of Civil Engineers, Journal of the Hydraulics Division*, **105**(HY4), 365–379.

Hey, R.D. 1988. Bar form resistance in gravel-bed rivers. *Proceedings of the American Society of Civil Engineers, Journal of Hydraulic Engineering*, **114**(12), 1498–1508.

Hey, R.D. and Thorne, C.R. 1983. Accuracy of surface samples from gravel bed material. *Proceedings of the American Society of Civil Engineers, Journal of Hydraulic Engineering*, **109**(6), 842–851.

Hooke, R.L. 1975. Distribution of sediment transport and shear stress in a meander bend. *Journal of Geology*, **83**(5), 543–565.

ISO. 1992. *Sampling and Analysis of Gravel Bed Material*. International Standards Organization Document 9195. Also *BS3680, Part 10E, Measurement of Liquid Flow in Open Channels*. BSI, Milton Keynes.

Itakura, T. and Kishi, T. 1980. Open-channel flow with suspended sediments. *Proceedings of the American Society of Civil Engineers, Journal of the Hydraulics Division*, **106**(HY8), 1325–1343.

Jackson, P.S. 1981. On the displacement height in the logarithmic profile. *Journal of Fluid Mechanics*, **111**, 15–25.

Jaeggi, M.N.R. 1984. Formation and effects of alternate bars. *Proceedings of the American Society of Civil Engineers, Journal of Hydraulic Engineering*, **110**(2), 142–156.

Jarrett, R.D. 1984. Hydraulics of high-gradient streams. *Proceedings of the American Society of Civil Engineers, Journal of Hydraulic Engineering*, **110**(11), 1519–1539.

Klaassen, G.J., Ogink, H.J.M. and van Rijn, L.C. 1986. *DHL-Research on Bedforms, Resistance to Flow and Sediment Transport*. Delft Hydraulics Communication No 362, Delft Hydraulics Lab., Emmeloord, The Netherlands, paper presented at 3rd International Symposium on River Sedimentation, Jackson, Mississippi.

Knight, D.W. and Demetriou, J.D. 1983. Flood plain and main channel flow interaction. *Proceedings of the American Society of Civil Engineers, Journal of Hydraulic Engineering*, **109**(8), 1073–1092.

Knight, D.W. and Shiono, K. 1990. Turbulence measurements in a shear layer region of a compound channel. *Journal of Hydraulic Research*, **28**(2), 175–196.

Kouwen, N. 1992. Modern approach to design of grassed channels. *Proceedings of the American Society of Civil Engineers, Journal of Irrigation and Drainage Engineering*, **118**(5), 733–743.

Lau, Y.L. 1983. Suspended sediment effect on flow resistance. *Proceedings of the American Society of Civil Engineers, Journal of Hydraulic Engineering*, **109**(5), 757–763.

Leopold, L.B., Wolman, M.G. and Miller, J.P. 1964. *Fluvial Processes in Geomorphology*. Freeman, San Francisco, 522pp.

Limerinos, J.T. 1970. *Determination of the Manning Coefficient from Measured Bed Roughness in Natural Channels*. Water Supply Paper 1898-B, US Geological Survey, Washington DC, 47pp.

Lisle, T.E. 1986. Effects of woody debris on anadromous salmonid habitat, Prince of Wales Island, Southeast Alaska. *North American Journal of Fisheries Management*, **6**, 538–550.

Marchand, J.P., Jarrett, R.D. and Jones, L.L. 1984. *Velocity Profile, Water-surface Slope, and Bed-material Size for Selected Streams in Colorado*. Open-File Report 84-733, US Geological Survey, Lakewood, Colorado, USA.

Markham, A.J. and Thorne, C.R. 1992. Geomorphology of gravel-bed river bends. In: Billi, P., Hey, R.D., Thorne, C.R. and Tacconi, P. (Eds), *Dynamics of Gravel-bed Rivers*, Wiley, Chichester, 433–450.

Masterman, R. and Thorne, C.R. 1992. Predicting influence of bank vegetation on channel capacity. *Proceedings of the American Society of Civil Engineers, Journal of Hydraulic Engineering*, **118**(7), 1052–1058.

Mosley, M.P. 1976. An experimental study of channel confluences. *Journal of Geology*, **84**, 535–562.

Myers, W.R.C. 1978. Momentum transfer in a compound channel. *Journal of Hydraulic Research*, **16**(2), 139–150.

Myers, W.R.C. and Brennan, E.K. 1990. Flow resistance in compound channels. *Journal of Hydraulic Research*, **28**(2), 141–155.

Newson, M.D. and Harrison, J.G. 1978. *Channel Studies in the Plynlimon Experimental Catchments*. Institute of Hydrology, Report 47, Wallingford, 61pp.

Nouh, M.A. and Townsend, R.D. 1979. Shear-stress distribution in stable channel bends. *Proceedings of the American Society of Civil Engineers, Journal of the Hydraulics Division*, **105**(HY10), 1233–1245.

Parker, G. and Peterson, A.W. 1980. Bar resistance of gravel-bed streams. *Proceedings of the American Society of Civil Engineers, Journal of the Hydraulics Division*, **106**(HY10), 1559–1575.

Pitlick, J.C. and Thorne, C.R. 1987. Sediment supply, movement and storage in an unstable gravel-bed river. In: Thorne, C.R., Bathurst, J.C. and Hey, R.D. (Eds), *Sediment Transport in Gravel-bed Rivers*. Wiley, Chichester, 151–178.

Prestegaard, K.L. 1983. Bar resistance in gravel bed streams at bankfull stage. *Water Resources Research*, **19**(2), 472–476.

Prus-Chacinski, T.M. 1954. Patterns of motion in open-channel bends. *Proceedings of the International Association for Scientific Hydrology Rome General Assembly*. IASH Publication No. 38, 311–318.

Rajaratnam, N. and Ahmadi, R. 1981. Hydraulics of channels with flood plains. *Journal of Hydraulic Research*, **19**(1), 43–60.

Rouse, H. 1965. Critical analysis of open-channel resistance. *Proceedings of the American Society of Civil Engineers, Journal of the Hydraulics Division*, **91**(HY4), 1–25.

Simons, D.B. and Şentürk, F. 1977. *Sediment Transport Technology*. Water Resources Publications, Fort Collins, Colorado, 807pp.

Smith, C.D. 1978. Effect of channel meanders on flood stage in valley. *Proceedings of the American Society of Civil Engineers, Journal of the Hydraulics Division*, **104**(HY1), 49–58.

Tamai, N., Ikeuchi, K., Yamazaki, A. and Mohamed, A.A. 1983. Experimental analysis on the open channel flow in rectangular continuous bends. *Journal of Hydroscience and Hydraulic Engineering*, **1**(2), 17–31.

Thorne, C.R. and Hey, R.D. 1979. Direct measurements of secondary currents at a river inflexion point. *Nature*, **280**(5719), 226–228.

Thorne, C.R. and Zevenbergen, L.W. 1985. Estimating mean velocity in mountain rivers. *Proceedings of the American Society of Civil Engineers, Journal of Hydraulic Engineering*, **111**(4), 612–624.

Thorne, C.R., Markham, A.J. and Oldfield, J. 1989. Secondary current measurements in a meandering gravel-bed river. In: Ranga Raju, K.G. (Ed.), *Proceedings of the Third International Workshop on Alluvial River Problems*, University of Roorkee, India, 219–230.

Toebes, G.H. and Sooky, A.A. 1967. Hydraulics of meandering rivers with flood plains. *Proceedings of the American Society of Civil Engineers, Journal of the Waterways and Harbors Division*, **93**(WW2), 213–236.

Vanoni, V.A. (Ed.) 1975. *Sedimentation Engineering*. American Society of Civil Engineers Manuals and Reports on Engineering Practice, No. 54, New York, 745pp.

Vanoni, V.A. and Nomicos, G.N. 1959. Resistance properties of sediment-laden streams. *Proceedings of the American Society of Civil Engineers, Journal of the Hydraulics Division*, **85**(HY5), 77–107.

White, W.R., Paris, E. and Bettess, R. 1979. *A New General Method for Predicting the Frictional Characteristics of Alluvial Streams*. Report IT 187, Hydraulics Research Station, Wallingford.

Whittaker, J.G. 1987. Sediment transport in step-pool streams. In: Thorne, C.R., Bathurst, J.C. and Hey, R.D. (Eds), *Sediment Transport in Gravel-bed Rivers*. Wiley, Chichester, 545–570.

Whittaker, J.G. and Jaeggi, M.N.R. 1982. Origin of step-pool systems in mountain streams. *Proceedings of the American Society of Civil Engineers, Journal of the Hydraulics Division*, **108**(HY6), 758–773.

Wilkinson, R.H. 1984. A method for evaluating statistical errors associated with logarithmic velocity profiles. *Geo-Marine Letters*, **3**, 49–52.

Willis, J.C. 1983. Flow resistance in large test channel. *Proceedings of the American Society of Civil Engineers, Journal of Hydraulic Engineering*, **109**(12), 1755–1770.

Wolman, M.G. 1954. A method of sampling coarse river-bed material. *Transactions of the American Geophysical Union*, **35**(6), 951–956.

Wormleaton, P.R. and Merrett, D.J. 1990. An improved method of calculation for steady uniform flow in prismatic main channel/flood plain sections. *Journal of Hydraulic Research*, **28**(2), 157–174.

Wormleaton, P.R., Allen, J. and Hadjipanos, P. 1982. Discharge assessment in compound channel flow. *Proceedings of the American Society of Civil Engineers, Journal of the Hydraulics Division*, **108**(HY9), 975–994.

Zevenbergen, L.W. 1984. Discussion of 'Accuracy of surface samples from gravel bed material' by R.D. Hey and C.R. Thorne. *Proceedings of the American Society of Civil Engineers, Journal of Hydraulic Engineering*, **110**(7), 1016–1018.

5 Sediment Erosion, Transport and Deposition

IAN REID[1], JAMES C. BATHURST[2], PAUL A. CARLING[3], DES. E. WALLING[4] AND BRUCE W. WEBB[4]
[1]*Department of Geography, Loughborough University, UK*
[2]*Department of Civil Engineering, University of Newcastle-upon-Tyne, UK*
[3]*Department of Geography, University of Lancaster, UK*
[4]*Department of Geography, University of Exeter, UK*

INTRODUCTION

Many problems of river management arise from the inadequate prediction of sediment behaviour during flood flows. Some of this uncertainty stems from the complex nature of the drainage system and our incomplete understanding of the links between runoff and sediment sources both within and outside the river channel. However, much uncertainty also arises from the limits placed on our ability to observe the processes that lead to sediment movement in real rivers. Conceptualisation is often based on observations made either in laboratory flumes, where conditions are inevitably artificially contrived, or on inferential analyses of post-flood channel bed characteristics and the grain size distribution of mobilised sediment in the field.

5.1 BED SEDIMENT CHARACTER

The entrainment and transport of sediment is governed, in large part, by the composition and arrangement of the particles that make up the channel bed. Bed composition and arrangement (i.e. 'fabric') have been shown to vary systematically in a downstream direction, the coarse sediments of headwaters giving way to progressively finer alluvium as base-level is approached. The downstream-fining 'law' was first defined succinctly in the mid-19th century with observations made on the bed sediments of the Alpine Rhine (Sternberg, 1875). Since then, the importance of fining as a diagnostic tool used by sedimentologists interested in establishing ancient drainage patterns and the provenance of rock constituents, together with the interest of geomorphologists in recognising order in Earth surface processes, has spawned a very extensive literature. However, despite the plethora of worked examples, there is still considerable controversy over the mechanisms that bring about downstream fining. This stems from both the long-term nature of the processes involved and our inability to establish records of suitable length, and our

Applied Fluvial Geomorphology for River Engineering and Management, Edited by C. R. Thorne, R. D. Hey and M. D. Newson.

incomplete knowledge of the ways in which sediment is transferred downstream. There is still uncertainty over the relative importance of size selection both at the time of entrainment and during transport and clast attrition, despite the elegance of studies which have attempted to separate them (Werritty, 1992). The problem is compounded by the fact that the pattern and the rate of downstream fining vary considerably between channel types and in different geomorphological environments (Figure 5.1). Notwithstanding this, the rate of change in bed material calibre has important implications for downstream changes in flow resistance (as outlined in Chapter 4) and sediment transport and, because of this, it cannot be ignored in river management.

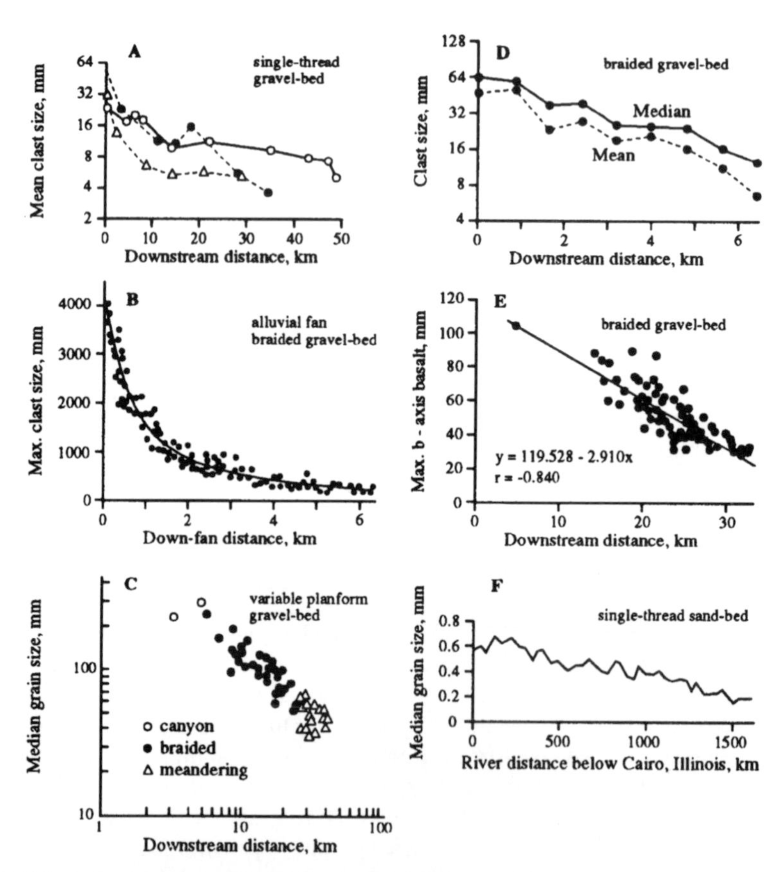

Figure 5.1 Downstream decline in clast size in single-thread and braided gravel- and sand-bed channels (Reid and Frostick, 1994). (A) Three tributaries of the Cheyenne River, Black Hills, South Dakota (after Plumley, 1948); (B) an alluvial fan of the Santa Catalina Mountains, Arizona (after Blissenbach, 1952. With permission of the Society for Sedimentary Geology); (C) the Squamish River, British Columbia (after Brierley and Hickin, 1985); (D) a braided reach of the Upper Kicking Horse River, British Columbia (after Smith, 1974); (E) the Koobi Fora Plateau Gravel, Northern Kenya (after Frostick and Reid, 1980. With permission of the Geological Society); (F) the Mississippi River below Cairo, Illinois (after Mississippi River Commission, 1935)

Of more local importance is the vertical segregation which is characteristic of gravel-bed channels that have perennial flow regimes. The armour layer clast size distribution is generally coarser than its substratum (Figure 5.2) such that the armour ratio (armour median/subarmour median) lies generally in the region of 1.5 to 2 but ranges not untypically up to a value of 6 (Andrews and Parker, 1987; Sutherland, 1987). The importance of armour development has long been recognised in river engineering, although an incomplete understanding of the processes and circumstances actually responsible for disruption of the layer has helped to fuel the debate about the potential mobility of clasts of differing size. It is becoming clear that this disruption allows fuller representation of the bed material size distribution in the bedload (Diplas, 1987; Shih and Komar, 1990). There is also some suggestion in recent work that seasonal or ephemeral streams do not develop strong armour layers because, among other factors,

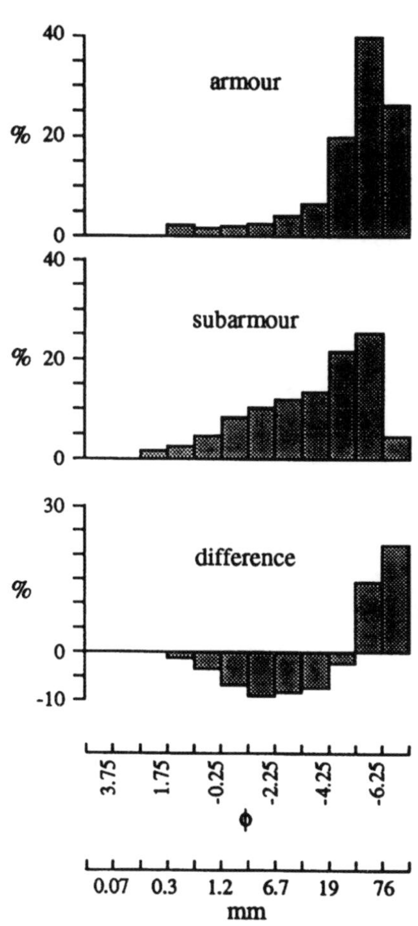

Figure 5.2 Bed material size distributions for Oak Creek, Oregon, and the difference by subtraction between the armour and subarmour layers (after Milhous, 1973; Shih and Komar, 1990)

ɩe finer fractions of the bed are not winnowed by prolonged low flows and sediment ɩupply from contributing hillslopes is not limited (Buffington et al., 1992; Laronne et al., 1992; Laronne and Reid, 1993). This has substantial implications for sediment transport in rivers that drain arid and semi-arid areas.

Another characteristic of the channel bed which has a bearing on flow resistance and sediment transport is the presence of microforms. In gravel-bed rivers, the most prevalent microform is the pebble cluster (Reid et al., 1992; Figure 5.3). These are oriented along the local streamline, range in length from decimetre to metre scale, and are known to delay both entrainment and transport of constituent clasts, and to reduce bedload transport rates by increasing flow resistance. Other repeating microforms that have been recognised are transverse ribs (Koster, 1978), clast dams (Bluck, 1987) and steps (Whittaker and Jaeggi, 1982). In sand-bed streams, the chief microforms are ripples and dunes. Their formation is dependent on both flow strength and grain size (Figure 5.4). Their impact on the flow is to increase form roughness (expressed as Manning's coefficient) to levels 1.75 times those of plane-bed conditions (Simons and Richardson, 1962).

Sand-beds usually consist of a single structural component – the framework. In contrast, more often than not, alluvial gravel-beds are composed of both a framework of coarse clasts and a matrix of finer material that fills, or partially fills, the interstices. These interstitial fines are not necessarily deposited at the same time as the framework. Indeed, there has been an increasing realisation that the matrix is an important source of, and sink for, both suspended and finer bedload sediment (Carling and Reader, 1982; Frostick et al., 1984; Lisle, 1989; Church et al., 1991), and that it has a potentially

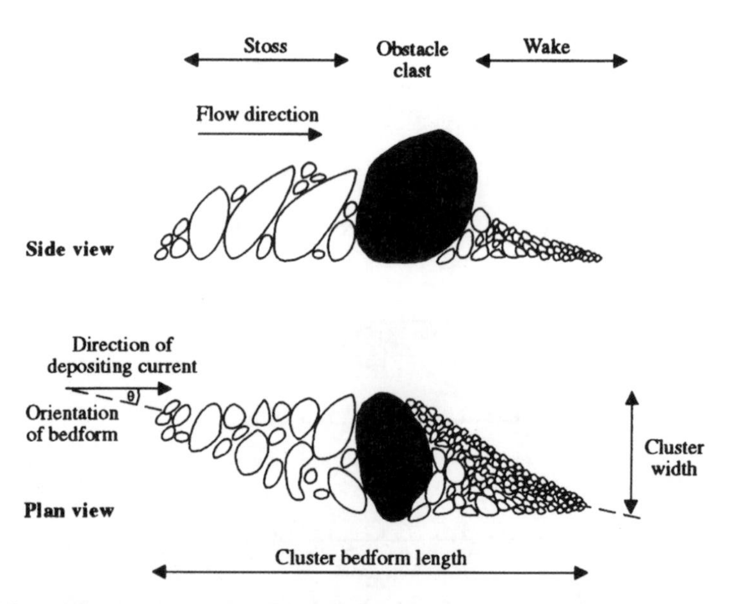

Figure 5.3 Archtype pebble cluster showing component parts, not all of which need be represented in all family members of this gravel-bed microform (after Reid et al., 1992)

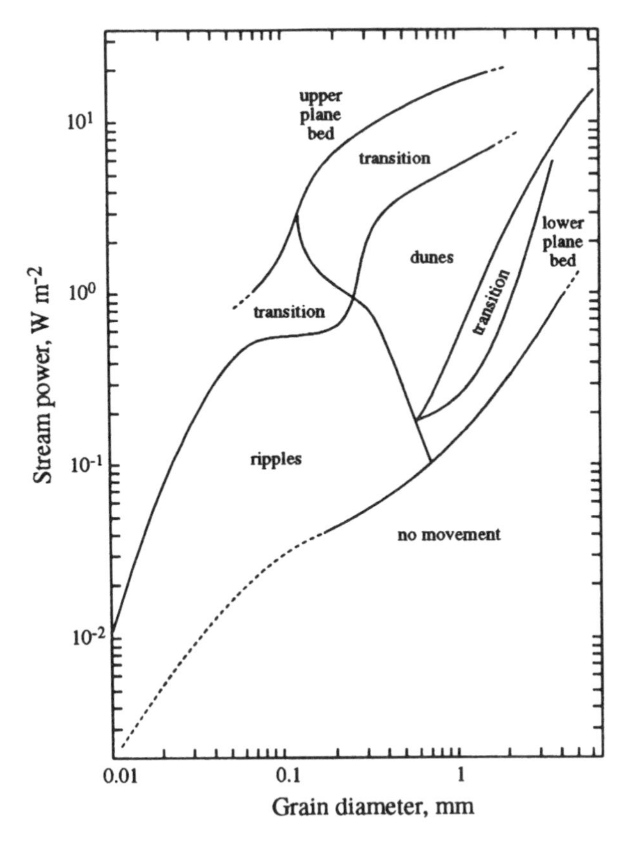

Figure 5.4 Stability fields of small-scale bedforms on sand-beds (after Allen, 1982). With kind permission of Elsevier Science

important role in strengthening the bed, so delaying the entrainment of the coarser framework at times of high shear (Reid et al., 1985).

The fact that the channel bed is structured and spatially inhomogeneous means that a careful sampling strategy must be adopted if size distributions are to be adequately representative. In the first instance, only single depositional units should be sampled. These may be easier to identify in sands than in gravels. Secondly, it is essential to recognise the need to sample the armour and subarmour layers separately in gravel-beds. However, there will be an equal need to take account of macrobedforms (bars, pools etc.) in deciding the number of sample sites if a reach is to be characterised thoroughly. Volumetric sampling gives the least biased estimates (Church et al., 1987; Diplas and Fripp, 1992), but there are practical difficulties in identifying a volumetric method for the surface layer. As a result, the surface should be sampled either by area and a correction factor applied to the size distribution (Kellerhals and Bray, 1971), or by grid (Wolman, 1954).

5.2 ENTRAINMENT THRESHOLDS

5.2.1 Concept

The initiation of sediment particle movement by flowing water is an important component of the sediment transport process. Many bedload transport equations incorporate a term that defines critical flow conditions (or entrainment thresholds) at which particles are just able to move, and reliable application of the equations therefore depends on appropriate specification of these conditions. Prediction of flow conditions at which bed material movement may be expected also has many practical applications, for example the effect of reservoir releases on downstream channel stability and the occurrence of scour around engineering structures. The conditions under which particle movement ceases are equally important (and are not necessarily the same as the conditions for initiation of motion). For example, the nature of sedimentary deposits and the conditions in which they formed are an essential ingredient of sedimentological studies.

The concept of a critical flow condition at which the bed material is set in motion, and below which there is not motion, involves a degree of approximation. This is particularly so for gravel sediments with non-uniform size distributions. As long ago as 1971, Paintal (1971) showed that for such sediments there is no single value of bed shear stress below which not a single particle will move and above which all the particles of the same size will move. The exposure of individual particles to the flow varies, the forces acting on a particle fluctuate about some average value as a result of turbulence, and the effects of bed material consolidation and particle cluster formation further complicate the picture. Consequently, the movement of particles is both unsteady and non-uniformly distributed over the bed. Nevertheless, Paintal showed that, for all practical purposes, a limiting shear stress for the bed sediment can be defined below which the bedload transport rate is insignificant. Some of the factors affecting this threshold are reviewed here. Formulae for determining the threshold are reviewed in Section 5.2.3.

5.2.2 Non-uniform Bed Material Size Distribution and Equal Mobility

For a level bed composed of uniformly sized particles, or of material with a narrow size distribution such as sand, all the particles may be expected to begin movement at approximately the same flow condition. For a bed material of non-uniform size distribution such as a sand/gravel or gravel/boulder mixture, though, the various particle sizes may be brought into motion over a range of discharges. If the individual size fractions had no influence on each other, the force required to initiate movement of a given size would be equal to that required to move the same size in a bed composed of uniform material of that size. Several studies have shown, however, that the stability of a particle is affected by the position it has within the overall size distribution relative to a reference size (Egiazaroff, 1965; Andrews, 1983; Bathurst, 1987a; Wiberg and Smith, 1987). Particles smaller than the reference size tend to be sheltered behind larger particles and require a stronger flow to set them in motion than would be necessary for uniform materials of the same size. Conversely, particles larger than the reference size tend to project into the flow and can be moved by flows weaker than would be necessary for uniform materials of the same size. Particles of the reference size are unaffected by the hiding/exposure effect and behave as if in a bed of uniform material. Emp reference size has been found to be of the order of D_{50}, the median distribution (e.g. Çeçen and Bayazit, 1973; Proffitt and Sutherland, 198 1987a).

The general pattern is illustrated in Figure 5.5 where the variation of critical discharge with particle size, determined as if there were no hiding/exposure effect (using a flume-based formula), is compared with the variation observed in a gravel/boulder-bed river with a hiding/exposure effect. In the latter case, smaller particles are relatively difficult to move and larger particles are relatively easy to move compared with the case represented by the flume-based formula. Overall, the range of discharges over which all particle sizes are brought into motion is decreased relative to the case without a hiding/exposure effect. In the extreme case, the hiding/exposure effect (acting to reduce the difference in critical conditions for motion between particles of different sizes) could exactly compensate for the effect of particle size and weight (acting to differentiate the critical conditions between particles). All sizes are then brought into motion at the same flow condition, the so-called condition of equal mobility. It should be noted, however, that the issue of equal, near-equal or unequal mobility continues to be the subject of considerable debate (Section 5.6.3).

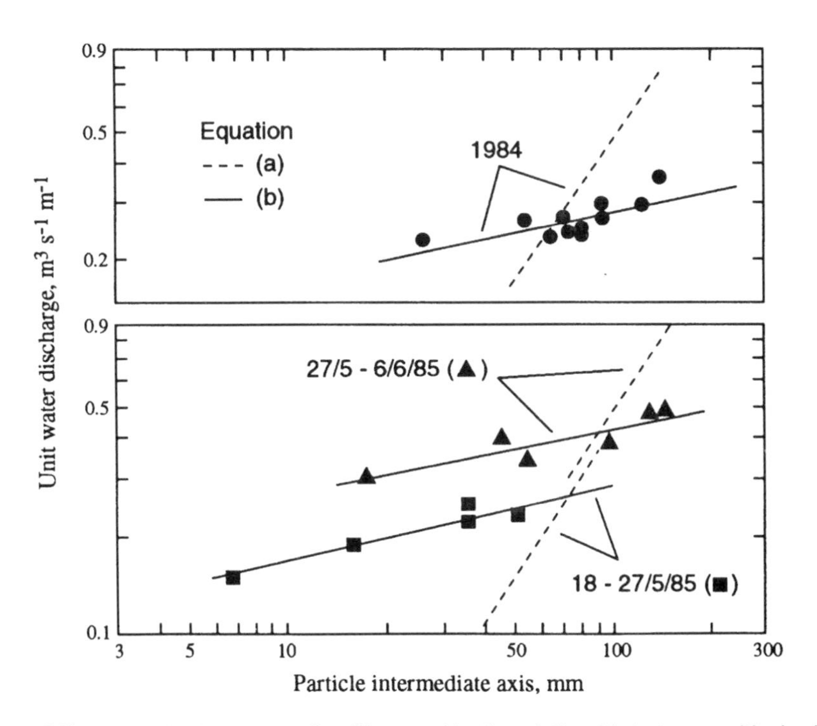

Figure 5.5 The hiding/exposure effect illustrated by the relationship between critical unit water discharge and size of particle intermediate axis for a site on the Roaring River, Colorado, USA. Equation (b) is fitted to the data and compared with Eqn (a) derived for uniform materials (Eqn 5.4). Symbols indicate period of validity (from Bathurst, 1987a)

5.2.3 Thresholds for Initiation and Cessation of Motion

As a result of the variety of particle sizes and shapes, gravel-bed material is often packed into an interlocking structure in which particles are tightly held by their neighbours. Similarly, particle clusters may form on the bed surface conferring, through mutual protection, a stability in excess of that which would be possessed by the constituent particles in isolation (as described in Section 5.1). Because of these bed structures, it is more difficult to dislodge individual particles than would otherwise be the case and several field studies have shown that the flow required to initiate bedload transport is consequently higher than would be expected with a uniform size distribution. This is especially noticeable after a prolonged period with no sediment transport, when the bed material has time to become consolidated. Infiltration of fine, particularly cohesive, sediments into the framework of coarser sizes can create a powerful cementation effect, for example. Field measurements by Reid and Frostick (1984) and Reid et al. (1985) show that, during the rising stage of the first flood after a long period of inactivity, relatively little transport may occur. Eventually, though, the bed structure is broken and, on the falling stage of the flood, transport occurs to a much greater degree (Figure 5.6). Measurements show that it may continue to a value of the Shields shear stress typically one-third of the value characterising initiation of motion. In other words, the conditions for initiation and cessation of motion can be substantially different (Figure 5.6).

On the other hand, when floods follow each other closely, the bed material remains comparatively loose and offers less resistance to entrainment; substantial amounts of bedload can then be transported on the rising limb of the hydrograph (Reid et al., 1985). The critical transport conditions are, therefore, to some extent dependent on previous high flow history.

5.3 BEDLOAD TRANSPORT

Many bedload transport equations are available for generating an estimate of sediment transport rates. However, although they appear to arise from several different deterministic

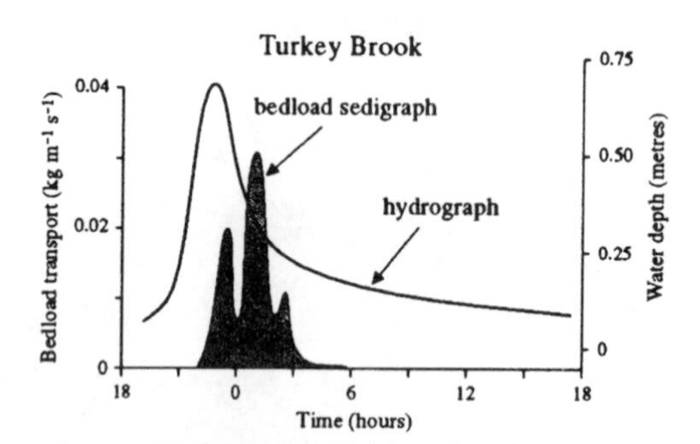

Figure 5.6 Example of flood hydrograph and bedload measured continuously on Turkey Brook, a small gravel-bed river in southern England. Bedload pulses can be seen as can the poor phase between the water and sediment transport waves that arises, in part, from protection that the armour layer with its interstitial fines gives to the bed on the rising limb of the flood (after Reid et al., 1985)

approaches, the fact that the problem they purport to solve is reasonably w inevitably means that there is considerable genetic similarity. This often me choice of one or other equation may be dictated by the availability within a databa requisite input parameters, or even by personal preference founded on ease of computa tion, rather than by some intrinsic hydraulic advantage. Some equations are uncalibrated or are calibrated only with flume data. As a consequence, there may be dangers in applying them in field situations. Indeed, even where transport equations have been calibrated using field data, experience suggests that reasonable results may only be obtained if they are applied to streams having similar character, as their predictive capability is known to be poor when they are used in conditions that differ from those of the test-bed. In fact, there is presently no universally applicable solution to the problem of predicting bedload sediment flux despite the apparent elegance of those functions that find a regular place in textbooks on hydraulics (Reid et al., 1996).

Normally, the determination of bulk bedload transport rate at a given river stage requires detailed information about cross-stream hydraulic parameters such as near-bed velocity or shear stress. This in itself raises logistical and, therefore, financial difficulties and it is one of the prime reasons why the computation of bedload transport rates is comparatively rare in river hydraulics. Even where an assessment is made, a significant problem lies in the fact that most transport equations assume that a capacity load is being carried, that is, the stream is transporting as much sediment as its competence will allow. For a variety of reasons, the actual flux may be much less than theoretical calculations indicate (Carling, 1992; Reid and Frostick, 1994).

The first reason for this is that, within a given reach, the supply of sediment may not be from the bed (invariably assumed as the source in all transport equations) but from tributaries, from bank erosion, or from mass-wasting on valley sideslopes. An extreme example, not particularly uncommon in tectonically active uplands, would be where bedrock channels are fed by occasional debris flows that originate either in headwaters or in chutes on adjacent hillslopes. For much of the time there would be a dearth of bedload whilst at other times transporting capacity might be equalled or even exceeded. But substantial changes in sediment supply are not confined to such dramatic river environments. Indeed, studies have shown that sediment availability in alluvial channels changes over a number of timescales. These can range from years to decades as large sediment waveforms pass downstream (Meade, 1985; Roberts and Church, 1986), to interflood and seasonal as the river bed becomes compacted or cemented during extended periods of lowflow or loosened during periods of floodflow (Reid et al., 1985; Carling and Hurley, 1987).

Even where the river bed is actually the immediate source of bedload, vertical and spatial sorting of the bed material may result in variable transport rates. This is especially important in gravel-bed rivers where a coarse surface layer – the armour – protects a finer substratum, preventing or delaying its entrainment. When and where the flow is unable to move the armour, bedload transport rates are likely to be small, to be under-capacity, and to consist of comparatively fine sediment moving over or between clasts that form the armour. This is an overpassing or 'throughput' load, and might be dubbed 'Phase I transport' (Bathurst, 1987b). The origin of this material is often either the winnowing of fines from the armour layer, or another source beyond the channel bed. Once the armour is disrupted, the prevailing force is usually greater than the threshold value required to move the finer subarmour sediment. Consequently, transport rates are often high once this happens, the grain size distribution of the bedload is much wider, and transport rates may approach capacity values. This is 'Phase II transport'. Figure 5.7 shows an example of the control exerted by a coarse surface layer on the maximum grain size of bedload. Two functions indicate considerable separation of the two data subsets. The lower one (E_1) represents Phase I transport, while the higher (E_2) indicates the relationship between bedload grain size and stream power during Phase II. It is evident that, in this stream, the threshold for disruption of the surface armour lies slightly in excess of $100\,\mathrm{W\,m^{-2}}$. It follows that bedload transport rates might be small on the rising limb of a flood hydrograph because the armour is intact, but higher on the falling limb should the armour have been disrupted perhaps at peak flow (Figure 5.6; Reid et al., 1985). Consequently, bedload transport rates might be expected to range widely at the same river stage, even during the same flood event (Figure 5.8), and it gives an explanation for both the wide scatter in the relationship between hydraulic parameters and bedload transport, and the poor prediction of sediment load by all transport functions since, inevitably, no allowance is made for such hysteresis.

In addition to the control of bedload by vertical segregation, it is now well established that transport rates vary at scales ranging from seconds through minutes to hours, even when flow appears to be steady and uniform (Gomez et al., 1989). This variation is often loosely termed 'pulsing'. It may be related in part to the periodic behaviour of turbulent flow – the sporadic and spasmodic impact of eddies on the bed (i.e. the burst-sweep phenomenon) – that makes sediment entrainment a discontinuous process. However, several other processes which operate at larger scales have been identified as also responsible. The destabilisation and break-up of small-scale bedforms, chiefly pebble clusters in gravel-bed channels, generates short periods of intense bed material motion

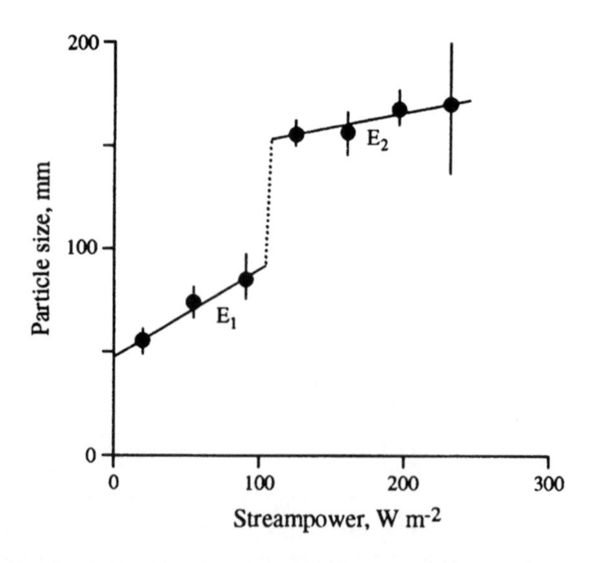

Figure 5.7 Example of how an armour layer controls the size of bedload. Curve E_1 represents fine bedload overpassing a static cobble bed. Curve E_2 represents the increase in bedload grain size subsequent to armour mobilisation as stream power increases beyond a threshold of *c.* $100\,\mathrm{W\,m^{-2}}$

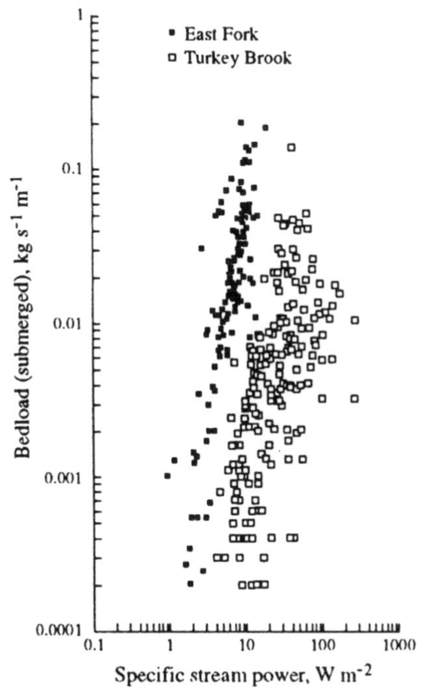

Figure 5.8 Bedload discharge as a function of stream power in the East Fork River, Wyoming (where the load is predominantly sand with a median diameter of 1.5 mm) and Turkey Brook, England (where the load is gravel with a median diameter of 11 mm). Note the degree of scatter – greater in the case of the gravel-bed – and the separation of the two envelopes, indicating greater efficiency on the gravely-sand bed of East Fork (after Leopold and Emmett, 1976; Reid and Frostick, 1986)

(Naden and Brayshaw, 1987). Longitudinal grain sorting processes operating within the mobile layer itself have been shown to produce discrete bedload sheets, the axes of which lie cross-channel (Iseya and Ikeda, 1987; Dietrich et al., 1989). The passage of each sheet is marked first by high transport rates and comparatively coarse sediment which are then followed by lower transport rates and finer material.

However, the interpretation of bedload pulses needs care (Hoey, 1992). For example, variation in bed roughness and topography means that bedload may move in discrete longitudinal ribbons which snake back and forth across the channel bed. This is especially possible where sand overpasses a static gravel bed (Leopold and Emmett, 1976; Ferguson et al., 1989). In this case, the sampling strategy might erroneously suggest unsteady transport when, in reality, a more or less continuous ribbon is traversing fixed sampling locations.

Bedload sheets may be the low-amplitude mobile bedforms typical of gravel-bed rivers, although the number of observations is as yet too low for confirmation, and a physical rationale relating them to the flow structure is still awaited. However, in sand-bed streams, equivalent bedforms have been described exhaustively and their relationship to the flow is well established. The passage of ripples and dunes produces a rhythmic fluctuation in bedload transport rate. In arriving at an estimate of sediment transport, acknowledgement has to be given to the fact that these small-scale bedforms are conservative of their volume, migrating as a mass. Sampling during the passage of a trough will underestimate bedload, while sampling only the avalanche faces will lead to overestimates.

In fact, where flow conditions dictate either ripple or dune bedforms, transport rates are often derived from a consideration of longitudinal bedform geometry and celerity. Because the different bedforms (ripple, dune, plane bed) exist under fairly well defined conditions of flow, and because flow conditions vary both spatially across and along the channel and temporally as a flood wave waxes and wanes, it follows that transport rates will vary in a complex fashion as one microform is replaced by another. As with gravel-beds, but for different reasons, there is no reason to expect a simple monotonic relationship between a hydraulic parameter such as stream power and sediment discharge. This can be illustrated by Figure 5.9 which has been drawn using data collected by Peters (1971) in the dune fields of the Zaire River. In general, higher transport rates are associated with the energetic flow conditions in which the dunes have been washed out and replaced by an upper-stage plane bed. There is, however, an overlap of the envelopes representing the two bed states because of form-lag. Dunes survive at a higher level of stream power during an upward transition phase than that at which they re-establish themselves during a downward phase. In the Zaire, this transition lies between 4 and 7 W m^{-2}, and the difference in transport rate between the bed states lies around 0.001 kg s^{-1} m^{-1}.

In these circumstances, sampling strategies have to be given special consideration if bedload transport is to be measured with any degree of accuracy. For example, after

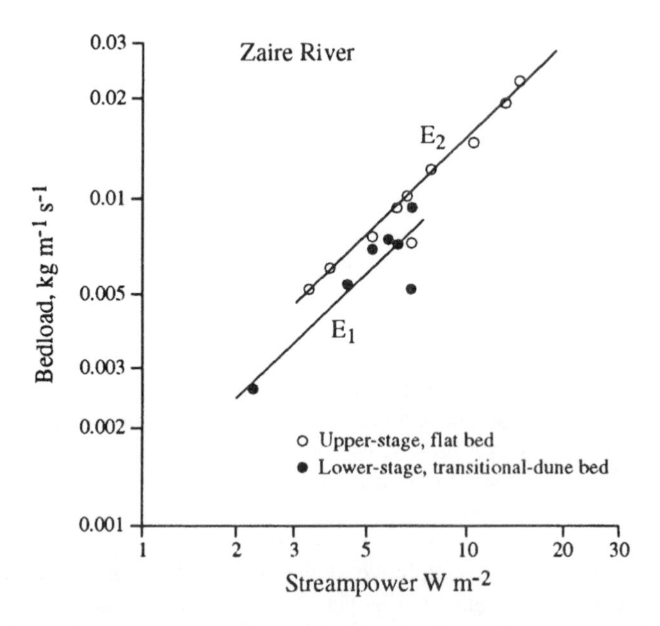

Figure 5.9 Example of the variability of bedload transport rates owing to lag effects in the adjustment of a sand-bed ($D_{50} = 0.6$ mm) from lower phase dunes to upper phase plane-bed conditions. Curve E_1 represents a predominantly dune-covered bed, whilst E_2 represents plane-bed conditions. Data from the Zaire River (Peters, 1971)

obtaining data for dune fields on the River Nile, Gaweesh and van Rijn (1992) considered that as many as 20 samples would be required to estimate transport rate to within $\pm 22\%$ of the true mean, while 50 would be needed to estimate to within $\pm 12\%$. This indicates, as much as anything else, the formidable practical problems of establishing robust estimates of bedload transport in natural rivers.

5.4 SUSPENDED SEDIMENT TRANSPORT

The suspended sediment load of a river is conventionally subdivided into two components, namely, the suspended bed material load and the wash load. This subdivision will to some degree be reflected in the grain size of the sediment, since the suspended bed material will commonly consist of sand-sized (>0.063 mm) particles picked up from the channel bed, whereas the wash load will generally be dominated by finer (<0.063 mm) material 'washed' into the river from the surrounding drainage basin. More importantly, perhaps, these two components of the suspended load have also been distinguished in terms of their hydraulic behaviour. Thus, whereas the suspended bed material load can, at least in theory, be treated as a capacity load, the wash load will be supply-dependent and must be viewed as a non-capacity load. In reality, the suspended bed material load will, like bedload, also evidence non-capacity behaviour (Section 5.3), but it is, nevertheless, more amenable to estimation using theoretical procedures. The supply of wash load to a channel will, in contrast, be influenced by a range of spatially and time-variant processes operating within the upstream drainage basin and is therefore very difficult to estimate theoretically. In most rivers, the wash load component will dominate the suspended sediment load and in the following discussion emphasis will be placed on its non-capacity behaviour and the importance of supply control, and on empirical approaches to characterising suspended sediment transport.

Figure 5.10A emphasises the importance of supply control on suspended sediment transport, by presenting an example of a plot of suspended sediment concentration versus water discharge at the time of sampling, based on data from the River Creedy in Devon, UK. In this example, the values of suspended sediment concentration sampled at specific levels of discharge range over more than two orders of magnitude in response to variations in the supply of fine-grained sediment to the river. There is clearly no simple, well defined relationship between suspended sediment concentration and discharge in this river. The positive trend exhibited by Figure 5.10A should not be seen as evidence of a transport function, but rather as indicating that periods characterised by high discharges are also periods of increased sediment supply. Furthermore, the non-capacity nature of suspended sediment transport can be further stressed by noting that this river would be capable of transporting concentrations at least an order of magnitude greater than those recorded. Under conditions of hyperconcentrated flow, such as those documented in China by Zhou et al. (1983) or discussed by Bradley and McCutcheon (1987), concentrations well in excess of 200 000 mg l^{-1} can exist.

The scatter evident in the plot of suspended sediment concentration versus water discharge presented in Figure 5.10A reflects the large number of factors, other than the instantaneous water discharge, which control the supply of suspended sediment to a river (Guy, 1964; Walling, 1974). These include factors such as rainfall intensity, antecedent

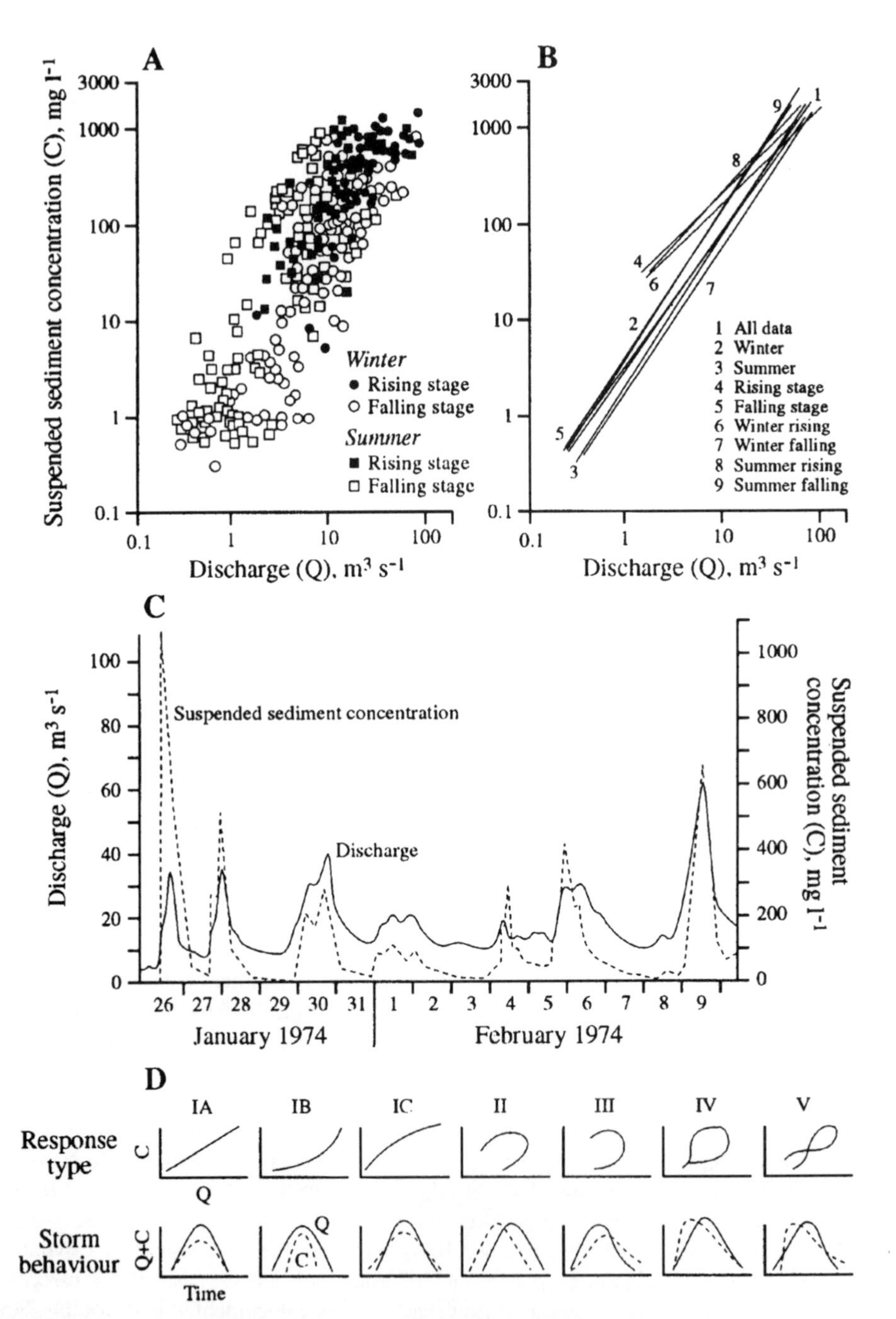

Figure 5.10 Characteristics of suspended sediment transport by rivers: (A) presents a plot of suspended sediment concentration versus water discharge for the River Creedy, Devon, UK, and (B) demonstrates how a series of rating relationships representing different seasons and rising and falling stage conditions can be fitted to the scatter plot; (C) illustrates a typical record of the variation of suspended sediment concentration during a sequence of storm runoff events for the same river; (D) illustrates the five characteristic response types associated with suspended sediment/discharge relationships proposed by Williams (1989)

moisture conditions, hydrograph shape and temporal variations in surface condition and vegetation cover, which will influence erosion processes operating on the catchment slopes, and antecedent temperature, moisture and discharge conditions, which will influence sediment supply by gully erosion and erosion of channel banks. Temporal variability in the storage and remobilisation of fine sediment within the channel (Bogen, 1980; Duijsings, 1986) will further complicate the relationship. Figure 5.10C illustrates a typical pattern of variation of suspended sediment concentrations during a series of storm events recorded on the River Creedy in Devon, UK, which demonstrates the influence of some of the these factors. In this record, there is no clear relation between the magnitude of the maximum suspended sediment concentration and water discharge associated with individual hydrographs; rather, there is evidence of an 'exhaustion' effect, wherein the highest concentrations are recorded in the first event and peak concentrations associated with the succeeding three events show a progressive decline. This pattern reflects both an exhaustion of the sediment supply and the progressive increase in baseflow discharges through the eight day period which result in an increased dilution of suspended sediment inputs delivered to the channel by surface runoff (Walling and Webb, 1982). There is apparently some recovery of sediment availability by the event of 4 February, but there is less evidence of exhaustion in the two succeeding hydrographs. Variations in the relative timing of the sediment concentration and water discharge peaks between individual hydrographs, and contrasts in the sediment concentration associated with a specific level of discharge between the rising and falling stages of the hydrograph, evident in Figure 5.10B, will also be reflected in the scatter apparent in Figure 5.10A. In the longer term, seasonal variations in catchment surface condition and vegetation cover, as well as in precipitation type and intensity, will introduce additional scatter into the concentration versus discharge plot.

The considerable scatter that is generally evidenced by plots of suspended sediment concentration versus discharge effectively precludes the establishment of a simple rating curve or functional relation between these two variables for most rivers. Many workers have, however, attempted to overcome this problem by subdividing the data set and deriving a series of relation or rating curves which take account of some of the major causes of scatter. Seasonal rating curves and relations which distinguish between rising and falling stages have frequently been produced (Figure 5.10B), but these still possess many limitations, particularly in terms of the assumption that fluctuations of concentration and discharge will occur in phase. In small drainage basins, the peak suspended sediment concentration will often precede the hydrograph peak, and it is well known that in larger basins the sediment concentration peak may progressively lag the hydrograph peak, as the flood wave travels downstream (Heidel, 1956). In reality, therefore, the relationship between suspended sediment concentration and discharge during individual flood events will frequently exhibit hysteresis, reflecting the relative timing and form of the water and sediment responses. Williams (1989) has identified five characteristic forms for such hysteretic loops, which in turn reflect the phasing and relative shapes of the responses, and these are portrayed schematically in Figure 5.10D.

Because suspended sediment transport is closely controlled by the supply of material to the channel from the surrounding drainage basin, it is also necessary to look beyond the river channel into the drainage basin in order to understand longer-term patterns of suspended sediment yield. The annual suspended sediment load of a river may be sensitive to critical thresholds governing that supply. For example, if a high-magnitude, low-frequency storm event occurs and triggers intensive sheet and rill erosion or mass movements within a drainage basin, the suspended sediment load associated with that period may be very substantially greater than that for periods without such events. Meade and Parker (1989) provide valuable examples of the potential significance of extreme events associated with hurricanes and very high-magnitude storms from the sediment records for the Juniata and Delaware Rivers in Pennsylvania and New Jersey, and the Eel River in California, USA. Their findings are summarised in Figure 5.11. In the case of the Delaware and Juniata Rivers, the suspended sediment loads transported by the floods generated by Hurricanes Agnes and Connie were equivalent to the total loads transported during three years of more average flows. The example from the Eel River is even more extreme, since in this case the suspended sediment load transported by a flood which occurred in December 1964, and which lasted about 10 days, was equivalent to the total load transported in 10 normal years.

For the same reasons, the suspended sediment loads of rivers will commonly be highly sensitive to changes in land use within a drainage basin, since, irrespective of whether or not the runoff regime is modified, these changes can influence sediment supply to the river. Table 5.1 provides several examples of the magnitude of the impact of land-use change on sediment yields based on experimental studies of small drainage basins. These examples relate to increases in sediment load, but reductions could occur where soil conservation measures reduce sediment supply.

Although most existing research on suspended sediment transport by rivers has focused on the total amounts of sediment moved and the concentrations involved, the recent growth of interest in the wider environmental significance of fine sediment, for example in the transport of nutrients and contaminants, has directed increasing attention to the physical

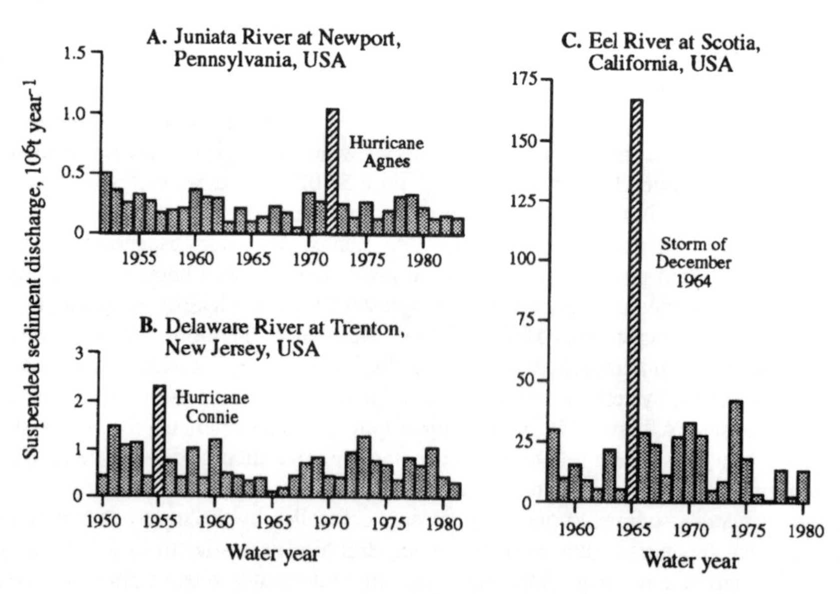

Figure 5.11 The importance of extreme events to the long-term suspended sediment yield of river basins. Based on records from three US river basins presented by Meade and Parker (1989)

Table 5.1 The impact of land-use change on suspended sediment yields, based on evidence from experimental catchment studies

Region	Land-use change	Increase in sediment yield	Study
Westland, New Zealand	Clearfelling	× 8	O'Loughlin et al. (1980)
Oregon, USA	Clearfelling	× 39	Frederiksen (1970)
Texas, USA	Forest clearance and cultivation	× 310	Chang et al. (1982)
Maryland, USA	Building construction	× 126–375	Wolman and Schick (1987)

and geochemical properties of the sediment involved (Allan, 1986; Walling, 1989; Krishnappan and Ongley, 1989; Horowitz, 1991). Information on grain size composition, organic matter content, mineralogy, chemical composition and particle surface chemistry is increasingly required. Detailed discussion of sediment geochemistry and particle characteristics lies outside the scope of this Chapter, but attention can usefully be directed to grain size composition and some geochemical properties.

By definition, the suspended sediment load of a stream will comprise fine-grained material capable of being transported in suspension. The grain size distribution will thus in part reflect hydraulic conditions within the channel, but in most instances the major control will again be exerted by the nature of the material supplied to the river. This in turn will reflect both the textural composition of the source material and the selectivity of the sediment mobilisation and transport processes. It is well known that eroded sediment is generally enriched in the finer fractions and depleted in the coarser fractions, relative to the source material, and Walling and Kane (1984) have shown how the proportion of clay (<0.002 mm) contained in the suspended sediment transported by two Devon rivers was typically c. 1.4–2.4 times greater than in the catchment soils. Figure 5.12A, based on the work of Walling and Moorehead (1989), presents typical particle size distributions for a selection of world rivers and emphasises the very considerable variability that exists. In most cases, this variability is primarily a reflection of catchment lithology.

The grain size composition of suspended sediment will vary through time in response to fluctuations in discharge and other environmental variables, and Figure 5.12B emphasises the potential complexity of this behaviour by presenting situations where the proportions of clay and sand increase, decrease and remain essentially constant with increasing discharge. The tendency for suspended sediment to become coarser as flow increases represents the traditional 'hydraulic' view of grain size behaviour (Horowitz, 1991), which can be accounted for in terms of increasing transport capacity and shear stress as flow increases. The reverse case, with the proportion of clay increasing as flow increases, can be explained in terms of an increased supply of fine sediment eroded from the slopes of the catchment during high flows, and therefore the interaction of runoff-contributing areas and sediment sources.

The grain size data discussed above represent the products of traditional laboratory analyses, which involve removal of the organic fraction and chemical dispersion. Such data are appropriate for considering relationships between the grain size composition of sediment and source materials, but they may be inappropriate for analysing the hydraulic

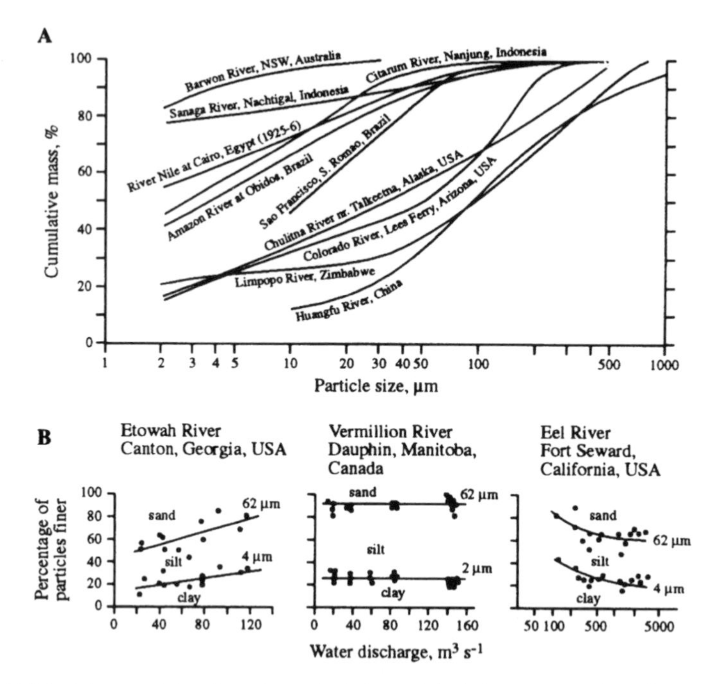

Figure 5.12 The particle size characteristics of fluvial suspended sediment: (A) presents examples of characteristic grain size distributions of suspended sediment for a selection of world rivers and (B) illustrates contrasting examples of the response of the particle size composition of suspended sediment to changing discharge (based on Walling and Moorehead, 1989)

behaviour of sediment particles during transport. There is a growing body of evidence which indicates that much of the suspended sediment transported by rivers exists as composite particles (i.e. flocs or aggregates) and that the *in situ* or effective grain size distribution may differ significantly from the ultimate size distribution, representing the individual discrete particles (Droppo and Ongley, 1989, 1992; Walling and Moorehead, 1989; Walling and Woodward, 1993; Woodward and Walling, 1992). Thus, for example, Walling and Moorehead (1989) have reported measurements undertaken on suspended sediment from rivers in Devon, UK, which indicate an order of magnitude difference between the median grain size of the effective and ultimate grain size distributions, and Walling and Woodward (1993) report results which suggest that more than 70% of the total suspended sediment load of the River Exe at Thorverton may exist as composite particles. Because of the non-linear relationship between particle size and fall velocity, an order of magnitude difference between the median particle size of the two distributions could result in a difference of two orders of magnitude in the equivalent fall velocities. More work is undoubtedly required to elucidate the mechanisms responsible for particle aggregation, and there is a need to develop reliable methods for *in situ* measurement of the grain size distribution of suspended sediment containing composite particles if their role and behaviour are to be more clearly understood. In the meantime, however, the implications of the existence of composite particles should be more clearly recognised.

In view of the close relationship between the geochemistry and grain size of suspended sediment particles (Horowitz, 1991), temporal variations in the grain size distribution of suspended sediment, such as those noted above, will also be reflected in variations in sediment geochemistry. Seasonal and shorter-term variations in sediment sources could also be expected to contribute additional variability in sediment geochemistry. Ongley et al. (1981), for example, report seasonal variations in the geochemistry of the suspended sediment transported by Wilton Creek, Ontario, Canada, which in part reflect the different sediment sources contributing during springmelt and summer storm events. Figure 5.13, based on the work of Walling et al. (1992), presents data relating to the temporal variation of the caesium-137 and total phosphorus (total-P) content of suspended sediment transported by the River Exe at Thorverton, Devon, UK. In this case, the caesium-137 content of suspended sediment shows only limited systematic variation in response to changes in discharge and with season, whereas total-P concentrations evidence much greater variability, decreasing with increasing flow and defining a clear seasonal cycle. This contrast can be related to the ultimate sources of the two substances. Caesium-137 is a fallout radionuclide and is associated primarily with the surface soils of the catchment. Since these represent an important source of suspended sediment at all levels of flow and at all times of the year, there is little systematic variation in the caesium-137 content of suspended sediment. In the case of total-P, however, a variety of sources are involved,

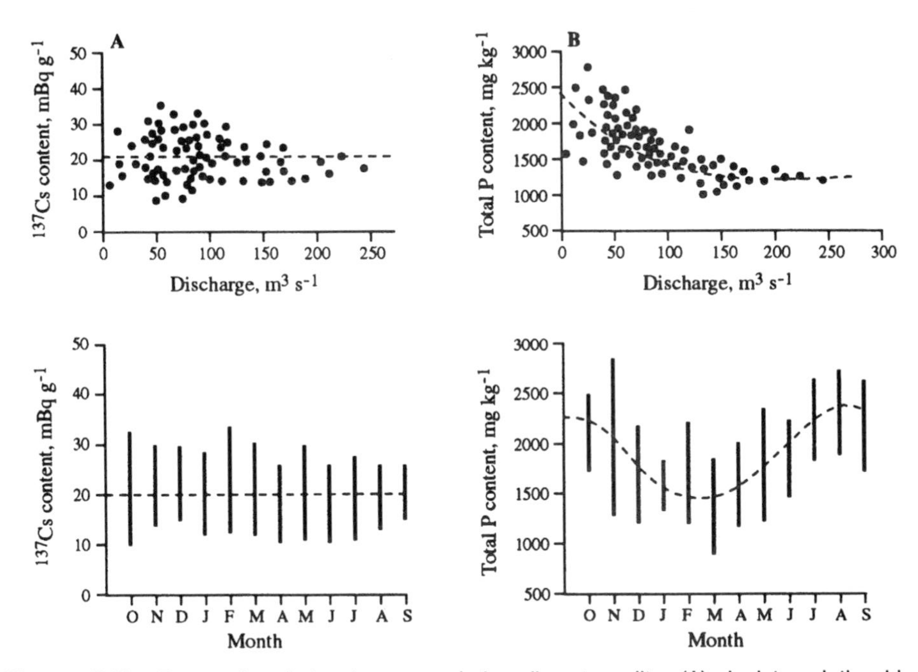

Figure 5.13 Temporal variation in suspended sediment quality: (A) depicts relationships between the caesium-137 and total phosphorus content of suspended sediment and water discharge for the River Exe at Thorverton, Devon, UK, and (B) illustrates seasonal variation in these two sediment properties at the same site. The bars represent the range of concentrations encountered in individual months (based on Walling et al., 1992)

ing eroded soil, land-use activities, point source effluent discharge and authigenic sses. Variations in the relative importance of these sources and in their behaviour g the year result in clear patterns of variability of the total-P content of suspended ment.

5.5 METHODS OF ESTABLISHING BEDLOAD AND SUSPENDED SEDIMENT FLUX

The veracity of sediment transport estimates, whether short or long term, depends fundamentally upon the quality of the data which have been collected. Sadly, despite the valour which often accompanies attempts to collect information in conditions which are frequently hostile, many river sediment databases are flawed if only because the variance that is inherent in sediment transport has not been allowed for. Several questions have to be addressed when establishing a sampling programme. Firstly, the efficiency of the sampling device should be known. Secondly, the likelihood of spatial variation in sediment transport rates, whether throughout a reach or across a single section, should be acknowledged. Thirdly, the short-, medium- and long-term temporal variation in transport rates should be allowed for when establishing not only single discrete samples, but also a time series.

5.5.1 Bedload Transport Rates

For a variety of reasons, the movement of bedload is more difficult to establish than that of suspended load. One of the problems is undoubtedly the fact that its incidence is not only highly discrete but, in many river environments, it is also unpredictable because our understanding of the transport mechanisms is far from perfect. Establishing a sampling programme requires an element of risk-taking unless the flow regime is guaranteed to induce bedload, as, for example, where snowmelt determines runoff. The more the researcher moves towards rain-fed arid and semi-arid regions, the greater the risk of not obtaining data. This is undoubtedly the prime reason why so little information is available in these areas (Schick et al., 1987).

One way of establishing bedload in small rivers, which avoids the problem of being on hand during transport, is to install a drop-structure or pit at a convenient location in the channel. Sediment collects in the pit over a period of time and can be excavated for weighing and for establishing size distributions. The method has been deployed successfully at a number of locations, for example, upland Wales (Newson and Leeks, 1987) and the Pennines, England (Carling, 1983). In the absence of a structure, repeated surveys of reservoir or lake deltas have been used, although the assumptions about the transport mode of the deposited material, the deposit density and thickness etc. are usually large so that only medium- and long-term generalisations about sediment movement are possible (Duck and McManus, 1993). The main disadvantage of the method of collecting sediment over a protracted period, whether this be a few weeks or half a century, is that hydraulics cannot be matched with transport except in extremely gross ways, for example maximum bedload clast size may be rated against the peak discharge of the sampling period.

Overcoming this problem inevitably involves investment of resources in employing people, in complex machinery, in data loggers etc. Three types of device have been developed and deployed successfully in the field at different locations. Logistical considerations dictate that only small- and medium-sized rivers have been tackled. All three devices employ a slot which lies in a plane that is conformable with the local bed surface and through which the bedload falls. Slot dimensions vary, but second slots deployed downstream of the primary receiver indicate efficiencies of around 100% at least during low and moderate flows (Milhous, 1973; Hayward, 1980). This is where similarity of the devices ends. The conveyor-belt sampler deployed on the East Fork River, Wyoming, has a set of moving belts that transport the trapped sediment to the stream bank where it is weighed and sampled (Leopold and Emmett, 1976; Klingeman and Emmett, 1982). Technicians are required to supervise the machinery and weigh the sediment. The vortex-tube sampler deployed on Oak Creek, Oregon (Milhous, 1973), on the Torlesse Stream, New Zealand (Hayward, 1980), and on Virginio Creek, Italy (Tacconi and Billi, 1987), uses the roller vortex that develops in an angled slot in the bed to convey sediment towards a measuring station (Figure 5.14). In the case of Oak Creek and Torlesse, sample weights are established manually; in Virginio Creek an impressive train of conveyors and weighing hoppers deals with the sediment, although supervision is required. The pressure-pillow sampler, deployed on Turkey Brook, England (Reid et al., 1980), Goodwin Creek, Mississippi (Kuhnle et al., 1988), Casper Creek, California (Lewis, 1991), River Kennett, England (Harris and Richards, 1995) and Nahal Yatir and Nahal Eshtemoa, Israel (Laronne et al., 1992), continuously weighs the bedload that falls through

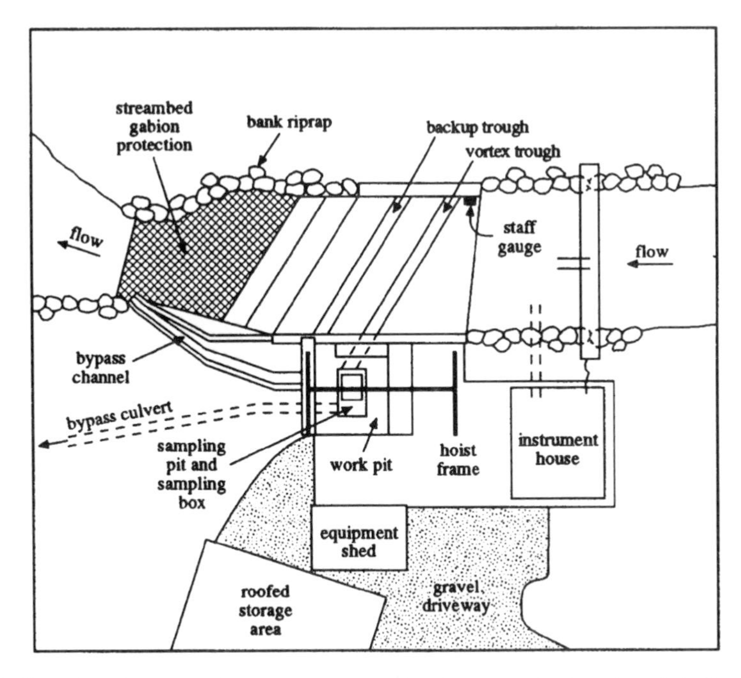

Figure 5.14 Plan diagram of the vortex-tube bedload sampler installed on Oak Creek, Oregon (after Milhous, 1973)

the bed slot (Figure 5.15). No technical supervision is required during a transport event, giving an advantage where rain-fed floods are unpredictable, but material for size analysis has to be obtained after the event by layer-sampling of the accumulated sediment.

The data obtained with such fixed installations are considered excellent, especially when information about sediment transport can be rated against contemporaneous hydrographic records. However, the expense of both the installation and the upkeep means that few rivers have been monitored, and these are only single-thread channels, deliberately avoiding the complications of braided streams (Reid and Laronne, 1995).

In order to overcome these limitations, several types of portable sampler or sensor have been devised and deployed with variable degrees of success. Hubbell (1964) compiled a list which was comprehensive at the time of publication and still gives an indication of the range of type-devices despite the passage of time. The device which has gained popularity is the pressure-difference sampler. Portability has meant that it has been deployed not only in small streams (Pitlick and Thorne, 1987), but also in larger rivers (Emmett, 1976). It has also proved useful in the complicated channel system of a braided stream (Ashworth et al., 1992). Early versions such as the so-called VUV had limited sampling efficiency (Novak, 1957). In contrast, the Helley–Smith version (Helley and Smith, 1971; Figure 5.16) has been shown to have near-perfect sampling efficiency when rated against the conveyor-belt sampler on the East Fork River, at least for grain sizes between 0.5 and 16 mm (Emmett, 1980). Subsequently, Hubbell (1987) has shown that efficiency falls off as transport rates rise above comparatively modest levels of around 0.05 to 0.1 $\mathrm{kg\,s^{-1}\,m^{-1}}$.

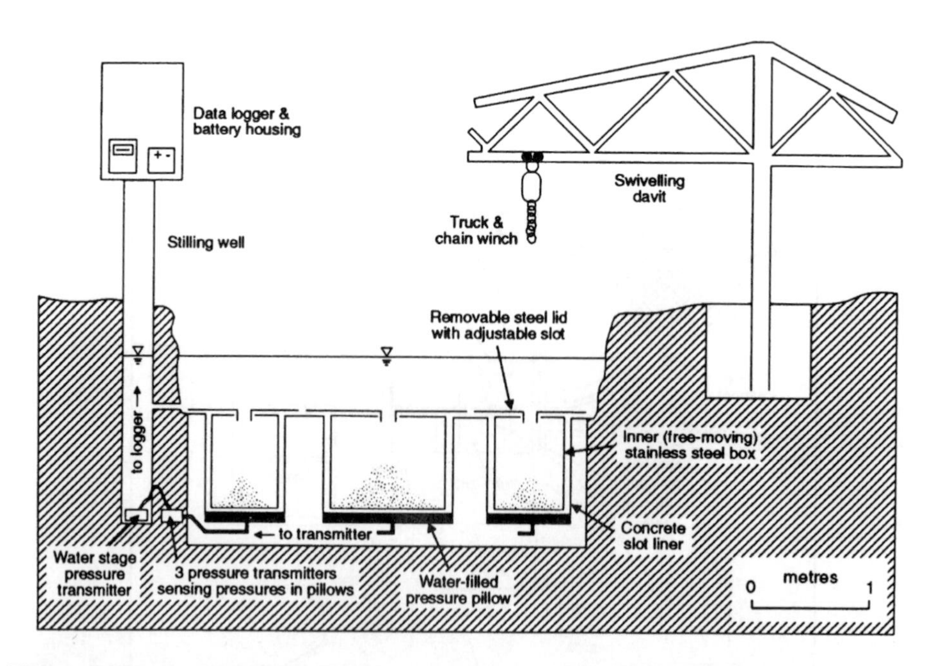

Figure 5.15 Cross-sectional diagram of the pressure-pillow samplers installed on the Nahal Yatir, Negev Desert, Israel (after Reid et al., 1980; Laronne et al., 1992. With permission of the Institute of Hydrology)

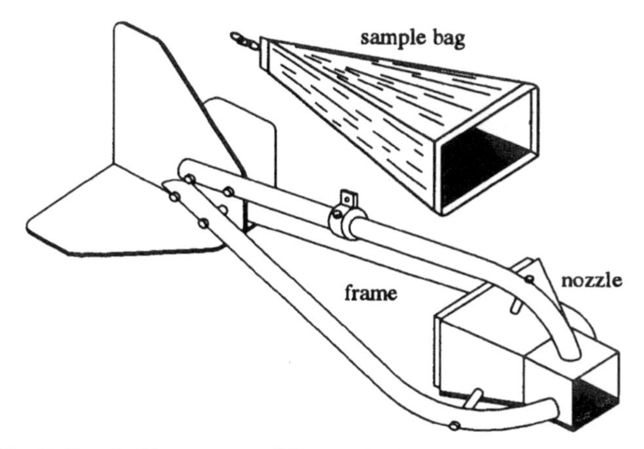

Figure 5.16 Helley-Smith pressure-difference bedload sampler (after Helley and Smith, 1971)

There are various problems that need consideration when using portable samplers. Firstly, their emplacement and maintenance on the stream bed requires skill to avoid digging the bed sediment and to ensure the passage of mobile particles into and not under the orifice. In addition, the drag exerted on the instrument by the flow is large, meaning that considerable assistance is needed to station any of these devices if bottom flow velocity approaches $1\ \mathrm{m\,s^{-1}}$. Secondly, their capacity usually means a short sampling period (of seconds to minutes), especially if transport rates are high, since the sample container fills and reduces efficiency. This poses problems for data interpretation if pulses characterise the bedload, because the sampling period may coincide with the trough or peak of the pulse. Thirdly, cross-channel variation in transport, either as a product of ribbons (Ferguson et al., 1989) or of sidewall drag (Reid et al., 1985; Pitlick, 1987), may be difficult to characterise because of large temporal and spatial variation in transport rates. Indeed, Emmett (1981) recommends 20 equally spaced traverse locations on each of two traverses to establish a single mean transport rate. Hubbell (1987) indicates that for moderately non-uniform transport in a channel cross-section, the maximum probable error can be reduced to 1% where 20 traverse positions are fixed, but only if more than 100 traverses are run!

5.5.2 Suspended Sediment Flux Rates

The measurement of suspended sediment fluxes is treated in a considerable number of manuals and technical reviews (ASCE, 1975; Allen, 1981; Guy and Norman, 1970; US Geological Survey, 1978; Ward, 1984; WMO, 1981, 1989), and there has been much greater standardisation of equipment and field sampling protocols than in the case of bedload. Specially designed suspended sediment samplers are commonly employed for measurements of suspended sediment concentration, and most of these incorporate a streamlined body and intake nozzle in order to provide the isokinetic sampling conditions essential for collection of representative samples (Guy and Norman, 1970). Both point-integrating and depth-integrating samplers have been developed (WMO, 1989). Use of simple dip samples and sampling equipment designed primarily for water quality monitoring can provide unrepresentative results (Yuzyk et al., 1992). The quest for automatic data collection from unmanned sites has also prompted the development of many types of automatic pump samplers capable of collecting discrete samples and storing them in individual bottles (Allen, 1981; Bogen, 1986; US Geological Survey, 1978; Walling, 1984; Ward, 1984). Such equipment does, however, possess limitations in terms of isokinetic sampling and the need to locate the sampler intake at a representative point in the channel cross-section. Optical turbidity meters (Brabben, 1981; Gippel, 1989; Grobler and Weaver, 1981; Jansson, 1992; Lawler et al., 1992; Truhlar, 1978) and nuclear sediment gauges (Berke and Rakoczi, 1981; Lu Zhi et al., 1981; Tazioli, 1981) have also been successfully employed to obtain continuous records of sediment concentration. Optical turbidity meters are limited to sediment concentrations below about 10 000 mg l^{-1}, whilst nuclear gauges are able to cope with higher concentrations, but with a minimum threshold of c. 500 mg l^{-1}. However, the reliability of such records depends heavily upon the nature and stability of the relationship between the absorption or scattering of the light or gamma radiation and the ambient suspended sediment concentration. In the case of optical turbidity meters, this relationship will be influenced by temporal variations in the grain size and other physical properties of the suspended sediment load (Fish, 1983; Foster et al., 1992; Gippel, 1989). Other devices which have been used to record suspended sediment concentrations include ultrasonic sensors (Flammer, 1962) and vibrating U-tube densimeters (Seely, 1982).

Whether a suspended sediment sampler or some other sensing device is used for measurement, it is important to recognise that suspended sediment concentrations may exhibit substantial variations within a river cross-section (Ongley et al., 1990). Sampling and measurement procedures must take such variations into account, if accurate estimates of instantaneous flux are required. In the case of the clay and silt components of suspended sediment (i.e. <0.063 mm), cross-sectional variations in concentration are frequently minimal in rivers with a reasonable degree of turbulence (Culbertson, 1977), but such variations may assume increasing importance in large rivers (Meade and Stevens, 1990) and in rivers where the suspended sediment load includes a significant proportion of sand. Depth-integrating samplers provide a means of obtaining velocity-weighted mean concentrations in specific verticals, but in the case of automatic pump samplers and optical turbidity meters and nuclear gauges, which collect data from a fixed point in the cross-section, there is a need to relate such fixed-point measurements to the overall cross-section (Horowitz et al., 1992).

Although existing manuals and technical reviews have clearly identified the problems associated with obtaining accurate measurements of instantaneous suspended sediment fluxes, much less attention has been given to the problems of obtaining reliable estimates of longer-term loads. In large rivers, where suspended sediment concentrations fluctuate relatively slowly, the need to define the continuous record of sediment concentration necessary to calculate longer-term loads may pose few problems (Porterfield, 1972). However, in smaller rivers, where concentrations may fluctuate rapidly (Figure 5.10C), it could prove extremely difficult to define the concentration record in the absence of automatic sampling or continuous recording equipment. Where only relatively infrequent sampling is possible, extrapolation and interpolation techniques are commonly used to estimate long-term loads (Walling and Webb, 1981; Walling et al., 1992). Interpolation techniques essentially involve the assumption that individual samples are representative of the time period between samples, whereas extrapolation techniques involve the use of sediment rating curves and similar relationships which enable the sediment concentration record to be synthesised from a continuous record of another variable (e.g. discharge). Walling et al. (1992) report results from a study of the 600 km^2 basin of the River Exe above Thorverton in Devon, UK, in which they used the continuous record from an optical turbidity meter to calculate the suspended sediment load for a two year period (1978–1980). This calculated load was then compared with the load estimates that would have been obtained by applying both a standard rating relationship and rating curves, corrected for bias using the parametric (I) and non-parametric (II) correction factors of Ferguson (1986) and Koch and Smillie (1986), to data collected from a programme of regular weekly sampling (i.e. 104 samples) and the same programme supplemented by samples collected during flood events (i.e. 174 samples). In each case, 50 replicate sets of sediment concentration data representing the different sampling strategies were generated by systematic and random sampling from the continuous record of sediment concentration provided by the turbidity monitor. The results presented in Figure 5.17 demonstrate that the use of rating curve procedures to estimate suspended sediment loads, even when bias correction factors are applied, can involve very substantial errors. These errors are likely to increase as catchment size decreases, and their potential magnitude indicates that when

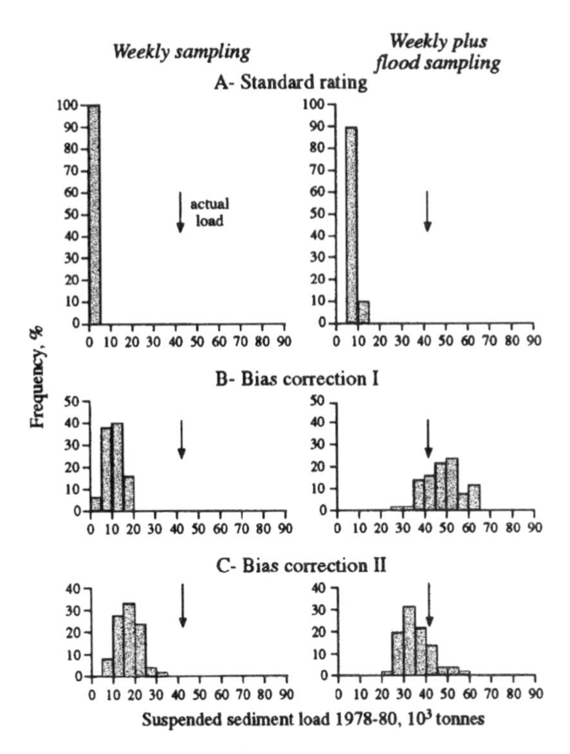

Figure 5.17 A comparison of the estimates of the suspended sediment load of the River Exe at Thorverton, Devon, UK, for the two year period 1978–1980 obtained using the specified sampling strategies and rating curve procedures with the actual load for the period (based on Walling et al., 1992)

long-term suspended sediment loads are being assessed, minor errors in the measurement of sediment concentration may be much less important than the problems associated with the use of a programme of infrequent sampling.

5.6 LIMITATIONS OF BEDLOAD TRANSPORT FORMULAE

5.6.1 General Difficulties

For several reasons it has not been possible to develop a universal bedload transport equation, or even a theoretical basis comparable with that developed for suspended load transport:

1. It is not easy to develop a theory because the basic principles and mechanisms are not easily studied;
2. Development of equations has tended to rely mainly on empirical and experimental work, especially in flumes and with uniform bed materials;
3. The various approaches can be tested only by comparison with data;
4. It is difficult to obtain accurate field data, so it is difficult to validate equations. None of the available bedload transport equations has been tested to the extent that it can be considered universal.

Consequently, a wide range of equations has been developed over the last half century or so. All of them involve some empirical elements, none is universally applicable and each should be used only under conditions similar to those for which it was derived.

5.6.2 General Form

Most bedload transport equations are based on hydraulic considerations and account only for the interaction of the flow with the channel bed material. In other words, they consider implicitly or explicitly the fluid forces acting on the bed material particles and the capacity of the flow to transport material. Their ability to account for inputs of material from other sources (e.g. the channel banks or outside the channel) is limited and depends on the continuous updating of information on bed material characteristics as these are affected by inputs.

Most bedload transport equations are based on the DuBoys (1879) concept that transport is a function of the excess of some flow quantity above the threshold value for initiation of transport:

$$q_b = f(X - X_c) \tag{5.1}$$

where q_b = bedload transport rate; X = a flow quantity (shear stress, discharge, velocity, stream power); and subscript c denotes the critical or threshold value. An important exception to this is Einstein's (1937,1942) formula which is a continuous relationship between bedload transport intensity and flow intensity and which does not therefore specify a sudden discontinuity in transport at the critical condition.

Most of the early development of bedload transport equations was aimed at sand-bed channels, for which the basic form of Eqn 5.1 is relatively easily adaptable, albeit empirically. For such channels, the size range of the bed material is narrow, a critical flow condition can be specified relatively easily and, for a given set of conditions, the onset of bedload transport is reasonably spatially uniform. Further up the channel system, though, where gravel and boulder materials predominate, consideration has to be given to the non-uniform particle size distribution. Not all particle sizes begin moving at the same condition and bedload transport is characterised by unsteadiness and non-uniformity as armour layers and other bed structures form and break up. The upper channel reaches are also more susceptible to supply effects, such as bank collapse, landslide inputs and other inputs from outside the channel. Although Eqn 5.1 is applied to gravel-bed rivers, it may be questioned whether the form is entirely appropriate. A number of new approaches for gravel-bed rivers have therefore been developed.

The following examines the available means of determining the critical conditions for initiation of motion and the bedload transport rate. It applies to non-cohesive materials only.

5.6.3 Initiation of Motion

The concept of a critical threshold for initiation (and cessation) of motion and the factors which determine the threshold are described in Section 5.2.

The standard means of calculating critical conditions for initiation of bed material movement is Shields' (1936) relationship:

$$\tau_{*\mathrm{c}} = f(Re_*) \qquad (5.2a)$$

where $\tau_{*\mathrm{c}}$ is the Shields parameter:

$$\tau_{*\mathrm{c}} = \frac{\tau_\mathrm{c}}{(\rho_\mathrm{s} - \rho) g D} \qquad (5.2b)$$

and τ_c = critical shear stress; ρ_s = particle density of bed material; ρ = fluid density of water; D = bed material particle diameter; Re_* = shear Reynolds number, $u_* D/\nu$; u_* = shear velocity; ν = kinematic viscosity; and g = acceleration due to gravity. The function in Eqn 5.2a has been evaluated experimentally and, for gravel and coarser materials (D greater than 5 mm for quartz-density sediment) with uniform size distributions, the Shields parameter is typically assigned a constant value in the range 0.04–0.06. Progress has also been made in extending the relationship to bed materials with wide size distributions, where hiding/exposure effects are important. Andrews and associates (Andrews, 1983; Andrews and Erman, 1986; Andrews and Sutherland, 1987) have proposed, on the basis of field measurements, the relationship:

$$\frac{\tau_{*\mathrm{ci}}}{\tau_{*\mathrm{c50}}} = \left(\frac{D_\mathrm{i}}{D_{50}}\right)^a \qquad (5.3)$$

where $\tau_{*\mathrm{ci}}$ = the average critical Shields parameter for particles of size D_i in the surface or armour layers of the bed; subscript i = a percentile within the overall sediment size distribution; $\tau_{*\mathrm{c50}}$ = the critical Shields parameter for particles of size D_{50} in the subsurface or parent bed material; and a = an exponent. Andrews (1983) originally evaluated a as -0.872 but subsequent analyses suggested that the exponent could achieve the value of -1. At this value all particle sizes are brought into motion at the same shear stress, the so-called condition of equal mobility. This is, however, the subject of considerable controversy. Komar (1987), using field data from several sources, declared that a value of about -0.7 appropriately reflects unequal mobility and this appears to have been confirmed by Ashworth and Ferguson (1989). Wilcock (1992) points to a progressive shift away from unequal and towards equal mobility as shear stress exceeds a value twice that at the critical condition for initiation of motion. Indeed, Komar and Shih (1992) acknowledge that a similar shift might have emerged in their analyses had the range in the field data which they used been more extensive. Here, researchers are faced with a familiar problem and one that has bedevilled successful prediction of bedload discharge: obtaining adequate field data poses major logistical difficulties. The upshot is that there are no databases that allow us to follow river behaviour over a wide range of flow conditions.

Several studies have shown that the Shields parameter for critical conditions achieves values of 0.1 and higher as channel slope increases above 1% and the ratio of depth to particle size falls below about 10 (Ashida and Bayazit, 1973; Bathurst et al., 1987). In addition, the rough flow conditions associated with such characteristics (typical of boulder-bed rivers) turn the measurement of depth (needed in determining shear stress) into a difficult practical problem. Often, water discharge is more readily available than depth. Bathurst et al. (1987) therefore adopted the Schoklitsch (1962) approach to predicting the critical flow conditions, based on water discharge per unit width rather than shear stress. Using flume data they developed the empirical relationship (Figure 5.18):

$$q_c = 0.15\, g^{0.5} D_{50}^{1.5} S^{-1.12} \tag{5.4}$$

where q_c = critical water discharge per unit width for the particle size D_{50}; D_{50} = that size of particle intermediate axis for which 50% of the particles are finer; S = slope; and g = the acceleration due to gravity. The equation was derived for the range of slopes $0.25 \leqslant S \leqslant 20\%$, the range of particle sizes $3 \leqslant D_{50} \leqslant 44$ mm and for ratios of depth to particle size as low as 1. It may be considered to apply to uniform bed materials. A particular problem with this approach is the definition of the width to be used in determining unit water discharge. Further research is needed to show whether it should be channel width (varying with flow stage) or the active width over which bedload transport occurs.

For non-uniform bed materials with a hiding/exposure effect, Bathurst (1987a) used limited river data to extend this approach in a manner similar to that used for the shear stress-based approach:

$$\frac{q_{ci}}{q_{cr}} = \left(\frac{D_i}{D_r}\right)^b \tag{5.5}$$

where q_{ci} = the critical unit water discharge for movement of particles of size D_i; q_{cr} = the critical unit water discharge for the reference particle size D_r which is unaffected by the hiding/exposure effect; and b = an exponent. An independent evaluation of q_{cr} is given by

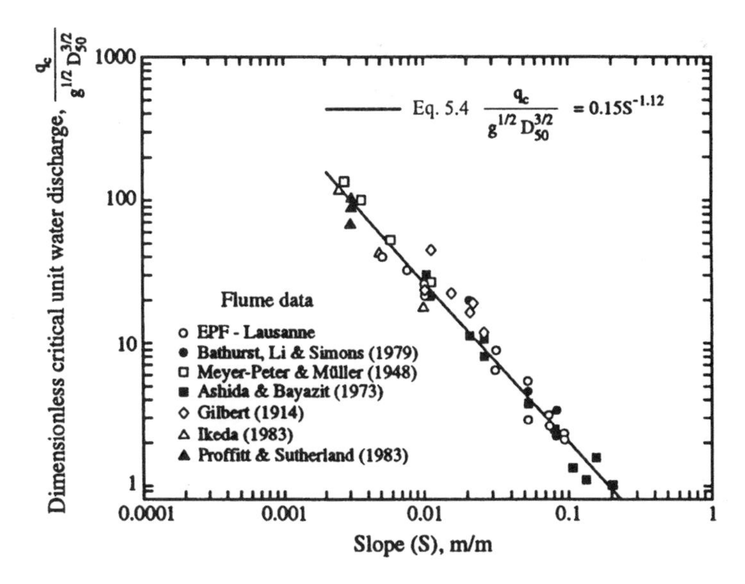

Figure 5.18 Basis of the empirical relationship Eqn. 5.4 (from Bathurst et al., 1987). Data sources given in the original reference

Eqn 5.4 replacing D_{50} by D_r (although typically D_r appears to be similar to D_{50}). The exponent b was tentatively calculated as:

$$b = 1.5\left(\frac{D_{84}}{D_{16}}\right)^{-1} \tag{5.6}$$

where D_{84} and D_{16} refer respectively to the particle sizes for which 84% and 16% of the particles in the surface layer are finer. However, further research into the form of this relationship was recommended. A value of zero for exponent b implies equal mobility.

Both the shear stress-based approach and the discharge-based approach to accounting for the hiding/exposure effect are still at the research stage and should be applied with due caution. No formulae are yet available for distinguishing between the critical conditions for initiation and cessation of motion.

5.6.4 Bedload Transport Formulae

The wide range of transport formulae which are available may be gauged from a review of such classic texts as Graf (1971), Vanoni (1975) and Simons and Şentürk (1977). There is little point in repeating the list here. A large proportion of these equations were developed with sand-bed rivers in mind and, noting their empirical element, they should not be applied outside such conditions. Fewer equations have been developed specifically for gravel- and boulder-bed rivers but relevant reviews are provided in White et al. (1975), Bathurst et al. (1987) and Gomez and Church (1989) (Box 5.1). With a few exceptions

Box 5.1 Choice of a bedload equation

It should be evident from the main text that predicting bedload discharge is fraught with difficulties. It is also evident that interaction between the flow and the bed material that forms the major source of mobile sediment is far from simple. In such a complex system, modelling has been, and continues to be, very approximate and predictive equations such as those bedload formulae that are repeated from text to text can only offer a guide to the levels of bedload transport to be expected in rivers where the only information is hydrologic or hydraulic. Gomez and Church (1989) offer a recent and excellent appraisal of the performance of a large number of formulae. Each is rated against a field bedload data set that has been carefully chosen and screened so that, for example, flow was steady at the time of establishing bedload, thereby removing or reducing the likelihood of fluctuations due to changes in hydraulic conditions. None of the formulae performs perfectly (see figure), and there is considerable over- and under-prediction. Indeed, formulae that are favourites among the engineering and geomorphological communities such as that of Meyer-Peter and Müller (1948) perform just as poorly as those that are less visited. Gomez and Church's conclusion is that the approach embodied in Bagnold's (1980) formula looks most promising, and, in their analysis, it appears to perform better than others at least in perennial armoured gravel-bed streams. However, in ephemeral desert streams', Reid et al. (1996) show the advantages of using the Meyer-Peter and Müller (1948) formula. Here, the channel bed is unarmoured due to the plentiful supply of sediment from the slopes of the water catchment.

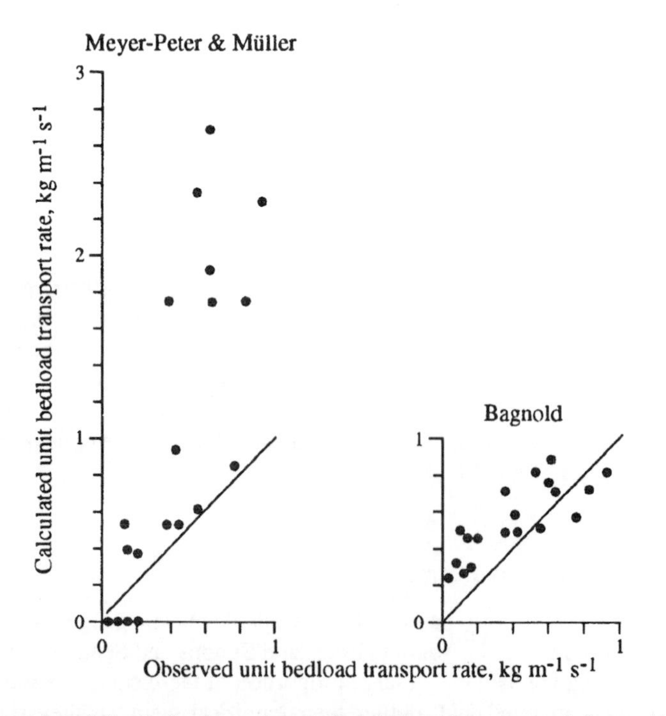

Performance of predictive bedload formulae rated against carefully screened observed bedload data obtained from the Elbow River, Canada (after Gomez and Church, *Water Resources Research*, **25**, *1161–1186, 1989, copyright by the American Geophysical Union)*

they tend to be modelled on the form of Eqn 5.1 and therefore have difficulties coping with non-uniform bed material size distributions and sediment supply/bed structure effects.

Where the bed material contains particles ranging from sand to cobbles or boulders, only the highest flows are capable of mobilising the entire range of sizes. During most moderate bedload transport events, only partial transport occurs, with some of the sizes in motion and others (the largest) stationary (Figure 5.19). Under these conditions, the observed transport rate for the moving size fraction is not directly comparable with the rate for a uniform bed material of the same mean size as those fractions. The former is always smaller than the latter because it is determined by the availability, or proportion, of those size fractions within the bed material. Consequently, bedload transport calculations based on the assumption of a uniform bed material size will overestimate the observed rate since they assume that, once any transport begins, all size fractions are in motion. This is particularly the case for 'Phase I transport' (Section 5.3).

One means of allowing for the effect of bed material size distribution is to apply the transport equation separately to each bed material size fraction (in proportion to its occurrence in the bed) and sum the resulting partial transport rates to give the total rate. However, it is then necessary to account also for the relative effects of the different size fractions in impeding or promoting each other's movement (in a manner similar to the hiding/exposure effect), and for the effect of bed armouring. Little progress has yet been

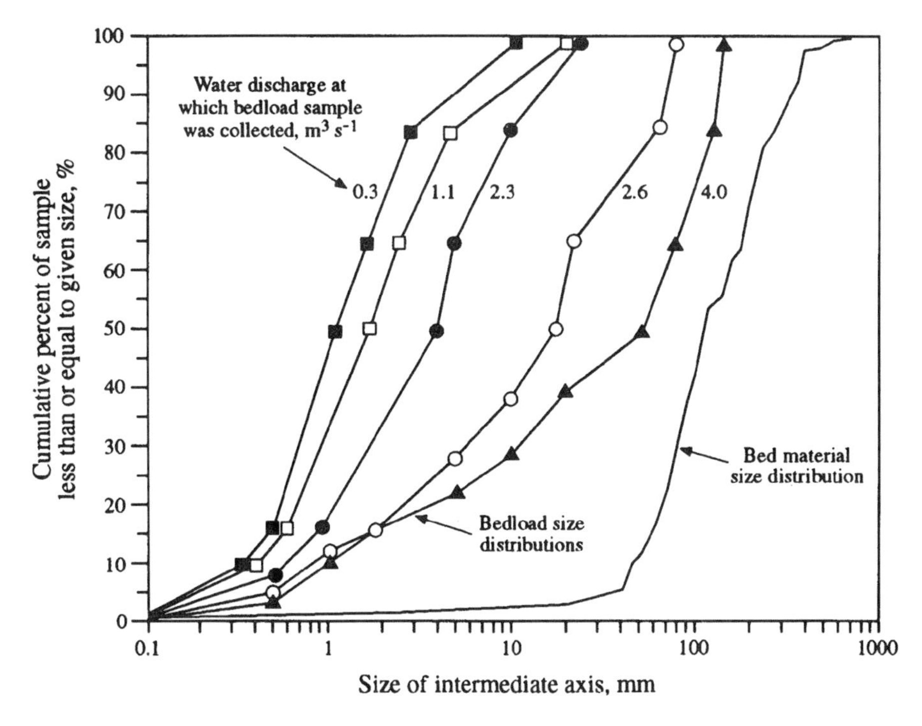

Figure 5.19 Variation of bedload size distribution with water discharge for the Roaring River, Colorado, USA, July 1984. The samples were collected with a Helley–Smith sampler and frequency analysis was by weight. Also shown is the bed material size distribution for the same site, obtained by grid sampling with frequency analysis by number (from Bathurst, 1987b with permission)

made in this direction, although flume studies with fine gravel by Misri et al. (1984) and Samaga et al. (1986) have demonstrated the presence of particle size interaction. Developments by Parker (1990) are also promising.

Available bedload transport equations are concerned only with the relationship between flow hydraulics and particle movement and do not allow for supply effects, changes in channel bed structure and unsteady or even steady particle wave movement. Where such effects are significant, bedload transport is non-uniform and unsteady and can fluctuate over an order of magnitude for given flow conditions. Hysteresis loops characterise the pattern of transport over the rise and fall of water discharge, both seasonally and for single events (Figure 5.20). No single rating curve relating bedload transport rate and water discharge (in essence, Eqn 5.1) can be applied to all events, and a full understanding of the relationship between flow and bedload discharge with time is achieved only when day-to-day changes are studied in chronological order (e.g. Reid et al., 1985).

It is important to understand what is meant by sediment supply as it affects bedload transport. In an alluvial channel there is always sediment available since the channel is

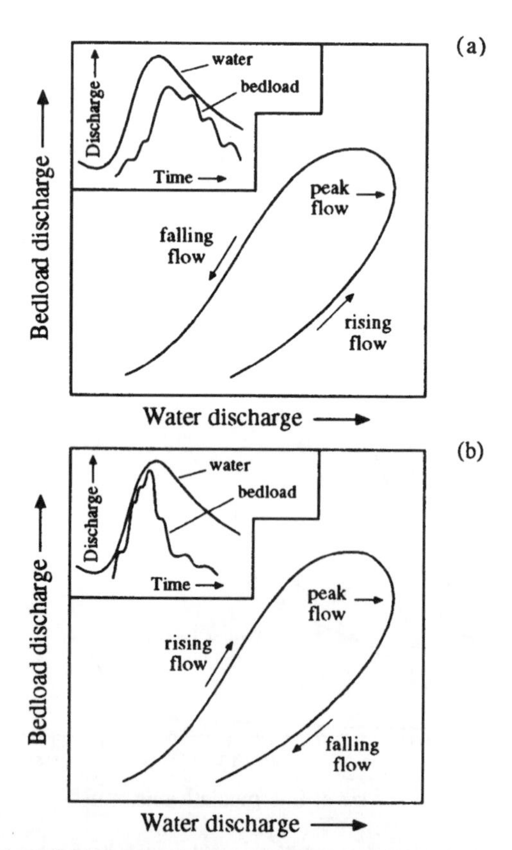

Figure 5.20 Variation of bedload discharge with water discharge through a flood hydrograph: (a) where sediment becomes available during the falling limb; (b) where sediment supplies are depleted during the rising limb. Water and bedload discharge hydrographs are inset (from Bathurst 1987b with permission)

formed in sediment. Consequently, supply limitation does not mean that, because the supply from outside the channel is limited, there is a strict absence of sediment available for transport within the channel and that this is the reason for low transport rates. Instead, it is related to the availability in the channel of particles fine enough to be moved by a given flow. Thus an injection of relatively fine material may increase that proportion, thereby enabling the bedload transport to increase. As the fine sediment is washed away, however, bedload transport may fall even though water discharge remains constant and despite the presence of an ample availability of coarser particles on the channel bed.

Probably the best means of accounting for supply effects in predicting bedload transport rate is to apply a catchment-scale erosion and sediment yield model (e.g. Wicks et al., 1992; Wicks and Bathurst, 1996). By simulating the erosion of material across the catchment and routing this material into the channel, such models provide a means for monitoring the availability of material (by size fraction) for channel transport. However, this is a complex approach not suitable for all cases. A simpler approach is to collect field data for the site of interest and develop an empirical relationship for that site. This will not necessarily be applicable elsewhere, nor will it necessarily represent all features of observed bedload transport variation. However, as a calibrated rating curve it is likely to be reasonably accurate on average and when applied over time periods of a month or so. As with suspended sediment (Section 5.4), though, different relationships may apply at different times of the year.

5.7 REFERENCES

Allan, R.J. 1986. *The Role of Particulate Matter in the Fate of Contaminants in Aquatic Ecosystems*. Inland Waters Directorate, Environment Canada, Scientific Series no. 107, 24pp.

Allen, J.R.L. 1982. *Sedimentary Structures*. Elsevier, Amsterdam.

Allen, J.R.L. 1983. River bedforms: progress and problems. In: Collinson, J.D. and Lewin, J. (Eds), *Modern and Ancient Fluvial Systems*. International Association of Sedimentologists Special Publication **6**, 19–33.

Allen, P.B. 1981. *Measurement and Prediction of Erosion and Sediment Yield*. US Department Agriculture Report, ARM-S-15, USDA, Washington DC.

Andrews, E.D. 1983. Entrainment of gravel from naturally sorted riverbed material. *Geological Society of America Bulletin*, **94**, 1225–1231.

Andrews, E.D. and Erman, D.C. 1986. Persistence in the size distribution of surficial bed material during an extreme snowmelt flood. *Water Resources Research*, **22**(2), 191–197.

Andrews, E.D. and Parker, G. 1987. Formation of a coarse surface layer as a response to gravel mobility. In: Thorne, C.R., Bathurst, J.C. and Hey, R.D. (Eds), *Sediment Transport in Gravel-bed Rivers*. Wiley, Chichester, 269–300.

ASCE. 1975. *Sedimentation Engineering*. ASCE, New York, 745pp.

Ashida, K. and Bayazit, M. (1973). Initiation of motion and roughness of flows in steep channels. *Proceedings of the 15th Congress of the International Association of Hydraulic Research*, Istanbul, Vol. 1, 475–484.

Ashworth, P.J. and Ferguson, R.I. 1989. Size-selective entrainment of bed load in gravel bed streams. *Water Resources Research*, **25**(4), 627–634.

Ashworth, P.J., Ferguson, R.I., Ashmore, P.E., Paola, C., Powell, D.M. and Prestegaard, K.L. 1992. Measurements in a braided river chute and lobe. 2. Sorting of bed load during entrainment, transport and deposition. *Water Resources Research*, **28**, 1887–1896.

Bagnold, R.A. 1980. An empirical correlation of bedload transport rates in flumes and natural rivers. *Proceedings of the Royal Society of London Series A*, **372**, 453–473.

Bathurst, J.C. 1987a. Critical conditions for bed material movement in steep, boulder-bed streams. In: *Erosion and Sedimentation in the Pacific Rim*. IAHS Publication no. 165, Institute of Hydrology, Wallingford, 309–318.

Bathurst, J.C. 1987b. Measuring and modelling bedload transport in channels with coarse bed materials. In: Richards, K. (Ed.), *River Channels: Environment and Process*. Blackwell, Oxford, 272–294.

Bathurst, J.C., Graf, W.H. and Cao, H.H. 1987. Bed load discharge equations for steep mountain rivers. In: Thorne, C.R., Bathurst, J.C. and Hey, R.D. (Eds), *Sediment Transport in Gravel-bed Rivers*. Wiley, Chichester, 453–477.

Berke, B. and Rakoczi, L. 1981. Latest achievements in the development of nuclear suspended sediment gauges. In: *Erosion and Sediment Transport Measurement*. IAHS Publication no. 133, 91–96.

Blissenbach, E. 1952. Relation of surface angle distribution to particle-size distribution on alluvial fans. *Journal of Sedimentary Petrology*, **22**, 25–28.

Bluck, B.J. 1987. Bedforms and clast size changes in gravel-bed rivers. In: Richards, K.S. (Ed.), *River Channels: Environment and Process*. Blackwell, Oxford, 159–178.

Bogen, J. 1980. Hysteresis effects in the sediment transport system. *Norsk Geografisk Tidsskrift*, **34**, 45–54.

Bogen, J. 1986. Sampling of suspended sediments in stream. In: Hashdt, B. (Ed.), *Partikulaert Bundet Stofftransport i Vann og Jorderosjon*. NHP report no. 14, 9–25.

Brabben, T. 1981. Use of turbidity monitors to assess sediment yield in East Java, Indonesia. In: *Erosion and sediment transport measurement*. IAHS Publication no. 133, 105–113.

Bradley, J.B. and McCutcheon, S.C. 1987. Influence of large suspended sediment concentrations in rivers. In: Thorne, C.R., Bathurst, J.C. and Hey, R.D. (Eds), *Sediment Transport in Gravel-bed Rivers*. Wiley, Chichester, 645–689.

Brierley, G.J. and Hickin, E.J. 1985. The downstream gradation of particle sizes in the Squamish River, British Columbia. *Earth Surface Processes and Landforms*, **10**, 597–606.

Buffington, J.M., Dietrich, W.E. and Kirchner, J.W. 1992. Friction angle measurements on a naturally formed gravel stream bed: implications for critical boundary shear stress. *Water Resources Research*, **28**, 411–425.

Carling, P.A. 1983. Threshold of coarse sediment transport in broad and narrow rivers. *Earth Surface Processes and Landforms*, **8**, 1–18.

Carling, P.A. 1992. In-stream hydraulics and sediment transport. In: Calow, P. and Petts, G.E. (Eds), *The Rivers Handbook, Vol. 1*. Blackwell, Oxford, 101–125.

Carling, P.A. and Hurley, M.A. 1987. A time-varying stochastic model of the frequency and magnitude of bed load transport events in two small trout streams. In: Thorne, C.R., Bathurst, J.C. and Hey, R.D. (Eds), *Sediment Transport in Gravel-bed Rivers*. Wiley, Chichester, 897–920.

Carling, P.A. and Reader, N.A. 1982. Structure, composition and bulk properties of upland stream gravels. *Earth Surface Processes and Landforms*, **7**, 349–366.

Çeçen, K. and Bayazit, M. 1973. Critical shear stress of armored beds. *Proceedings of the 15th Congress of the International Association of Hydraulic Research*, Istanbul, Vol. 1, 493–500.

Chang, M., Roth, F.A. and Hunt, E.V. 1982. Sediment production under various forest site conditions. In: *Recent Developments in the Explanation and Prediction of Erosion and Sediment Yield*. IAHS Publication no. 137, 13–22.

Church, M., McLean, D.G. and Wolcott, J.F. 1987. River bed gravels: sampling and analysis. In: Thorne, C.R., Bathurst, J.C. and Hey, R.D. (Eds), *Sediment Transport in Gravel-bed Rivers*. Wiley, Chichester, 43–79.

Church, M., Wolcott, J.F. and Fletcher, W.K. 1991. A test of equal mobility in fluvial sediment transport: behaviour of the sand fraction. *Water Resources Research*, **27**, 2941–2951.

Culbertson, J. 1977. Influence of flow characteristics on sediment transport with emphasis on grain size and mineralogy. In: Shear, H. and Watson, A.E.P. (Eds), *The Fluvial Transport of Sediment-Associated Nutrients and Contaminants*. International Joint Commission, Windsor, Ontario, 117–133.

Dietrich, W.E., Kirchner, J.W., Ikeda, H. and Iseya, F. 1989. Sediment supply and the development of the coarse surface layer in gravel bedded rivers. *Nature*, **340**, 215–217.

Diplas, P. 1987. Bedload transport in gravel-bed streams. *American Society of Civil Engineers, Journal of Hydraulic Engineering*, **113**, 277–292.

Diplas, P. and Fripp, J.B. 1992. Properties of various sediment sampling procedures. *American Society of Civil Engineers, Journal of Hydraulic Engineering*, **118**, 955–970.

Droppo, I.G. and Ongley, E.D. 1989. Flocculation of suspended solids in southern Ontario rivers. In: Hadley, R.F. and Ongley, E.D. (Eds), *Sediment and the Environment*. IAHS Publication no. 184, 95–103.

Droppo, I.G. and Ongley, E.D. 1992. The state of suspended sediment in the freshwater fluvial environment: a method of analysis. *Water Research*, **26**, 65–72.

DuBoys, M.P. 1879. Etudes du régime et l'action exercée par les eaux sur un lit à fond de graviers indéfiniment affouilable. *Annales de Ponts et Chaussées*, Ser. 5, **18**, 141–195.

Duck, R.W. and McManus, J. 1993. Sedimentation in natural and artificial impoundments: an indicator of evolving climate, landuse and dynamic conditions. In: McManus, J. and Duck, R.W. (Eds), *Geomorphology and Sedimentology of Lakes and Reservoirs*. Wiley, Chichester, 1–3.

Duijsings, J.J.H.M. 1986. Seasonal variation in sediment delivery ration of a forested drainage basin in Luxembourg. In: Hadley, R.F. (Ed.), *Drainage Basin Sediment Delivery*. IAHS Publication no. 159, 153–164.

Egiazaroff, I.V. 1965. Calculation of nonuniform sediment concentrations. *Proceedings of the American Society of Civil Engineers, Journal of the Hydraulics Division*, **91**(HY4), 225–247.

Einstein, H.A. 1937. *Der Geschiebetrieb als Wahrscheinlichkeits Problem*. Verlag Rascher, Zürich, 110pp.

Einstein, H.A. 1942. Formulas for the transportation of bedload. *Transactions of the American Society of Civil Engineers*, **107**, 561–577.

Emmett, W.W. 1976. Bedload transport in two large gravel-bed rivers, Idaho and Washington. *Proceedings 3rd Federal Inter-agency Sedimentation Conference*, Denver, Colorado, 15pp.

Emmett, W.W. 1980. *A Field Calibration of the Sediment-trapping Characteristics of the Helley-Smith Bedload Sampler*. United States Geological Survey Professional Paper 1139.

Emmett, W.W. 1981. Measurement of bedload in rivers. In: Tacconi, P. and Billi, P. (Eds), *Erosion and Sediment Transport Measurement*. IAHS Publication no. 133, 3–15.

Ferguson, R.I. 1986. River loads underestimated by rating curves. *Water Resources Research*, **22**, 74–76.

Ferguson, R.I., Prestegaard, K.L. and Ashworth, P.J. 1989. Influence of sand on hydraulics and gravel transport in a braided gravel bed river. *Water Resources Research*, **25**, 643–653.

Fish, I.L. 1983. *Partech Turbidity Monitors. Calibration with Silt and the Effects of Sand*. Technical Note OD/TNR, Hydraulics Research, Wallingford.

Flammer, G.H. 1962. *Ultrasonic Measurement of Suspended Sediment*. US Geological Survey Bulletin, 1141-A.

Foster, I.D.L., Millington, R. and Grew, R.G. 1992. The impact of particle size controls on stream turbidity measurement; some implications for suspended sediment yield estimation. In: Bogen, J., Walling, D.E. and Day, T.J. (Eds), *Erosion and Sediment Transport Monitoring Programmes in River Basins*. IAHS Publication no. 210, 51–62.

Frederiksen, R.L. 1970. *Erosion and Sedimentation Following Road Construction and Timber Harvest on Unstable Soils in Three Small Western Oregon Watersheds*. US Forest Service Research Paper no. PNW 104.

Frostick, L.E., Lucas, P.M. and Reid, I. 1984. The infiltration of fine matrices into coarse-grained alluvial sediments and its implications for stratigraphical interpretation. *Journal of the Geological Society, London*, **141**, 955–965.

Gaweesh, M.T.K. and van Rijn, L.C. 1992. Laboratory and field investigation of a new bedload sampler for rivers. In: Falconer, R.E., Shiono, K. and Mathews, R.G.S. (Eds), *Hydraulic and Environmental Modelling: Estuarine and River Waters. Proceedings 2nd International Conference on Hydraulic and Environmental Modelling of Coastal and River Waters*, Vol. 2. 479–488.

Gippel, C.J. 1989. *The Use of Turbidity Instruments to Measure Stream Water Suspended Sediment Concentration*. Monograph Series no. 4, Department of Geography and Oceanography, ADFA, University NSW, Australia.

Gomez, B. and Church, M. 1989. An assessment of bed load sediment transport formulae for gravel bed rivers. *Water Resources Research*, **25**(6), 1161–1186.

Gomez, B., Naff, R.L. and Hubbell, D.W. 1989. Temporal variations in bedload transport rates associated with the migration of bedforms. *Earth Surface Processes and Landforms*, **14**, 135–156.

Graf, W.H. 1971. *Hydraulics of Sediment Transport*. McGraw-Hill, New York, 513pp.

Grobler, D.C. and Weaver, A. van B. 1981. Continuous measurement of suspended sediment in rivers by means of a double beam turbidity meter. In: *Erosion and Sediment Transport Measurement*. IAHS Publication no. 133, 97–103.

Guy, H.P. 1964. *An Analysis of Some Storm-period Variables Affecting Stream Sediment Transport*. US Geological Survey Professional Paper 462E.

Guy, H.P. and Norman, V.W. 1970. Field methods for measurement of fluvial sediment. In: *Techniques of Water Resources Investigations of the US Geological Survey* (Book 5, Chapter C1). US Geological Survey, Reston, Virginia.

Harris, T. and Richards, K.S. Design and calibration of a recording bedload trap. *Earth Surface Processes and Landforms*, **20**, 711–720.

Hayward, J.A. 1980. *Hydrology and Stream Sediment from Torlesse Stream Catchment*. Tussock Grasslands and Mountain Lands Institute, Lincoln College, Canterbury, Special Publication, 17.

Heidel, S.G. 1956. The progressive lag of sediment concentration with flood waves. *Transactions of the American Geophysical Union*, **37**, 56–66.

Helley, E.J. and Smith, W. 1971. *Development and Calibration of a Pressure-difference Bedload Sampler*. United States Geological Survey Open-file Report, 18.

Hoey, T. 1992. Temporal variations in bedload transport rates and sediment storage in gravel-bed rivers. *Progress in Physical Geography*, **16**, 319–338.

Horowitz, A.J. 1991. *A Primer on Sediment Trace-Element Chemistry*. Lewis Publishers, Chelsea, Michigan, 136pp.

Horowitz, A.J., Elrick, K.A., Von Guerard, P.B., Young, N.O., Buell, G.R. and Miller, T.L. 1992. The use of automatically collected point samples to estimate suspended sediment and associated trace element concentrations for determining annual mass transport. In: Bogen, J., Walling, D.E. and Day, T.J. (Eds), *Erosion and Sediment Transport Monitoring Programmes in River Basins*. IAHS Publication no. 210, 209–218.

Hubbell, D.W. 1964. Apparatus and techniques for measuring bedload. *United States Geological Survey Professional Paper*, **1748**.

Hubbell, D.W. 1987. Bedload sampling and analysis. In: Thorne, C.R., Bathurst, J.C. and Hey, R.D. (Eds), *Sediment Transport in Gravel-bed Rivers*. Wiley, Chichester, 89–106.

Iseya, H. and Ikeda, H. 1987. Pulsations in bedload transport rates induced by longitudinal sediment sorting: a flume study using sand and gravel mixtures. *Geografiska Annaler*, **69A**, 15–27.

Jansson, M. 1992. Suspended sediment inflow to the Cachi reservoir. In: Jansson, M.B. and Rodriguez, A. (Eds), *Sedimentological Studies of the Cachi Reservoir, Costa Rica. Sediment Inflow, Reservoir Sedimentation and Effects of Flushing*. UNGI Report 81, Department of Physical Geography, University of Uppsala, Sweden.

Kellerhals, R. and Bray, D.I. 1971. Sampling procedures for coarse fluvial sediments. *Journal Hydraulics Division ASCE*, **97**, 1165–1180.

Klingeman, P.C. and Emmett, W.W. 1982. Gravel bedload transport processes. In: Hey, R.D., Bathurst, J.C. and Thorne, C.R. (Eds), *Gravel-bed Rivers*. Wiley, Chichester, 141–169.

Koch, R.W. and Smillie, G.M. 1986. Comment on 'River loads underestimated by rating curves' by R.I. Ferguson. *Water Resources Research*, **22**, 2121–2122.

Komar, P.D. 1987. Selective grain entrainment by a current from a bed of mixed sizes: a reanalysis. *Journal of Sedimentary Petrology*, **57**(2) 203–211.

Komar, P.D. and Shih, S-M. 1992. Equal mobility versus changing bedload grain sizes in gravel bed streams. In: Billi, P., Hey, R.D., Thorne, C.R. and Tacconi, P. (Eds), *Dynamics of Gravel-bed Rivers*. Wiley, Chichester, 73–93.

Koster, E.H. 1978. Transverse ribs: their characteristics, origin and palaeohydraulic significance. In: Miall, A.D. (Ed.), *Fluvial Sedimentology*. Memoir Canadian Society Petroleum Geologists **5**, 161–186.

Krishnappan, B.G. and Ongley, E.D. 1989. River sediment and contaminant transport – changing needs in research. In: *Proceedings Fourth International Symposium on River Sedimentation*, IRTCES, Beijing, 530–538.

Kuhnle, R.A., Willis, J.C. and Bowie, A.J. 1988. Measurement of bed load transport on Goodwin Creek, northern Mississippi. *Proceedings 18th Mississippi Water Resources Conference*, Jackson, Mississippi, 57–60.

Laronne, J.B. and Reid, I. 1993. Very high rates of bedload sediment transport in desert ephemeral rivers. *Nature*, **366**, 148–150.

Laronne, J.B., Reid, I., Yitshak, Y. and Frostick, L.E. 1992. Recording bedload discharge in a semiarid channel, Nahal Yatir, Israel. In: Bogen, J., Walling, D.E. and Day, T.J. (Eds), *Erosion and Sediment Monitoring Programmes in River Basins*. IAHS Publication no. 210, 79–86.

Lawler, D.M., Dolan, M., Tomasson, H. and Zophoniasson, S. 1992. Temporal variability of suspended sediment flux from a subarctic glacial river, southern Iceland. In: Bogen, J., Walling, D.E. and Day, T.J. (Eds), *Erosion and Sediment Transport Monitoring Programmes in River Basins*. IAHS Publication no. 210, 233–243.

Leopold, L.B. and Emmett, W.W. 1976. Bedload measurements, East Fork River, Wyoming. *Proceedings of the National Academy Science, USA*, **73**, 1000–1004.

Lewis, J. 1991. An improved bedload sampler. In: *Proceedings of the Fifth Federal Interagency Sedimentation Conference*. Las Vegas, Nevada, 6.1–6.8.

Lisle, T.E. 1989. Sediment transport and resulting deposition in spawning gravels, north coastal California. *Water Resources Research*, **26**, 1303–1319.

Lu Zhi, Yuren Lui, Leling Sun, Xianglin Xu, Yuging Yang and Lingqui Kong 1981. The development of nuclear sediment concentration gauges for use on the Yellow River. In: *Erosion and Sediment Transport Measurement*. IAHS Publication no. 133, 83–90.

Meade, R.H. 1985. Wave-like movement of bedload sediment, East Fork River, Wyoming. *Environmental Geology and Water Science*, **7**, 215–225.

Meade, R.H. and Parker, R.S. 1989. Sediment in rivers of the United States. In: *National Water Summary 1984*. US Geological Survey Water-Supply Paper, 2275, 49–60.

Meade, R.H. and Stevens, H.H. 1990. Strategies and equipment for sampling suspended sediment and associated toxic chemicals in large rivers – with emphasis on the Mississippi River. *Science of the Total Environment*, **97/8**, 125–135.

Meyer-Peter, E. and Müller, R. 1948. Formulas for bed-load transport. *International Association for Hydraulic Structures Research, Second Meeting, Stockholm*. pp. 39–64.

Milhous, R.T. 1973. *Sediment Transport in a Gravel-bottomed Stream*. Unpublished PhD thesis, Oregon State University, 232pp.

Misri, R.L., Garde, R.J. and Ranga Raju, K.G. 1984. Bed load transport of coarse nonuniform sediment. *Proceedings of the American Society of Civil Engineers, Journal of Hydraulic Engineering*, **110**(3), 312–328.

Mississippi River Commission. 1935. *Studies of River Bed Materials and their Movement with Special Reference to the Lower Mississippi River*. United States Waterways Experiment Station Paper 17.

Naden, P.S. and Brayshaw, A.C. 1987. Small and medium-scale bedforms in gravel-bed rivers. In: Richards, K.S. (Ed.), *River Channels: Environment and Process*. Blackwell, Oxford, 249–271.

Newson, M.D and Leeks, G.J. 1987. Transport processes at catchment scale. In: Thorne, C.R., Bathurst, J.C. and Hey, R.D. (Eds), *Sediment Transport in Gravel-bed Rivers*. Wiley, Chichester, 187–218.

Novak, P. 1957. Bed load meters – development of a new type and determination of their efficiency with the aid of scale models. *Transactions of the International Association Hydraulic Research 7th General Meeting, Lisbon*, Vol. 1, A9-1–A9-11.

O'Loughlin, C.L., Rowe, L.K. and Pearce, A.J. 1980. Sediment yield and water quality responses to clear felling of evergreen mixed forests in western New Zealand. In: *The Influence of Man on the Hydrological Regime with Special Reference to Representative and Experimental Basins*. IAHS Publication no. 130, 285–292.

Ongley, E.D., Bynoe, M.C. and Percival, J.B. 1981. Physical and geochemical characteristics of suspended solids, Wilton Creek, Ontario, *Canadian Journal of Earth Science*, **18**, 1365–1379.

Ongley, E.D., Yuzyk, T.R. and Krishnappan, B.G. 1990. Vertical and lateral distribution of fine grained particulates in Prairie and Cordilleran rivers: sampling implications for water quality programs. *Water Research*, **24**, 303–312.

Paintal, A.S. 1971. Concept of critical shear stress in loose boundary open channels. *Journal of Hydraulic Research*, **9**(1), 91–113.

Parker, G. 1990. Surface-based transport relation for gravel rivers. *Journal of Hydraulic Research*, **28**, 417–436.

Peters, J.J. 1971. *La dynamique de la sedimentation de la region divangante du bief maritime du fleuve Congo*. Report Internal, Laboratoire due Researches Hydrauliques (Borgerhout), Ministere des Travaux Publiques, Belgium.

Pitlick, J.C. 1987. Discussion of Hubbell, D.W. Bed load sampling and analysis. In: Thorne, C.R., Bathurst, J.C. and Hey, R.D. (Eds), *Sediment Transport in Gravel-bed Rivers*. Wiley, Chichester, 106–108.

Pitlick, J.C. and Thorne, C.R. 1987. Sediment supply, movement and storage in an unstable gravel-bed river. In: Thorne, C.R., Bathurst, J.C. and Hey, R.D. (Eds). *Sediment Transport in Gravel-bed Rivers*. Wiley, Chichester, 151–178.

Plumley, W.J. 1948. Black Hills terrace gravels: a study in sediment transport. *Journal of Geology*, **56**, 526–577.

Porterfield, G. 1972. Computation of fluvial sediment discharge. In: *Techniques of Water Resources Investigations of the US Geological Survey* (Book 3, Chapter C3). US Geological Survey, Reston, Virginia.

Proffit, G.T. and Sutherland, A.J. 1983. Transport of non-uniform sediments. *Journal of Hydraulic Research*, **21**(1), 33–43.

Reid, I. and Frostick, L.E. 1984. Particle interaction and its effect on the thresholds of initial and final bedload motion in coarse alluvial channels. In: Koster, E.H. and Steel, R.J. (Eds), *Sedimentology of Gravels and Conglomerates*. Canadian Society of Petroleum Geologists, Memoir 10, 61–68.

Reid, I. and Frostick, L.E. 1986. Dynamics of bedload transport in Turkey Brook, a coarse-grained alluvial channel. *Earth Surface Processes and Landforms*, **11**, 143–155.

Reid, I. and Frostick, L.E. 1993. Fluvial sediment transport and deposition. In: Pye, K. (Ed.), *Sediment Transport and Depositional Processes*. Blackwell, Oxford.

Reid, I. and Frostick, L.E. 1984. Particle interaction and its effect on the thresholds of initial and final bedload motion in coarse alluvial channels. In: Koster, E.H. and Steel, R.J. (Eds), *Sedimentology of Gravels and Conglomerates*, Canadian So. Petroleum Geologists, memoir 10, 61–68.

Reid, I. and Laronne, J.B. 1995. Bed load sediment transport in an ephemeral stream and a comparison with seasonal and perennial counterparts. *Water Resources Research*, **31**, 773–781.

Reid, I., Layman, J.T. and Frostick, L.E. 1980. The continuous measurement of bedload discharge. *Journal of Hydraulic Research*, **18**, 243–249.

Reid, I., Frostick, L.E. and Layman, J.T. 1985. The incidence and nature of bedload transport during flood flows in coarse-grained alluvial channels. *Earth Surface Processes and Landforms*, **10**, 33–44.

Reid, I., Frostick, L.E. and Brayshaw, A.C. 1992. Microform roughness elements and the selective entrainment and entrapment of particles in gravel-bed rivers. In: Billi, P., Thorne, C.R., Hey, R.D. and Tacconi, P. (Eds), *Dynamics of Gravel-bed Rivers*. Wiley, Chichester, 251–272.

Reid, I., Powell, D.M. and Laronne, J.B. 1996. Prediction of bedload transport by desert flash-floods. *Journal of Hydraulic Engineering ASCE*, **122**, 170–173.

Roberts, R.G. and Church, M. 1986. The sediment budget of severely disturbed watersheds, Queen Charlotte Ranges, British Columbia. *Canadian Journal of Forest Research*, **1**, 1092–1096.

Samaga, B.R., Ranga Raju, K.G. and Garde, R.J. 1986. Bed load transport of sediment mixtures. *Proceedings of the American Society of Civil Engineers, Journal of Hydraulic Engineering*, **112**(11), 1003–1018.

Schick, A.P., Lekach, J. and Hassan, M.A. 1987. Bedload transport in desert floods: observations in the Negev. In: Thorne, C.R., Bathurst, J.C. and Hey, R.D. (Eds). *Sediment Transport in Gravel-bed Rivers*. Wiley, Chichester, 617–636.

Schoklitsch, A. 1962. *Handbuch des Wasserbaues*, 3rd edn. Springer-Verlag, Vienna.

Seely, E.H. 1982. The Goodwin Creek Research Catchment: Part IV. Field data acquisition. In: *Hydrological Research Basins and their Use in Water Resources Planning, Proceedings Berne Symposium*, September 1982, Landeshydrologie, Berne, Switzerland, 173–182.

Shields, A. 1936. *Anwendung der Ähnlichkeitsmechanik und Turbulenzforschung auf die Geschiebebewegung*. Report 26, Mitteil. Preuss. Versuchsant. Wasserbau und Schiffsbau, Berlin, 24pp.

Shih, S.-M. and Komar, P.D. 1990. Differential bedload transport rates in a gravel-bed stream: a grain-size distribution approach. *Earth Surface Processes and Landforms*, **15**, 539–552.

Simons, D.B. and Richardson, E.V. 1962. Resistance to flow in alluvial channels. *Transactions of the American Society of Civil Engineers*, **127**, 927–953.

Simons, D.B. and Şentürk, F. 1977. *Sediment Transport Technology*. Water Resources Publications, Fort Collins, Colorado, 807pp.

Smith, N.D. 1974. Sedimentology and bar formation in the Upper Kicking Horse River, a braided outwash stream. *Journal of Geology*, **82**, 205–223.

Sternberg, H. 1875. Untersuchungen über Langen und Querprofil deschiebeführende Flüsse. *Z. Bauwesen*, **25**, 483–506.

Sutherland, A.J. 1987. Static armour layers by selective erosion. In: Thorne, C.R., Bathurst, J.C. and Hey, R.D. (Eds), *Sediment Transport in Gravel-bed Rivers*. Wiley, Chichester, 243–260.

Tacconi, P. and Billi, P. 1987. Bedload transport measurements by the vortex-tube trap on Virginio Creek, Italy. In: Thorne, C.R., Bathurst, J.C. and Hey, R.D. (Eds), *Sediment Transport in Gravel-bed Rivers*. Wiley, Chichester, 583–606.

Tazioli, G.S. 1981. Nuclear techniques for measuring sediment transport in natural streams – examples from instrumented basins. In: *Erosion and Sediment Transport Measurement*. IAHS Publication no. 133, 63–81.

Truhlar, J.F. 1978. Determining suspended sediment loads from turbidity records. *Hydrological Sciences Bulletin*, **23**, 409–417.

US Geological Survey. 1978. Sediment. In: *National Handbook of Recommended Methods for Water-data Acquisition*. USGS, Reston, Virginia.

Vanoni, V.A. (Ed.) 1975. *Sedimentation Engineering*. American Society of Civil Engineers Manuals and Reports on Engineering Practice, No 54, New York, 745pp.

Walling, D.E. 1974. Suspended sediment and solute yields from a small catchment prior to urbanization. In: Gregory, K.J. and Walling, D.E. (Eds), *Fluvial Processes in Instrumented Watersheds*. Institute of British Geographers Special Publication no. 6, 169–192.

Walling, D.E. 1984. Dissolved loads and their measurement. In: Hadley, R.F. and Walling, D.E. (Eds), *Erosion and Sediment Yields: Some Methods of Measurement and Modelling*, Geobooks, Norwich, 111–177.

Walling, D.E. 1989. Physical and chemical properties of sediment: The quality dimension. *International Journal of Sediment Research*, **4**, 27–39.

Walling, D.E. and Kane, P. 1984. Suspended sediment properties and their geomorphological significance. In: Burt, T.P. and Walling, D.E. (Eds), *Catchment Experiments in Fluvial Geomorphology*. Geobooks, Norwich, 311–334.

Walling, D.E. and Moorehead, P.M. 1989. The particle size characteristics of fluvial suspended sediment: an overview. *Hydrobiologia*, **176/177**, 125–149.

Walling, D.E. and Webb, B.W. 1981. The reliability of suspended sediment load data. In: *Erosion and Sediment Transport Measurement*. IAHS Publication no. 133, 177–194.

Walling, D.E. and Webb, B.W. 1982. Sediment availability and the prediction of storm-period sediment yields. In: Walling, D.E. (Ed.), *Recent Developments in the Explanation and Prediction of Erosion and Sediment Yield*. IAHS Publication no. 137, 327–337.

Walling, D.E. and Woodward, J.C. 1993. Use of a field-based water elutriation system for monitoring the *in situ* particle size characteristics of fluvial suspended sediment. *Water Research*, **27**, 1413–1421.

Walling, D.E., Webb, B.W. and Woodward, J.C. 1992. Some sampling considerations in the design of effective strategies for monitoring sediment-associated transport. In: Bogen, J., Walling, D.E. and Day, T.J. (Eds), *Erosion and Sediment Transport Monitoring Programmes in River Basins*. IAHS Publication no. 210, 279–288.

Ward, P.R.B. 1984. Measurement of sediment yields. In: Hadley, R.F. and Walling, D.E. (Eds), *Erosion and Sediment Yield: Some Methods of Measurement and Modelling*. Geobooks, Norwich, 67–70.

Werritty, A. 1992. Downstream fining in a gravel-bed river in southern Poland: lithologic controls and the role of abrasion. In: Billi, P., Thorne, C.R., Hey, R.D. and Tacconi, P. (Eds), *Dynamics of Gravel-bed Rivers*. Wiley, Chichester, 251–272.

White, W.R., Milli, H. and Crabbe, A.D. 1975. Sediment transport theories: a review. *Proceedings of the Institution of Civil Engineers*, Part 2, **59**, 265–292.

Whittaker, J.G. and Jaeggi, M.N.R. 1982. Origin of step-pool systems in mountain streams. *American Society of Civil Engineers, Journal of the Hydraulics Division*, **108**, 758–773.

Wiberg, P.L. and Smith, J.D. 1987. Calculations of the critical shear stress for motion of uniform and heterogeneous sediments. *Water Resources Research*, **23**(8), 1471–1480.

Wicks, J.M. and Bathurst, J.C. 1996. SHESED: A physically-based, distributed erosion and sediment yield component for the SHE hydrological modelling system. *Journal of Hydrology*, **175**(1–4), 213–238.

Wicks, J.M., Bathurst, J.C. and Johnson, C.W. 1992. Calibrating SHE soil-erosion model for different land covers. *Proceedings of the American Society of Civil Engineers, Journal of Irrigation and Drainage Engineering*, **118**(5), 708–723.

Wilcock, P.R. 1992. Experimental investigation of the effect of mixture properties on transport dynamics. In: Billi, P., Hey, R.D., Thorne, C.R. and Tacconi, P. (Eds), *Dynamics of Gravel-bed Rivers*. Wiley, Chichester, 110–131.

Williams, G.P. 1989. Sediment concentration versus water discharge during single hydrologic events in rivers. *Journal of Hydrology*, **111**, 89–106.

WMO. 1981. *Measurement of River Sediment*. WMO Operational Hydrology Report no. 16, Geneva, Switzerland.

WMO. 1989. *Manual on Operational Methods for the Measurement of Sediment Transport*. WMO Operational Hydrology Report no. 29, Geneva, Switzerland.

Wolman, M.G. 1954. A method of sampling coarse river-bed gravel. *Transactions of American Geophysical Union*, **35**, 951–956.

Wolman, M.G. and Schick, A.P. 1967. Effects of construction on fluvial sediment, urban and suburban areas of Maryland. *Water Resources Research*, **3**, 451–464.

Woodward, J.C. and Walling, D.E. 1992. A field sampling method to obtain representative samples of composite fluvial suspended sediment particles for SEM analysis. *Journal of Sedimentary Petrology*, **64**, 742–744.

Yuzyk, T.R., Gummer, W.D. and Churchland, L.M. 1992. A comparison of methods used to measure suspended sediment in Canada's federal monitoring programs. In: Bogen, J., Walling, D.E. and

Day, T.J. (Eds), *Erosion and Sediment Transport Monitoring Programmes in River Basins*. IAHS Publication no. 210, 289–297.

Zhou, W., Zeng, Q., Fang, Z., Pan, G. and Fan, Z. 1983. Characteristics of fluvial processes for the flow with hyperconcentration in the Yellow River. In: *Proceedings of the Second International Symposium on River Sedimentation*. Water Resources and Electric Power Press, Beijing, 618–626.

6 Bank Erosion and Instability

D. M. LAWLER[1], C. R. THORNE[2] AND J. M. HOOKE[3]
[1]School of Geography, University of Birmingham, UK
[2]Department of Geography, University of Nottingham, UK
[3]Department of Geography, University of Portsmouth, UK

6.1 INTRODUCTION

River bank erosion can present serious problems to river engineers, environmental managers and farmers through loss of agricultural land, danger to riparian and floodplain structures, increased downstream sedimentation, and occasional riverine boundary disputes. It may also signify the presence of a more general, system-wide, channel instability. Bank erosion processes are also key components in the evolution of meandering and braided river systems, floodplain renewal, the dynamics of basin sediment systems and, sometimes, hillslope instability. Sustained bank erosion and lateral channel shift can also help to maintain high biological diversity on floodplains and continually creates new opportunities for pioneering species of flora (e.g. Salo et al., 1986). In view of these many impacts, it is no surprise that a wide-ranging and sometimes contradictory literature base exists on the subject (Lawler, 1992a, 1993a; Thorne, 1991).

6.2 AIMS

Bank erosion rates and processes were relatively neglected research fields until the mid-1970s. Notable progress, however, in elucidating the many complexities and combinations of bank processes has been made recently. Nevertheless, uncertainties remain, and this is reflected in Hasegawa's (1989) assertion that the 'major controls on bank erosion remain unclear at present' and Odgaard's (1991, p. 1092) statement with regard to meander development models that 'the basic mechanisms of bank erosion and sediment transport are critical, and also poorly understood, ingredients in the stability analysis'. Against this background, then, the aims of this chapter are to:

(i) present a synthesis of our present knowledge of bank erosion rates, patterns and processes;
(ii) review briefly the main techniques available for the monitoring of bank retreat and instability;
(iii) illustrate the potential of fusing geomorphological and engineering approaches in channel stability studies;

Applied Fluvial Geomorphology for River Engineering and Management, Edited by C. R. Thorne, R. D. Hey and M. D. Newson.

(iv) outline the application of some contemporary bank stability models;
(v) highlight the key contributions in the rich scientific and applied literature dealing with bank erosion and channel instability problems.

Bank stabilisation measures are not discussed here; excellent accounts, though, can be found in Chang (1988), Hemphill and Bramley (1989) and US Army Corps of Engineers (1981), and further examples can be found elsewhere in this volume.

6.3 MEASUREMENT OF BANK RETREAT

In many river management or fluvial research situations there is a need for information on bank erosion rates through time and/or across space. A comprehensive review of bank erosion monitoring techniques appears in Lawler (1993a), and only a short summary is given here. It is important to remember that 'linear rates of bank erosion are commonly quoted ... but these rates have no meaning unless the method of measurement is given' (Brice, 1977, p. 43). Monitoring techniques, sampling frameworks and sources and magnitudes of errors should always be declared in reports.

A large range of methods has evolved, reflecting, amongst other things, different project aims, logistical constraints, site characteristics, disciplinary backgrounds of workers and, especially, timescale of interest (Lawler, 1993a). It is the timescale which provides the framework for discussion in the following sections. Within three appropriate timescales of application, eight major methods can be identified (Figure 6.1):

(a) long-timescale techniques (sedimentological evidence, botanical evidence and historical sources);
(b) medium-timescale techniques (planimetric resurvey and repeated cross-profiling);
(c) short-timescale techniques (erosion pins, terrestrial photogrammetry and the Photo-Electronic Erosion Pin (PEEP) system).

6.3.1 Long-timescale Techniques

Discussion of methods of inferring very long-term channel change from sedimentological and botanical evidence is reserved for Section II (see also Bluck, 1971; Hickin and Nanson, 1975; Nanson, 1986), and only the use of historical sources, possibly a more common method in applied studies, is dealt with here.

Historical sources include serial cartography, aerial photography, archived photogrammetric materials and, more recently, satellite imagery. They have been widely used to unravel lateral channel change over timescales of the order of 10–250 years (Figures 6.1 and 6.2). Hooke and Redmond (1989a), Lewin and Hughes (1976) and Lawler (1993a) provide more detailed accounts of available methods, advantages and problems. The technique can yield useful channel stability/instability data for larger rivers at the reach scale. A typical procedure involves: (a) obtaining maps (or air photographs) at a sufficiently large scale (e.g. 1:2500 or 1:10 000 in the UK) of the river course for different, known, dates; (b) performing cartometric or photogrammetric checks to ascertain source accuracy, and correcting where necessary (e.g. Hooke and Perry, 1976; Maling,

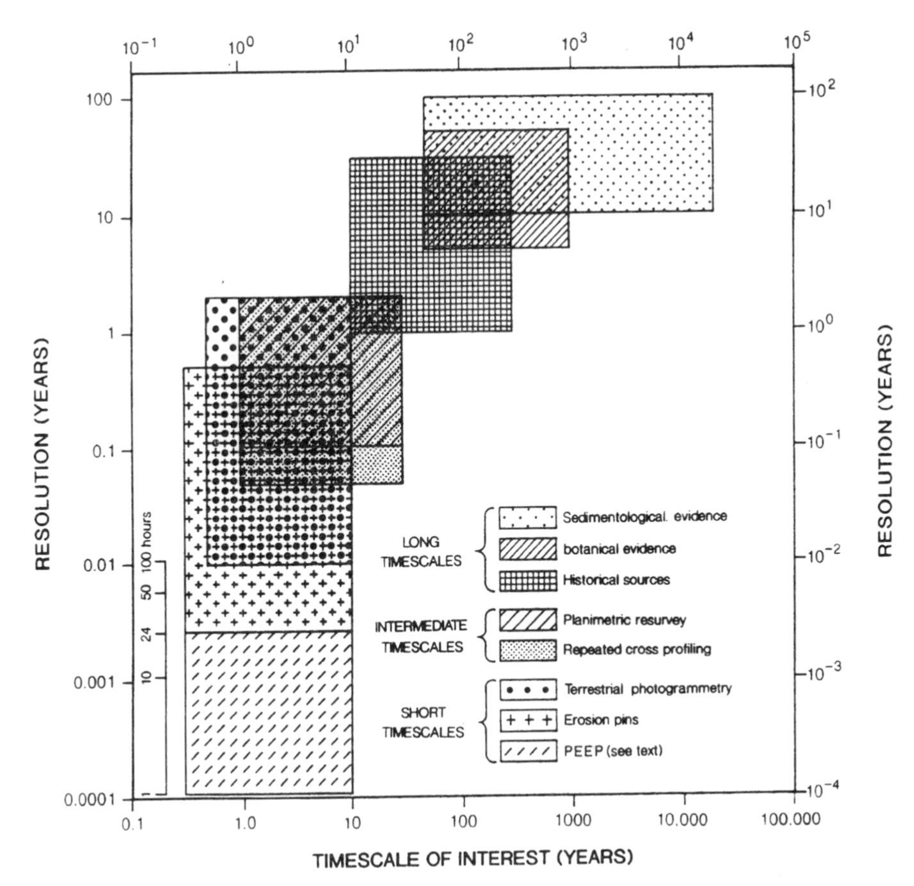

Figure 6.1 Appropriate and applicable timescales for techniques to measure bank erosion and lateral channel change rates (from Lawler, 1993a)

1989); (c) superimposing the maps at a common scale (e.g. Figure 6.2) by co-registering fixed detail (e.g. floodplain buildings); (d) measuring the linear or areal lateral shift of banklines between given dates; and (e) dividing by the intervening period to provide a time-averaged bank erosion rate. There is no guarantee, of course, that this rate will continue into the future, nor that past channel change has been regular and uniform, but the technique does provide a valuable longer-term perspective. Good illustrations of the method can be found in Hooke (1977a, 1980, 1989b) and Lewin (1987). For example, work on the Brahmaputra River, Bangladesh, has been used to establish the average rate and variability of bank erosion, both of which are useful inputs to floodplain management policy and the optimisation of resource development (Thorne et al., 1993). The recent development of analytical photogrammetry has improved the rigour and efficiency of studies based on aerial photograph superimpositions (see Lane et al., 1993). More recently still, Geographic Information System (GIS) methods have been applied to determine planform change in rivers and quantify error involved (Downward et al., 1994).

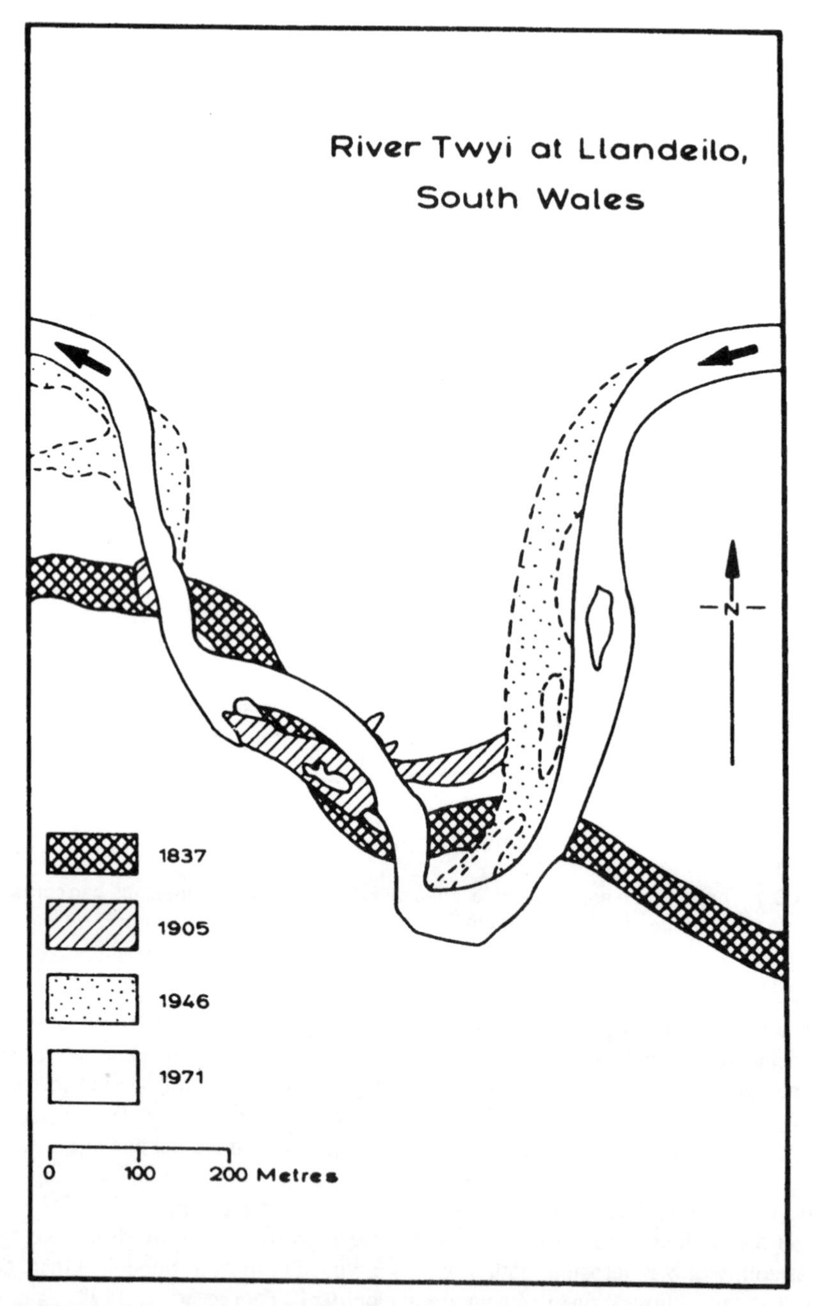

Figure 6.2 An example of superimposition of serial historical sources to document lateral channel change rates and patterns: the River Twyi, South Wales, UK (after Lewin and Manton, 1975)

6.3.2 Medium-timescale Techniques

The two techniques in this group are appropriate for the investigation of channel change over periods of between about one and 30 years, although usually at a fairly low temporal resolution (Figure 6.1).

Planimetric Resurvey

The planimetric resurvey method involves a periodic survey of the position of the bank edge using (a) simple offsets from a monumented baseline, (b) a compass-and-tape technique (Thorne, 1981), or (c) a theodolite or Electronic Distance Measurer (EDM) (Lawler, 1993a), as illustrated in Figure 6.3. A recent development is the use of Global Positioning Systems (GPS) for channel surveys. All these methods allow bankline migration rates to be determined from the superimposition of successive field surveys plotted at the same scale. Problems in the identification of the banktop may lead to significant errors, particularly if different personnel are used for successive field surveys. An agreed criterion of bankfull stage may be helpful; Schumm (1961) and Wolman (1955), for example, suggest maximum break of slope or minimum width–depth ratio, although vegetational criteria are occasionally employed (e.g. Leopold and Skibitske, 1967). The minimum width–depth ratio has become very popular. It may be helpful simply to think in

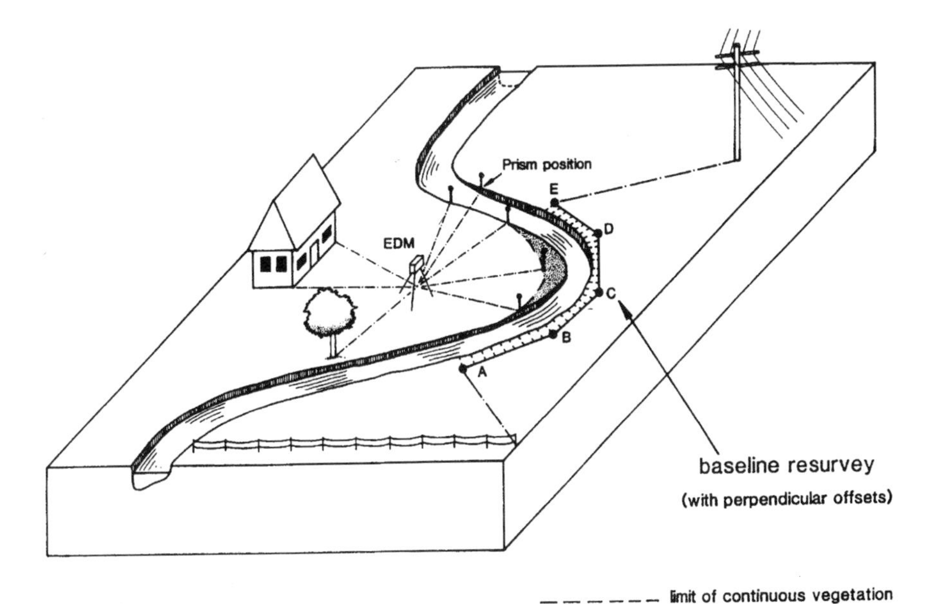

Figure 6.3 Typical methods of surveying river planform for lateral channel change investigations. Measurements can be taken with the EDM arrangement shown, compass-and-tape methods, or level-and-staff tacheometry. Alternatively, a baseline (e.g. A–E) can be established around the bend from which orthogonal offset measurements to the bank edge can be taken (from Lawler, 1993a)

terms of a 'bank edge', which is a topographic/vegetative break identified in the field. Definitions should be clearly declared in reports. It is stressed, however, that planimetric resurvey will only reveal changes to the bankline position, and not progressive under-cutting of the bank below floodplain level.

Repeated Cross-section Surveys

To overcome this problem of insensitivity to 'in-channel' bank changes, a network of monumented and repeatedly surveyed cross-profiles should also be established to identify changes within the whole cross-section (Figure 6.4). Profiles can be resurveyed with hydrographic sidescan sonar, echo-sounders (e.g. May, 1982), transit level-and-staff, EDM, or purpose-built profilers (e.g. Hudson, 1982; Pizzuto and Meckelnburg, 1989). On small streams a simple horizontal bar or survey staff can be laid across the channel, from which vertical distances to the channel boundary are measured (e.g. Park, 1975). The 'Tausendfuessler' of Ergenzinger (1992) is a more precise device with which to accomplish this, and can be used even during flood events. Ritchie et al. (1993) have recently applied airborne laser altimetry to surveys of streambank and gully erosion in Oklahoma and Mississippi.

Whatever the method used, surveys can be carried out more frequently than is usually meaningful for planform resurveys, to obtain additional data on seasonal and subseasonal changes to channel form. These temporal variations in rate are often crucial to correct identification of the processes responsible for bank erosion and the critical geometry for bank failure. Cross-sections should be established orthogonally to the channel, although subsequent channel evolution may cause a loss of orthogonality over time. Durable pegs marking each end of the profile (or baseline) should be positioned to allow for subsequent bank retreat. Lawler (1993a) discusses problems, precautions, procedures and recommendations in more detail (see also Leopold, 1973; Hughes, 1977; Lewin, 1990).

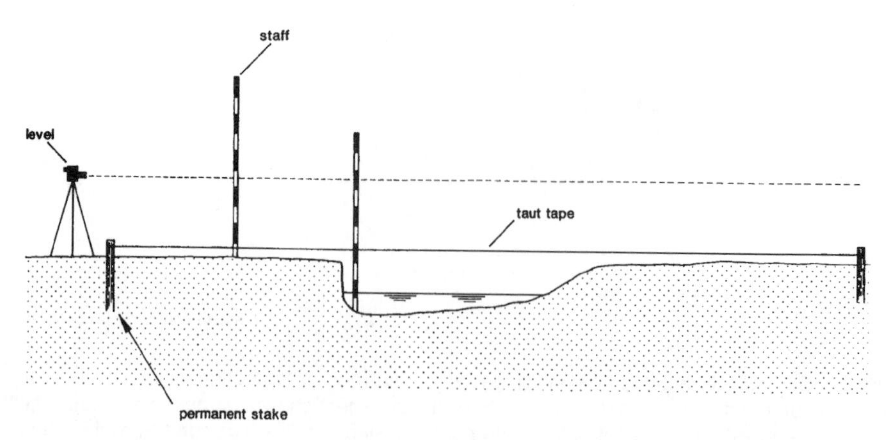

Figure 6.4 Cross-section survey of streams with the standard levelling technique (from Lawler, 1993a)

6.3.3 Short-timescale Techniques

Short-timescale techniques have usually been applied in those investigations which aim to identify causes, processes and mechanisms of bank retreat, rather than simply to establish recession rates. Figure 6.1 suggests appropriate timescales from a few months to a few years with an achievable time resolution of minutes to days. With this time resolution it becomes possible to relate erosional and depositional activity directly to hydraulic, hydrological and meteorological events or event combinations, and the capability to infer processes is thereby enhanced. Although a variety of miscellaneous or semi-quantitative methods have emerged (e.g. painted sections and tagged sediment (Thorne and Lewin, 1979), repeated photography (Harvey, 1977) and sediment traps (Lawler, 1993a)), the three main measurement techniques remain those separately discussed below.

Erosion Pins

Erosion pin deployment is the most popular technique of this category, because pins are simple and cheap to use, suitable for a reasonably wide range of fluvial contexts, and pick up spatial patterns of bank erosion important to process inference (Lawler, 1993a). Usually, a length of steel or brass rod is simply inserted horizontally into the bank face, leaving a known, marked amount protruding. Marking with heat-shrinkable coloured sleeving (as used in the electronics industry) is often better than paint because it is more durable and provides a 'cleaner' measurement datum (Lawler, 1989, 1993a). As retreat proceeds, more unmarked pin is exposed, which can be measured with vernier callipers at appropriate intervals (weekly or monthly recordings are common, with additional readings after major events). Problems include direct or indirect disturbance or reinforcement effects (Thorne, 1981), and movement, loss or burial of pins in dynamic situations, especially in large systems (Lawler, 1993a). The technique works reasonably well, especially on banks formed in alluvium which are not undergoing mass failure (e.g. sliding blocks (see Section 6.8) simply rip pins out) or severe erosion rates (i.e. where the amount of bank retreat between site visits exceeds the length of erosion pin, leading to pin loss). An initial reconnaissance investigation of bank erosion rates using serial historical sources (see above) may sometimes be useful in deciding on the suitability of pin (and other) methods.

Terrestrial Photogrammetry

Terrestrial photogrammetry has been used by Collins and Moon (1979) and Painter et al. (1974) specifically for bank retreat monitoring, and by Lane et al. (1994) for channel change studies in a proglacial environment. Stereoscopic images of areas of bank face are traditionally obtained using a phototheodolite mounted above stations established on the opposite bank (Collins and Moon, 1979; Lawler, 1989, 1993a). A three-dimensional model of the site can then be reconstructed using photogrammetric plotting machines (Collins and Moon, 1979). The method works best when camera stations can be placed reasonably close to sites which are relatively dynamic. 'Subtraction' of serial bank-surface images provides retreat/accretion rate data at, potentially, a very large number of points.

Lawler (1989, 1993a) notes the following advantages: minimisation of spatial sampling problems; negligible contact with the bank face; exchange of expensive fieldwork time for more economic office time; and the simultaneous recording of other, qualitative data relevant to erosion process interpretation (e.g. seasonal vegetation growth, soil moisture changes, progressive slippage and degradation of failed blocks). In short, terrestrial photogrammetry has the potential to yield much more information on the spatial structure of bank retreat – which can be important in process identification and river management decisions – with a minimum of site disturbance. More recently, the emergence of analytical photogrammetry has considerably refined the traditional analogue techniques (see Lane et al. (1993) for a detailed discussion), including the generation of serial Digital Terrain Models (DTMs) for use in channel evolution studies. Terrestrial photogrammetry, however, is inapplicable (with current technology) to subaqueous portions of bank profiles. Also, the capital costs involved in analogue photogrammetric approaches (e.g. for purchase of phototheodolite and plotting/analytical machines) are high. Renting equipment or sub-contracting these services is usually possible, however, and, with analytical photogrammetry, the need for expensive metric cameras may be relaxed if a network of control points is established in the field of view (Lane et al., 1993). Terrestrial photogrammetry, in general, may become more popular with the advent of computer-aided image processing techniques.

Photo-Electronic Erosion Pin (PEEP) System

Cross-section resurveys, erosion pins and repeat photogrammetry simply reveal net changes in bank form or position occurring between field visits. A new technique, however, recently developed at the University of Birmingham, UK, is the Photo-Electronic Erosion Pin (PEEP) system (Lawler, 1991, 1992a, 1992b, 1994). This method, for the first time, allows quasi-continuous erosion rate data to be collected automatically. The PEEP sensor is a simple and inexpensive optoelectronic device composed of an array of photosensitive cells enclosed within a clear acrylic tube. It is inserted into an eroding feature much like the conventional erosion pin (Figure 6.5). The sensor outputs an analogue millivolt signal directly proportional to its protruding length (i.e. that which is exposed to natural daylight). It is, in effect, a self-reading erosion pin. Once connected to a datalogger by cable, the PEEP instrument can provide a near-continuous record of bank retreat and advance at the measurement point. This allows the magnitude, timing and frequency of erosional and depositional activity to be ascertained with much greater precision than is available with traditional manual methods. It can also be applied to gully walls, drainage or irrigation channels and tidal creeks.

An example of how the precise timing of a falling-stage erosion event can be revealed by the PEEP system is given in Figure 6.6. Such erosional events can therefore be matched up much more rigorously with temporal fluctuations in the bank stresses applied by suspected controlling variables (e.g. as measured by continuous stage, discharge, flow velocity, soil moisture or temperature records); this enhances process inference capabilities and helps to disentangle competing hypotheses concerning the dominant erosion mechanisms. Figure 6.6, for instance, clearly demonstrates that material removal took place 43 hours after the flood peak on the Upper River Severn, and points to the importance of mass-instability processes here rather than excess fluid shear stresses (which might be

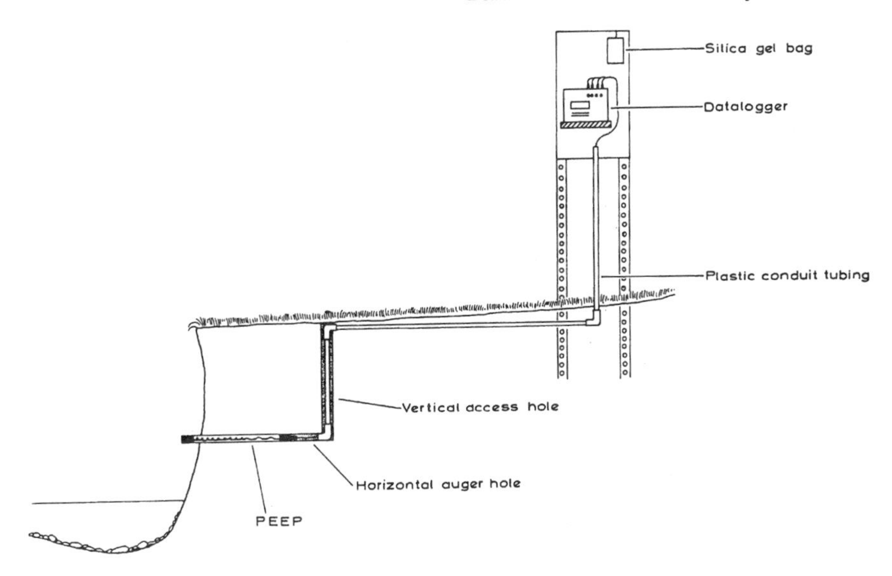

Figure 6.5 Typical installation of a Photo-Electronic Erosion Pin (PEEP) system at a bank site for the automatic, unattended collection of quasi-continuous data on bank erosion rates and the magnitude, frequency and timing of individual erosion events (after Lawler, 1992b)

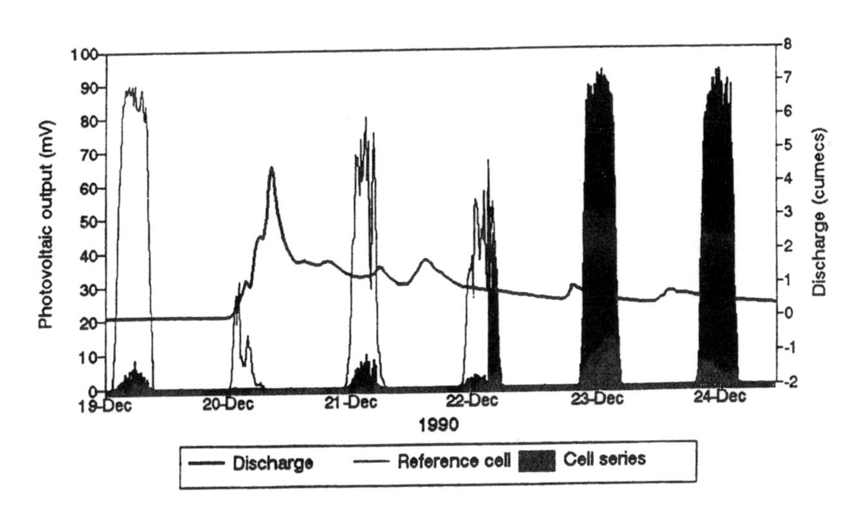

Figure 6.6 The precise timing of a bank erosion event (at 1330 GMT on 22 December 1990), revealed by the automated Photo-Electronic Erosion Pin monitoring system installed on a bend of the Upper River Severn, mid-Wales, UK. Note that the bank retreat event follows some 43 hours after the flow peak, indicating that excess boundary shear stresses were probably not strongly causative in this instance (from Lawler and Leeks, 1992)

expected to act synchronously with the flow peak). The method has also allowed a truer picture of bank erosion and deposition dynamics and temporal variability to be developed, and revealed that bank surfaces can be much more active than previously supposed (e.g. Lawler, 1994). PEEP sensors are compatible with almost all dataloggers, require no power in the field, and their low cost means that detailed sensor networks can be installed (to pick up spatial patterns) within reasonable budgets. Applications of the technique within the UK since 1989 on the lowland River Arrow, Warwickshire, and the upland River Severn, mid-Wales, have been extremely encouraging (Lawler, 1991, 1992a, 1994; Lawler and Leeks, 1992; Lawler et al., 1997). Real-time monitoring of bank erosion/deposition is also possible with telemetry systems.

6.3.4 Field Indicators of Erosion Problems

It is not always possible actually to monitor bank retreat rates. Frequently, an assessment of erosion problems and processes has to be made from visual, qualitative evidence alone (see the stream reconnaissance sheets and notes of Thorne (1991)). Failed blocks of bank material at the base of the bank and/or tension cracking on the floodplain surface indicate that mass failure is important, and retreat may be rapid. Overhang development indicates that flows well below bankfull level are effective in producing erosion. If a notch feature is evident, accentuated on the inside of a bend or narrow parts of the channel, then boat wash may be the cause on navigable waterways. Notches can also be produced by a combination of fluvial entrainment and freeze–thaw action on banks (Lawler, 1984, 1993b) and by piping processes (Hagerty, 1991; Thorne, 1991). Freeze–thaw action can lead to substantial accumulations of spalled aggregates (crumbs) at the base of the bank in winter. Similar accumulations in summer indicate the presence of desiccation (Section 6.6.2). If the banks are steep and free of vegetation, then they tend to be susceptible to fluvial erosion. However, bare banks in themselves do not necessarily indicate that severe erosion is occurring. Other bank features may indicate erosion, and possibly the longer-term rate of erosion (e.g. the pattern of 'embayments' between riparian trees; exposed tree roots; undermined fence posts).

6.4 BANK EROSION RATES

Even within the same hydroclimatic environment, bank erosion rates vary widely with river system scale, channel geometry, bank material characteristics and in response to other channel changes taking place. Variability has been noted at all spatial and temporal scales (Hooke, 1977a, 1980; Lawler, 1989, 1994; Thorne et al., 1993), which has to be assessed before confidence limits can be placed on derived mean rates, and as an aid to strong process inference.

Hooke (1980) and Lawler (1993a) have presented comprehensive tabulations of world lateral channel change rates of around $0.01-1000\,\mathrm{m\,a^{-1}}$, and a loose positive relationship between drainage basin area and mean erosion rate can be identified (Figure 6.7). Within some basins, however, maximum erosion rates have been found to occur in the middle reaches (Lewin, 1987) and this may relate to a possible peaking in stream power in these piedmont regions (Graf, 1982; Lawler, 1992a). Hooke (1980) and Lewin (1987), amongst

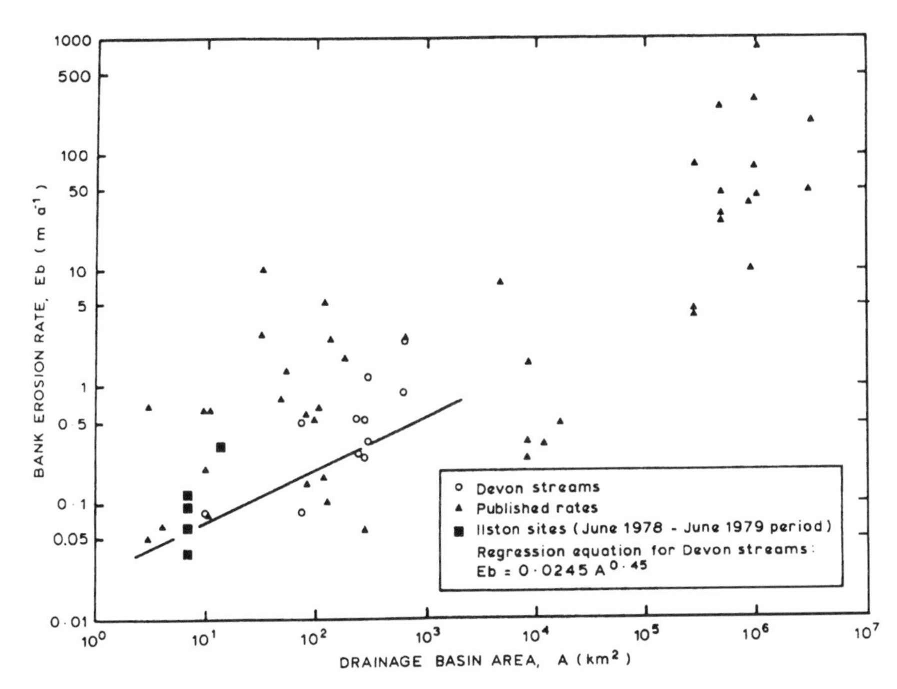

Figure 6.7 The relationship between drainage basin area and bank erosion rates for selected world rivers (from Hooke (1980), with additional data for the River Ilston, South Wales, UK, from Lawler (1984))

others, also found that, at the reach scale, change in one part of the system appeared to instigate or accelerate change in adjacent reaches; this illustrates the importance of adopting an integrated approach to river management problems – one that recognises the feedbacks and downstream effects in fluvial systems which can render undesirable purely site-specific engineering investigations and projects (e.g. Leeks et al., 1988). Finally, some researchers (e.g. Hooke, 1980) have noted a tendency for bank erosion rates to increase over the last 100 years, and the possibility of longer-term fluctuations in change rates, perhaps linked to wider environmental changes or system evolutionary behaviour (e.g. Hickin, 1983), should be borne in mind when managing river systems or extrapolating rates obtained over short monitoring periods. Spatial and temporal variability needs to be placed in the longer-term context of channel change, especially where changes in the hydrological and sediment regime of a river are known to be occurring in response to changing basin land use, climate or river management practices.

6.5 RIVER BANK RETREAT PROCESSES: INTRODUCTION

Box 6.1 defines the differences between bank erosion, failure, retreat and erodibility. There is a large variety of bank erosion mechanisms and causes, and usually processes act in concert to produce bank retreat (Wolman, 1959; Hooke, 1979; Thorne, 1991; Lawler, 1984, 1992a). The combinations themselves are likely to vary at seasonal and subseasonal

Box 6.1 Definition of terms

Bank erosion

Detachment, entrainment and removal of bank material as individual grains or aggregates by fluvial and subaerial processes

Bank failure

Collapse of all or part of the bank *en masse*, in response to geotechnical instability processes

Bank retreat

Net linear recession of bank as a result of erosion and/or failure

Bank advance

The opposite of bank retreat, i.e. net linear streamwise change in bank surface position, as a result of deposition of sediment or *in situ* swelling of bank materials

Bank erodibility

The ease with which bank material particles and aggregates can be detached, entrained and removed (normally by flow processes)

timescales. Establishing the dominant mechanisms can be very difficult, but a detailed consideration of the geomorphology and hydrology of the bank, the river reach and drainage basin contexts, as well as the geotechnical and hydraulic characteristics of the site, usually provides the basis for significant insights which can then be fed into the engineering aspects of the problem.

Twidale (1964) provided an early list of bank erosion processes, following Wolman's (1959) statement on the importance of erosive forces and erodibility changes. A comprehensive list of possible bank retreat processes that the river manager may have to contend with has been produced by the US Army Corps of Engineers (1981). These may be grouped into the following categories: weakening processes, direct fluid entrainment, and mass failure (Lawler, 1992a). Any one of these process groups can emerge as dominant, depending on site characteristics, especially near-bank hydraulic fields, bank height and material properties.

Clearly, it is unduly restrictive to interpret all bank erosion simply as a function of excess boundary shear stress. Although such approaches work well enough in rivers of sufficiently high energy to overcome the entrainment resistance of the sediments, and may be adequate for cohesionless boundary materials (e.g. Hickin and Nanson, 1984), they are less useful in the case of cohesive banks which can be very resistant to fluid entrainment. The considerable scatter in simple bivariate plots of flow indices against at-a-site erosion rates indicates that other variables are also significant in such contexts (see Figure 6.8). In fact, most recent work emphasises the interplay of a variety of weakening, fluvial erosion and mass failure processes in producing serious bank erosion, instability and/or retreat (e.g. Hooke, 1979; Thorne, 1982, Lawler, 1992a; Beatty, 1984). Box 6.2 contains a list of

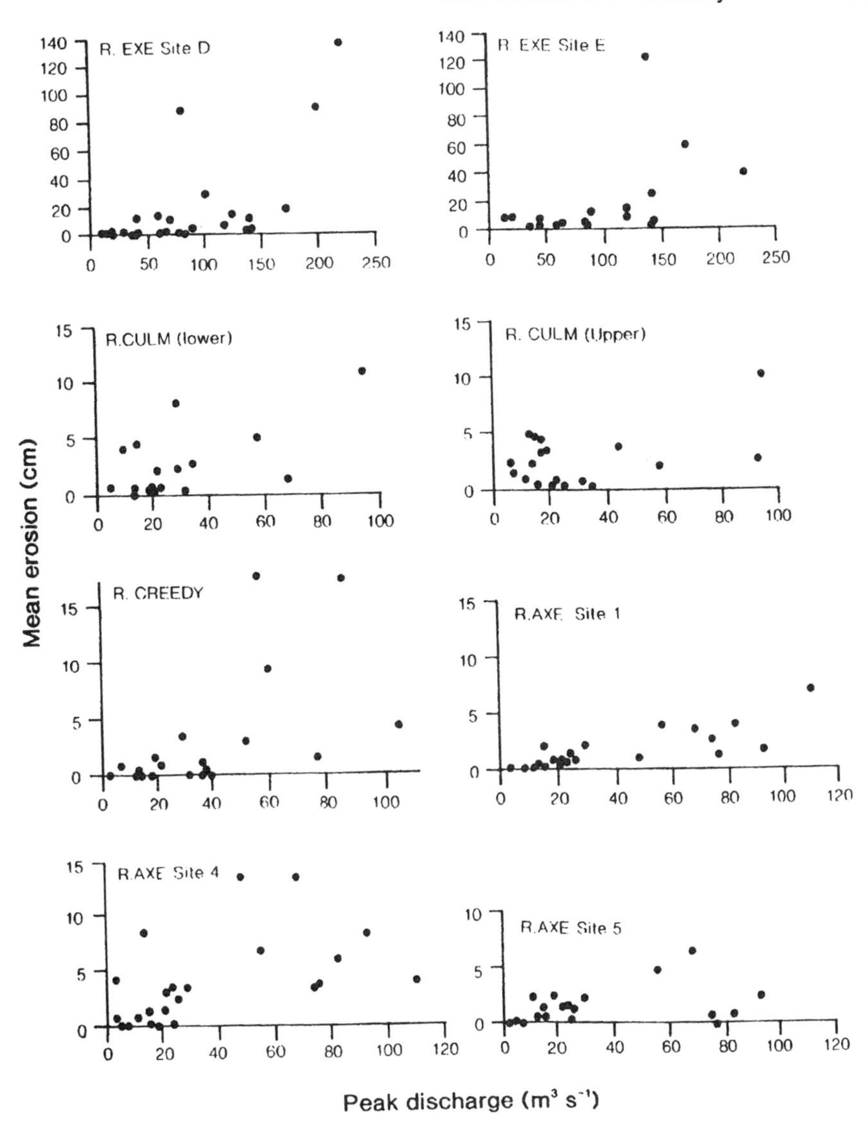

Figure 6.8 Relationships between spatially averaged amounts of bank erosion for various river sites in Devon, UK, and peak discharge in the measurement interval (from Hooke, 1977b)

general factors which influence the various bank retreat processes and mechanisms, and some of these are briefly discussed in Neill and Yaremko (1989).

Each major group (weakening, fluvial erosion and mass failure) is discussed separately in the succeeding sections. A suite of miscellaneous processes and mechanisms involved in bank retreat is then summarised in Section 6.9.

Box 6.2 Influential factors in bank erosion systems

Subaerial processes	Microclimate, especially temperature Bank composition, especially silt/clay percentage
Fluvial processes	Stream power Shear stress – but actual distribution influenced by position of primary currents Secondary currents Local slope Bend morphology – cs, curvature Bank composition Vegetation Bank moisture content
Mass failure	Bank height Bank angle Bank composition Bank moisture content or pore water pressure/tension

6.6 BANK WEAKENING PROCESSES

Unlike cohesionless sediments, cohesive fine-grained bank material can undergo substantial temporal change in its resistance to erosion, or erodibility. Although many possible agents can alter erodibility (US Army Corps of Engineers, 1981), the three most important weakening processes are usually pre-wetting, desiccation and freeze–thaw activity.

6.6.1 Pre-wetting

A number of studies have revealed that banks become more erodible when wet (Hooke, 1979; Knighton, 1973; Wolman, 1959), though few quantitative data exist on the patterns, controls and precise effects of higher moisture contents. Banks may be wetted by descent of the wetting front from above and inwards from the bank face as a result of precipitation. Bank storage processes (e.g. Sharp, 1977) during the passage of high-flow events will also increase bank moisture contents, as will the rise of the ground water table. Wolman (1959) and Hooke (1979) showed that late-winter flows, acting on thoroughly wetted banks, are more erosive than comparable events earlier in the season. Indeed, a multiple regression analysis of erosion rates revealed that bank moisture content, as represented by a simple Antecedent Precipitation Index (Gregory and Walling, 1973, p. 187), was as important as peak flow magnitude (Hooke, 1979). When very wet, seepage from banks can cause sapping of localised areas of bank face (Twidale, 1964). Similarly, piping processes (e.g. Hagerty, 1991) can be very important in bank erosion systems.

6.6.2 Desiccation

Even in humid temperate environments, substantial soil moisture deficits can be built up within river banks. However, there is almost a complete absence of published data on temporal variations in river bank material moisture contents, although some instantaneous moisture profiles have been published (e.g. Lawler, 1987) and further data sets exist in Hasan (1991) and Masterman (1994). Similarly, river bank thermal regime information is virtually non-existent, although the data that are available, for humid temperate environments, show that bank surfaces, especially near the banktop, can undergo substantial cooling in winter (Lawler, 1993b) and intense heating in summer (Lawler, 1992a). For example, early-morning heating rates monitored continuously on east-facing banks of the River Arrow, Warwickshire, UK, rose to $7^{\circ}C\,h^{-1}$, with peak temperatures well in excess of 30°C and diurnal temperature ranges of 20°C. Such strong heating was related to spalling (peeling away of micropeds and slabs of bank material as desiccation progressed (Lawler, 1992a).

There are number of references to the apparent effects of desiccational activity on bank surfaces (e.g. Oxley, 1974; Bello et al., 1978; Thorne, 1982; Lawler, 1992a), but contrasting views emerge. Many workers suggest that hard, dry banks are more resistant to fluid entrainment (e.g. Thorne, 1982). Others have found that desiccational activity can encourage higher retreat rates due to (a) the direct spalling of particle aggregates from the dryer upper bank surfaces which collect at the bank-foot and become available for entrainment during subsequent stage rises (e.g. Oxley, 1974; Lawler, 1992a), and (b) the cracking up and incipient exfoliation of the bank surface which, subsequently, allows flood water access around and behind unstable crumbs and ped structures (Lawler, 1992a). Reconciliation of these opposing views probably lies in the degree of bank cracking experienced, which is related to clay fabrics. The related process of slaking has been outlined by Thorne and Osman (1988b). Slaking is the subsequent 'bursting' of bank crumbs and peds during saturation because of a build-up intracrumb air pressures, generated by the influx of water into the soil pores during rapid immersion.

6.6.3 Freeze–Thaw Processes

A number of researchers in cool and temperate environments have pointed to the influence of freeze–thaw processes in conditioning cohesive bank material for later fluid entrainment. However, Lawler (1992a) showed that all the quantitative studies which have identified freeze–thaw activity as the dominant control on bank erosion rates (e.g. Hill, 1973; Leopold, 1973; McGreal and Gardiner, 1977; Blacknell, 1981; Gardiner, 1983; Lawler 1986, 1987; Stott et al., 1986) have been for relatively small river systems (basin areas $< 85\ km^2$). This is probably because frost action, being largely independent of the feedback controls in a typical fluvial system (Lawler, 1995), can inflict only a limited amount of 'damage' to bank surfaces, and this finite effect is, of course, relatively more important in smaller rivers where total erosion rates are usually low (see Figure 6.7).

Although different types of freeze–thaw activity are identifiable, and vertical zonations on individual banks common (e.g. Lawler, 1987), the effects of needle ice are probably the most significant. Needle ice is a globally widespread form of ice segregation in which elongated crystals of ice grow in the direction of nocturnal cooling (i.e. orthogonal to the bank surface), during moderately subzero air temperature depressions (Lawler, 1988). The crystals often lift and/or incorporate material in the process, depending on the nature and frequency of heat- and moisture-flux perturbations in the growing environment (Outcalt, 1971; Branson et al., 1992). For example, Lawler (1993b) estimated that sediment yields from river banks due to needle ice represented between 32% and 43% of total bank retreat over a two-year monitoring period. During ablation, the incorporated sediment is transported downslope, and sometimes directly to the stream below. Much, however, remains on the bank surface as a skin of disrupted, highly erodible material. Such sediment drapes are readily removed by subsequent stage rises, even if of modest magnitude (e.g. Leopold, 1973; Lawler, 1987, 1993b). Crumbs and larger aggregations of particles (e.g. in the size range 2–20 mm (Lawler, 1987)) are frequently transported downslope by freeze–thaw action in winter.

At larger scales, ice lenses can reduce cohesion by wedging peds apart and, in rivers that freeze over in the winter, serious damage can be done to the banks by cantilevers of ice attached to the banks and by ice rafts during the spring thaw (Church and Miles, 1982; Lawson, 1983; Walker et al., 1987). Bank instability caused by the prior cutting of thermoerosional niches is another important retreat mechanism where permafrost exists in the bank stratigraphy (Church and Miles, 1982). Niches are cut near the level of the river surface, in response to melting by the relatively warmer river water (Lawson, 1983).

6.7 FLUVIAL EROSION PROCESSES

Bank erosion consists of the detachment of grains or assemblages of grains from the bank surface, followed by fluvial entrainment (Box 6.1). Flow forces of lift and drag may be entirely responsible for both detachment and entrainment but, on cohesive materials especially, particles, aggregates and micropeds are loosened and partially or completely detached by weakening processes prior to entrainment (see Section 6.6). Entrainment occurs when the motivating forces due to the flow and to the downslope component of particle mass overcome the forces tending to hold the grain or aggregate in place. With regards to entrainment capabilities, a number of workers have demonstrated reasonable relationships between bank erosion rates and near-bank velocity perturbations or velocity distribution asymmetry (Knighton, 1973; Hasegawa, 1989; Pizzuto and Meckelnburg, 1989). The nature of the grain or grain assemblage entrained depends on the engineering properties of the bank material and, in particular, on whether the material is non-cohesive or cohesive (Thorne, 1982).

6.7.1 Erosion of Cohesive Banks

Cohesive, fine-grained bank material is usually eroded by the entrainment of aggregates or crumbs of soil rather than individual particles which are bound tightly together by electromechanical cohesive forces. No complete theory for this type of erosion exists, but empirical studies show that resistance to entrainment is not strongly controlled by soil mechanical properties such as compressive or shear strength (Arulanandan et al., 1980). Instead, erodibility is more a function of a complex combination of physico-chemical, intergranular bonding forces which act to control the resistance of a particle or aggregate to detachment by fluid shear applied at the bank surface. These properties are themselves heavily dependent on (a) the mineralogy, dispersivity, moisture content and particle size distribution of the bank material, and (b) the temperature, pH and electrical conductivity of the pore and eroding fluids (Thorne, 1978, pp. 67–81; Grissinger, 1982;

Kamphuis and Hall, 1983; Nickel, 1983; Osman and Thorne, 1988; Thorne and Osman, 1988b). Entrainment occurs when the motivating forces overcome the resistive forces of friction and cohesion. Field and flume experiments show that intact, undisturbed cohesive banks are much more resistant to fluvial entrainment than are non-cohesive banks (Thorne, 1982).

In addition, cohesive soils are often poorly drained and rapid drawdown in the channel following a high-flow event may generate positive porewater pressures in the bank that act to reduce friction and effective cohesion which, in extreme cases, can lead to liquefaction (a complete loss of strength and flow-type failures).

Lift forces are now known to be very significant in sediment transport processes (see Chapter 5). These are a function of the velocity differential that exists in the flow between the upper and lower surfaces of a particle or aggregate. If lift forces exceed the submerged weight of a particle, then it moves up off the channel boundary and into the flow. Aggregates have relatively low submerged weights (being around half the bulk density of quartz particles); their critical lift forces are therefore quite low, and subsequent fluid entrainment is likely to be relatively easily effected. Near-bank flows can be extremely complex, however, and difficult to measure without the aid of electromagnetic flow meters (Bathurst et al., 1979; Thorne et al., 1985).

6.7.2 Erosion of Non-cohesive Banks

Non-cohesive bank material is usually detached and entrained grain by grain, and may leave a pronounced notch marking peak stage achieved. Stability depends on the balance of forces acting on individual particles (Figure 6.9). The motivating forces are the downslope component of submerged weight, and fluid forces of lift and drag. Resisting these forces are the interparticle forces of friction and interlocking. The fluid forces can be represented by boundary shear stress when calculating the stability of non-cohesive

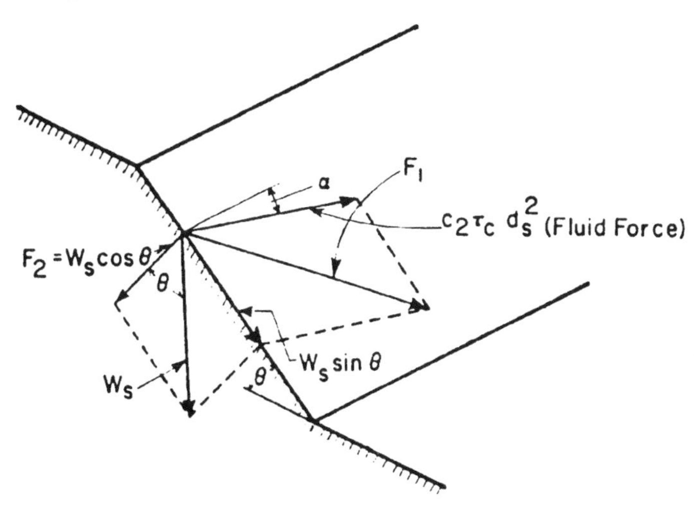

Figure 6.9 Forces on a particle at the surface of a submerged non-cohesive bank (from Thorne, 1982)

grains, because this scales on the same parameters of flow intensity as lift and drag forces (Wiberg and Smith, 1987). Non-cohesive bank materials in alluvial channels are usually derived from in-channel deposits. Consequently, they may exhibit close packing and, in the case of non-spherical grains, imbrication. This can greatly enhance their resistance to fluid entrainment, and can allow non-cohesive banks to stand at steep bank angles. If the degree of packing and imbrication is reduced (e.g. by the removal of fine interstitial material by wind or water, frost heave and ice segregation, and weakening by positive porewater pressures) then restoring forces are decreased; this greatly increases susceptibility to entrainment and can decrease the angle of repose by as much as 70% (Carson and Kirkby, 1972). However, truly non-cohesive banks are rare in natural channels. Usually, even banks formed in coarse materials exhibit some operational cohesion due to the binding effects of vegetation roots and effective cohesion introduced by porewater suction.

6.7.3 Vegetation and Bank Erosion Processes

Bank vegetation affects practically all aspects of bank erosion and stability, although the net results of these effects may not always be significant in engineering terms. The effect of vegetation is probably greatest on small rivers. Vegetation impacts are complex and their overall impact may be beneficial, neutral or deleterious to bank erodibility and stability. Vegetation influences on near-bank flow hydraulics and bank erodibility are considered first, while rooting effects on soil strength, bank hydrology and bank morphology are dealt with in Section 6.8.4.

Flourishing vegetation has been found to reduce the effectiveness of fluid erosion by between one and two orders of magnitude (Kirkby and Morgan, 1980; Smith, 1976). It has long been recognised that the stems and trunks of bank vegetation materially alter the distributions of near-bank velocity and boundary shear stress (Kouwen, 1987; Kouwen and Li, 1979). The results of recent research indicate that the potential impact of bank vegetation on the overall flow capacity of the channel is strongly related to the width–depth ratio. Masterman and Thorne (1992, 1994) developed a semi-theoretical analysis of composite channel roughness. Their simulations show that vegetation resistance on the banks is only significant in channels with width–depth ratios less than about 12. Most natural rivers have higher width–depth ratios than this. In the near-bank zone, flexible vegetation reduces the velocities and shear stresses experienced at the soil surface, primarily by shifting the virtual origin of the velocity profile upwards away from the soil boundary, and secondarily by damping turbulence. However, if velocities and stresses are sufficiently large, flexible vegetation becomes prone and its effectiveness in protecting the bank from erosion is diminished.

The stiff stems of trees act on the flow in a quite different fashion from the flexible stems of grasses and shrubs. Tree trunks act as large-scale roughness elements, reducing near-bank average velocities primarily through form drag. Masterman and Thorne (1992, 1994) synthesise the results of threoretical and empirical studies to suggest that it is the spacing and pattern of trees which determines their impact on flow resistance (Li and Shen, 1973). While trees do reduce mean velocities in the bank zone, they may also produce areas of accelerated flow and heavy turbulence associated with their wake zones. Hence, local pockets of erosion may be promoted by tree trunks, leading to possible destabilisation of themselves as well as the bank. Pizzuto and Meckelnburg (1989) show how different tree types and densities affect bank erosion rates on Brandywine Creek, Pennsylvania. Trees may also shade and suppress shorter riparian vegetation that helps to bind bank materials, leading to increased channel widths in some forested basins (Murgatroyd and Ternan, 1983).

6.8 BANK FAILURE MECHANISMS

When a section of bankline fails and collapses, blocks of bank material slide or fall towards the toe of the bank. They may remain there until broken down *in situ* or entrained by the flow. Mass failures can be analysed in geotechnical slope stability terms, or as the result of fluvial and gravitational forces which overcome resisting forces of friction, interlocking and cohesion (Schofield and Wroth, 1968; Thomson, 1970; Thorne and Tovey, 1981; Little et al., 1982; Osman and Thorne, 1988). However, much of the underlying theory was originally developed for hillslopes, dams, embankments and canal banks, and its applicability to streambanks can be questionable. Such analyses consider the balance of forces which promote and resist the downward motion of a given block of material. Blocks can be brought to failure by over-deepening due to bed scour, or undercutting which increases the bank angle. An increase in degree of bank material saturation can occur as a result of seepage into the bank during high-flow events, heavy precipitation or snowmelt. This substantially increases the mass of material above a potential failure surface and can be enough to instigate failure, even in the absence of over-deepening or over-steepening. The mechanics of the resulting failures depend on the size and geometry of the bank and the engineering properties of the bank material. Banks may be broadly classified on this basis as cohesive, non-cohesive or composite. Many of the data on the variables required for bank stability analyses (including those presented below) can be effectively gathered using the Stream Reconnaissance Record Sheets of Thorne (1991). A useful summary of bank failure modes and characteristics was compiled by Hey et al. (1991) (Figure 6.10).

6.8.1 Stability of Cohesive Banks

Cohesive banks often fail along a discrete surface deep within the bank. This is because in cohesive materials the shear strength increases less quickly than shear stress with depth. The stability of the bank with regard to mass failure depends on the balance of motivating and resisting forces on the most critical potential failure surface. The motivating force is the tangential component of the weight of the potential failure block. An increase in this force occurs when fluvial erosion leads to an increase in the bank height or bank angle and catastrophic failure occurs when the critical value of height or angle is reached. Block mass also increases when the moisture content of the soil increases. Of particular significance is the switch from submerged to saturated conditions during hydrograph recessions following high flows. This can cause the bulk unit weight of the soil to double, and can trigger drawdown failures, even without the generation of significant excess porewater pressures. A decrease in resisting force can arise because of a reduction in the cohesion or friction angle, or a decrease in the length of the intact potential failure surface

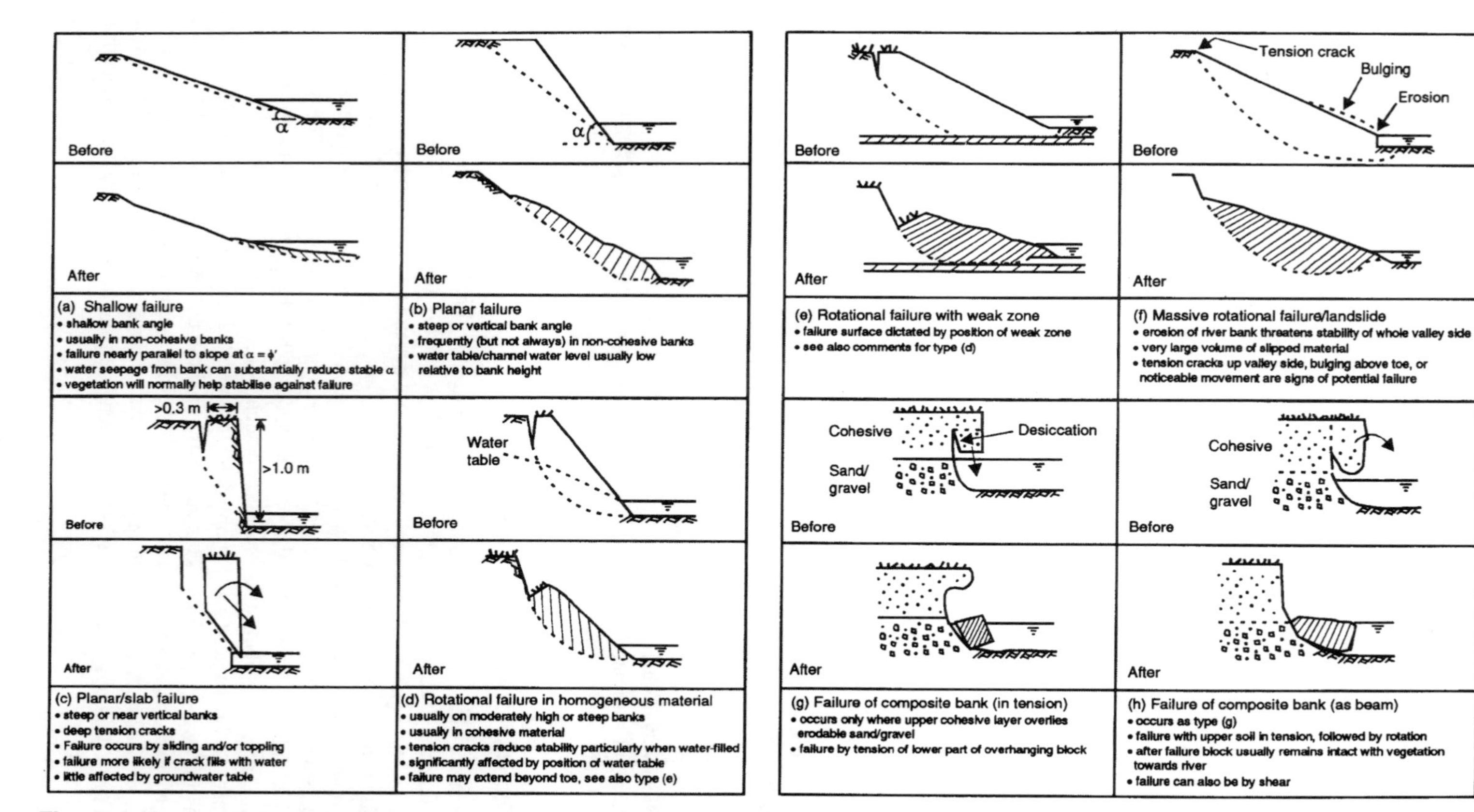

Figure 6.10 Modes and characteristics of bank failure (from Hey et al., 1991)

brought about by tension cracking (Figure 6.10c). Cohesion is reduced by softening of the soil due to high moisture content or to leaching of minerals from the soil. Friction is reduced by *in situ* weathering or by porewater pressure changes during the passage of a flood. The length of the intact surface – and hence the restoring force on the potential failure block – is reduced by tension cracking behind steep banks.

Little hard, quantitative, data on the timing of bank failures exist (see Section 6.3.3). However, many mass failures of cohesive banks occur a few hours or days after, rather than during, high-flow events (e.g. Figure 6.6). This is related to the time-lag in the development of high porewater pressures, as well as to the removal of the buttressing effect of the flood waters themselves. Also, failures in weakly cohesive soils may occur on the rising stage of flow events due to the reduction of effective strength that accompanies wetting of partially saturated soils. This is because soil suction related to moisture deficits actually contributes substantially to the strength of weakly cohesive soils. As the bank is wetted during the rising stage, its strength is reduced and this can trigger mass failure ahead of the flood peak. Consequently, the nature and timing of failure may only be weakly related to discharge, velocity or boundary shear stress peaks.

The shape of the most critical failure surface depends mostly on the bank height. In low, steep banks this surface is almost planar, and a slab or block of soil slides downwards and outwards before toppling forwards into the channel (Figure 6.10c). This is the most common type of failure for eroding river banks. One simple method of analysing slab failure is with the Culmann formula (cited in Selby, 1982, p. 138), which is based on total, rather than effective, stress principles (ignoring porewater pressure phenomena) and assumes a planar shear surface along which slab or wedge failure takes place (Figure 6.11). The critical bank height for planar failure is given by:

$$H_c = \frac{4c}{\gamma} \cdot \frac{\sin a \cos \phi}{\{1 - \cos(a - \acute{Y})\}} \tag{6.1}$$

where H_c = critical height of bank (m), c = bank material cohesion (kPa), γ = unit weight of bank material (kN m^{-3}), a = bank slope angle (°), ϕ = bank material friction angle (°).

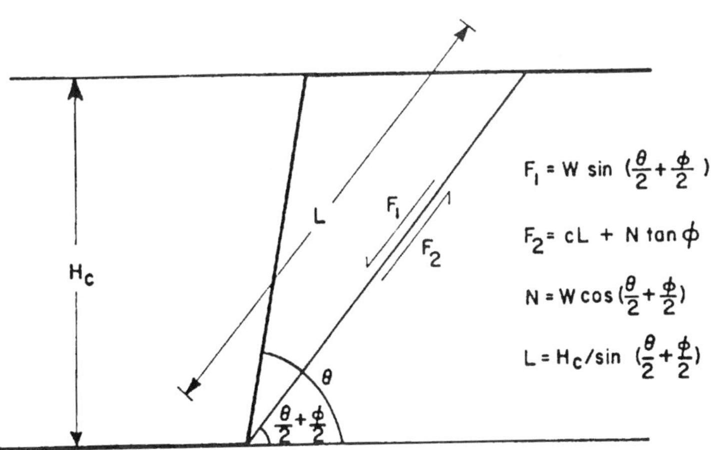

Figure 6.11 Culmann stability analysis for plane slip failure on a river bank (from Thorne, 1982)

Lohnes and Handy (1968), albeit in a highly simplified analysis, found good agreement between observed and predicted maximum heights of cuts in friable loess. Thorne (1978, pp. 57–60) notes that some of the assumptions of a Culmann analysis, however, may not be justified, particularly the exclusion of the possibility of tension cracking (Figure 6.10c), and presents a related equation for the critical height of a vertical bank, H'_c, in which a tension crack may extend to around one-half of total bank height (Figure 6.12) (Thorne, 1982):

$$H'_c = (2c/\gamma)\tan(45 + \phi/2) \tag{6.2}$$

Using Eqn 6.2, Lawler (1992a) constructed a family of curves which predict, for a range of saturated (i.e. worst-case condition) bulk unit weights, cohesions and friction angles, the critical height of river bank needed for wedge or slab failure to take place (Figure 6.13).

Given that bank height generally increases in a downstream direction, Lawler (1992a, 1995) argued that there should be a zone within each drainage basin where bank heights first exceed critical stability values for the boundary materials in question (which will themselves change downstream). Figure 6.14 shows, for example, how, by combining Culmann-type geotechnical theory with hydraulic geometry relationships (Leopold and Maddock, 1953), a rough prediction can be made of the location within a given basin of the scale threshold for wedge failure (Lawler, 1992a). For example, for a typical material where $\phi = 20°$, $c = 10$ kPa, and $\gamma = 19$ kN m^{-3}, vertical banks should be stable up to 1.5 m in height ($H'_c = 1.5$) (Eqn 6.2; Figure 6.14a). To calculate the threshold drainage area 'required' for bank height to exceed this critical value, H'_c can be inserted into the relevant hydraulic geometry equation expressing maximum channel depth (a simple and imperfect bank height surrogate, if detailed information on bank geometry is unavailable) as a function of drainage basin area. For the River Onny, Shropshire, UK, this is (Figure 6.14b):

$$D_{max} = 0.271\,A^{0.386} \qquad (r^2 = 0.869,\ n = 12,\ p < 0.001) \tag{6.3}$$

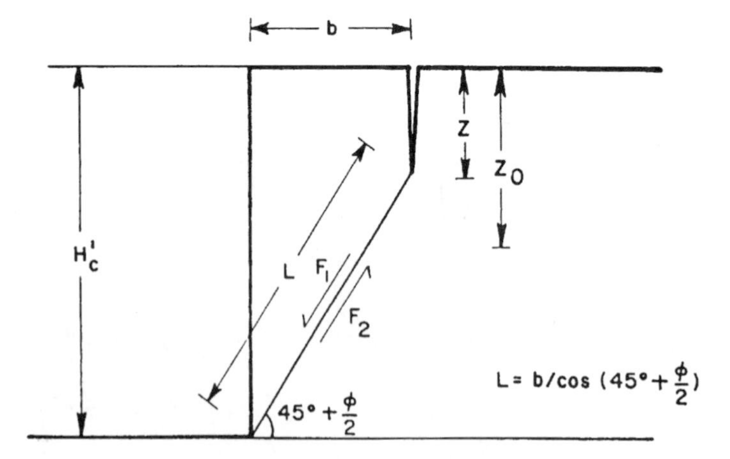

Figure 6.12 Culmann stability analysis for plane slip failure on a river bank, modified to incorporate effects of tension cracking behind the bank surface (from Thorne, 1982)

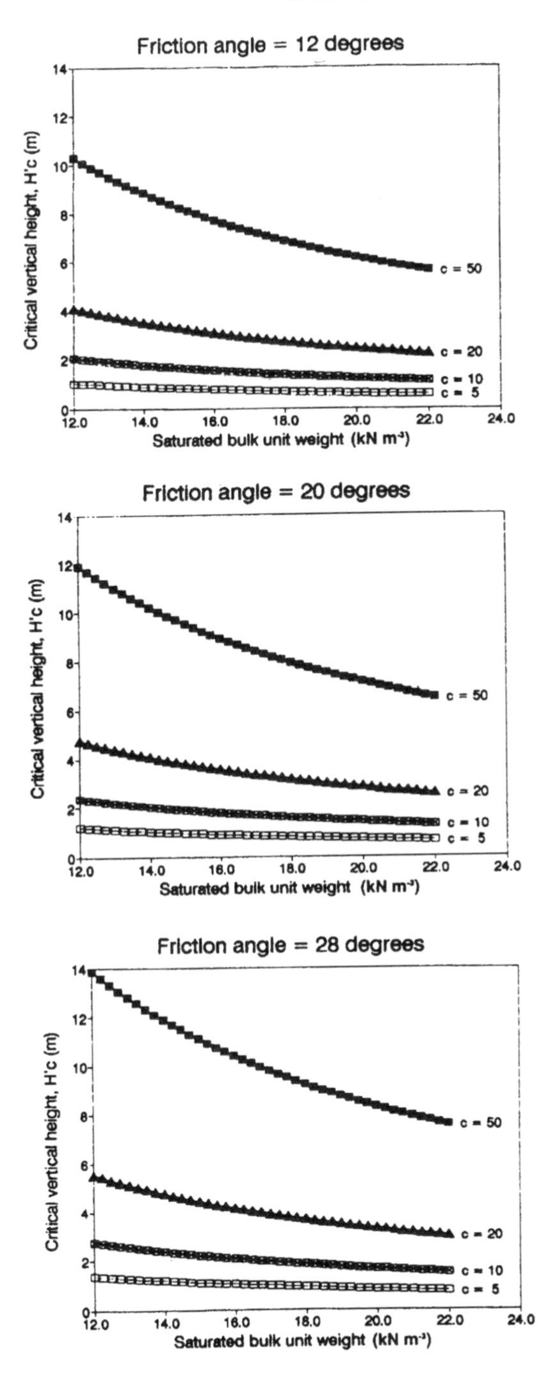

Figure 6.13 Family of Culmann-type bank stability curves, which allow prediction of maximum vertical heights for banks of given values of cohesion (*c*), friction angle and saturated bulk unit weight (from Lawler, 1992a)

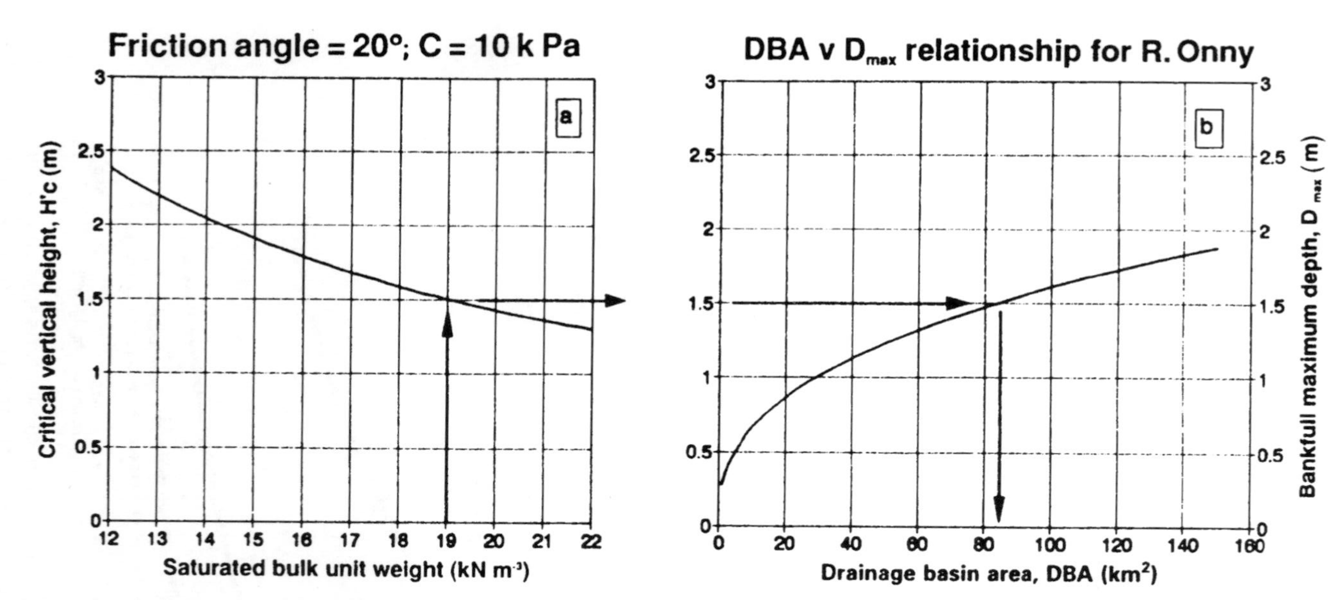

Figure 6.14 Combined use of (a) Culmann-type stable bank-height prediction for a hypothetical material with (b) hydraulic geometry relationship ($D_{max} = 0.271A^{0.386}$) to obtain approximate scale and drainage area thresholds for wedge failure on river banks (from Lawler, 1992a)

where D_{max} = maximum bankfull depth (m) and A = drainage basin area (km^2) (Lawler, 1992a). Then, solving Eqn 6.3 with $H_c' \sim D_{max} = 1.5$, we obtain a predicted threshold drainage area, A_t, of 84 km^2. This would represent a minimum drainage basin area required before wedge failure became a quantitatively significant mechanism responsible for bank retreat (Figure 6.14). Upstream of this zone, bank retreat should be mainly attributable to other mechanisms. It is for this reason that mass failure is likely to be the dominant bank retreat process only in large systems or at the downstream end of small systems (Lawler, 1995). This type of analysis has been developed more generally by Lawler (1995) into the concept of overlapping spatial zoning of dominant process groups.

Osman and Thorne (1988) and Thorne and Osman (1988b) developed a stability analysis for low, cohesive banks subject to slab-type failure. They showed that it is possible to predict the critical height for mass failure as a function of the initial geometry of the bank, the amount of toe scour and over-steepening, and the geotechnical properties of the bank material under average and 'worst case' conditions. Thorne (1988) tested the analysis using data from the Long Creek catchment in north Mississippi, and Thorne and Abt (1993) presented a spreadsheet program in LOTUS 1-2-3 (first written by Lyle Zevenbergen) to facilitate the application of the factor of safety and failure block dimensions. The Osman–Thorne approach represents a simple tool in the analysis of bank stability for reaches of channel in response to degradation and/or over-steepening. It is not intended for site-specific use, but gives a fair indication of the overall stability of the banks with regard to slab-type failure. For example, application to Mimmshall Brook, Hertfordshire, UK, showed how dredging for land drainage and flood control can trigger bank instability which leads to complex response when critical bank heights are exceeded (Darby and Thorne, 1992). Subsequently, Darby and Thorne (1996) further developed the Osman–Thorne model to include the effects of confining pressures during high-flow stages, and positive porewater pressures during rapid drawdown in banks vulnerable to slab-type failure.

In high, gently sloping banks the failure surface is curved and the failure block tends to rotate backwards as its toe slides outwards into the channel (Figure 6.10e). This occurs because the orientation of the principal stresses in gently sloping banks changes significantly with depth. Rotational slips may be analysed using well established procedures developed in geotechnical engineering (e.g. Schofield and Wroth, 1968; Fredlund, 1987), and are relatively less common than planar failures. Theoretical research at Colorado State University coupled with field observation of failure geometries in unstable banks indicates that rotational slips are mostly found in cohesive banks with angles less than about 60° (Osman, 1986; Thorne 1988). The STABL2 slope stability program was used by Osman (1986) and by Thorne (1988) to assess bank stability with respect to rotational slip, and gave reasonable results when applied to a variety of undercut and over-steepened banks.

6.8.2 Stability of Non-cohesive Banks

Non-cohesive banks fail by the dislodgement of individual clasts or by shear failure along shallow, very sightly curved slip surfaces. Deep-seated failures are rare because in non-cohesive banks shear strength usually increases more quickly than shear stress with depth. In drained banks, failures are associated with friction angle reduction due to loss of close packing and imbrication, and with bank angle increase due to basal scour and over-steepening. Such processes can be analysed using Shields dimensionless shear stress, but with account taken of the side-slope angle (Thorne, 1978).

6.8.3 Stability of Composite Banks

Thorne and Lewin (1979) and Thorne and Tovey (1981) present a conceptual model of the failure processes on composite river banks (i.e. banks with clearly differentiated layers of material of very different erodibility). These studies were based on observations of the Upper River Severn in Wales, UK, where fine-grained units overlie coarse basal gravel layers in the bank profiles. The lower layers were preferentially eroded (at rates an order of magnitude higher than the silty units above) to produce overhangs or cantilevers (Figure 6.10g, h). These overhangs subsequently failed in shear, beam or tensile modes

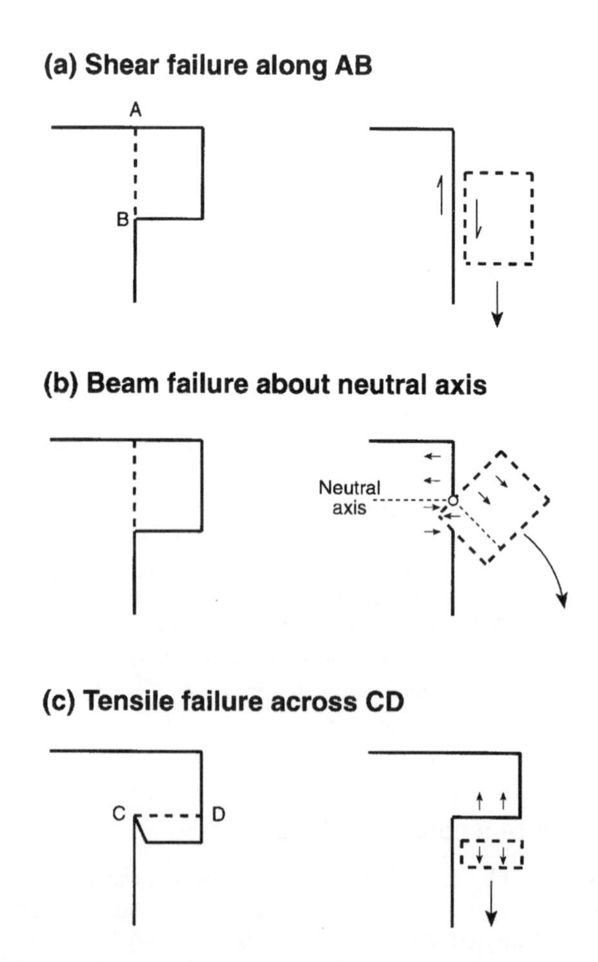

Figure 6.15 Modes of cantilever failure on composite river banks (from Thorne and Tovey, 1981)

(Figure 6.15) to deliver sizeable blocks of material to the lower bank zone. Blocks degraded subsequently at the bank toe. Thus, bank retreat here was seen as a combination of fluid entrainment of basal gravels and, later, mechanical failure of the upper overhanging sections, the precise mode of failure being dependent upon the geotechnical properties of the upper bank material. Fluid entrainment of basal material is, of course, vital to the whole process (Thorne, 1978; Pizzuto, 1984). Hence it is the stability of the lower bank and toe zone which is crucial to the retreat rate of composite banks in the medium to long term.

6.8.4 Vegetation and Bank Stability

Thorne (1990) and Thorne and Osman (1988a) argue that vegetation can either increase or decrease bank stability, depending on the type of vegetation, bank geometry and bank materials. The roots and rhizomes of bank vegetation have significant impacts on the mechanical properties of the bank material. Soil is strong in compression but weak in tension: plant roots are the reverse. Roots can therefore provide reinforcement to soils through the addition of tensile strength. For example, Thorne et al. (1981) showed that the tensile strength of root-permeated Mississippi bank materials was 10 times greater than root-free samples. Waldron (1977) reports that the roots of mature willow can increase the shear strength of soils 100%. The net result, however, may be only a marginal increase in bank stability. For example, Masterman (1994) and Amarasinghe (1993) both failed to find any significant effect of root reinforcement on operational bank material strength despite extensive and intensive testing. This is because bank strengthening is often offset against an increase in loading due to the surcharging effects of the biomass and wind throw forces of trees on the bank, and because root reinforcement is ineffective below the rooting depth of plants. Especially in large-scale cohesive banks, critical failure surfaces may well be located deep inside the bank, passing below the root zone. Roots are often cited as providing lines of weakness in a bank, particularly if the plants are dying or dead. However, while this may be a problem on levees, there are few documented cases of failures particularly associated with root castes. The role of root reinforcement remains equivocal and further research is necessary to delineate the conditions under which roots do and do not materially strengthen a river bank.

Often, bank failures are thought to be associated with saturated conditions and, particularly, with rapid drawdown in the channel. However, bank hydrology is a relatively under-researched topic, with a few notable exceptions such as Sharp (1977). In the case of hydrology-related failures, bank vegetation plays a beneficial role in suppressing failure, because saturated conditions occur less frequently in a vegetated bank where the plants operate as 'wicks' drawing water from the soil and generating negative pore pressures (Masterman, 1994). Also, vegetated banks have more open soil fabric and tend to be better drained. Often, engineers cite plant roots as providing preferential drainage paths which can lead to piping, especially in levees. Hard evidence of such effects is rare, however. In studies of tropical slopes, Collison and Anderson (1996) show that vegetation can either decrease or increase slope stability depending on the permeability of the soil. The effect of vegetation on streambank stability certainly requires further research.

At present there is no procedure to take the effects of bank vegetation into account explicitly when analysing mass failure processes in river banks. Until such a procedure is developed, the best approach is to incorporate vegetation effects into the parameters used to represent bank material unit weight, effective friction angle and effective cohesion. Further examination of these biomechanical issues can be found in Waldron (1977), Gray and Leiser (1983), Gray and MacDonald (1989), Thorne (1990) and Masterman (1994).

6.8.5 Basal Endpoint Control

The concept of basal endpoint control (Carson and Kirkby, 1972; Richards and Lorriman, 1987) can be used to link bank processes into the sedimentary system of the channel as a whole. The flow supplies sediment to the basal area from upstream by longstream sediment transport and removes sediment downstream. There may also be a net input or output of sediment laterally from cross-stream transport associated with transverse velocity components and secondary circulation, especially in bends. The various bank processes discussed in Sections 6.5–6.9 can also supply sediment to the basal area (Figure 6.16). Mechanical failures involve the movement of blocks of bank material directly from the intact bank to the basal area. The removal of failed material from the basal area depends

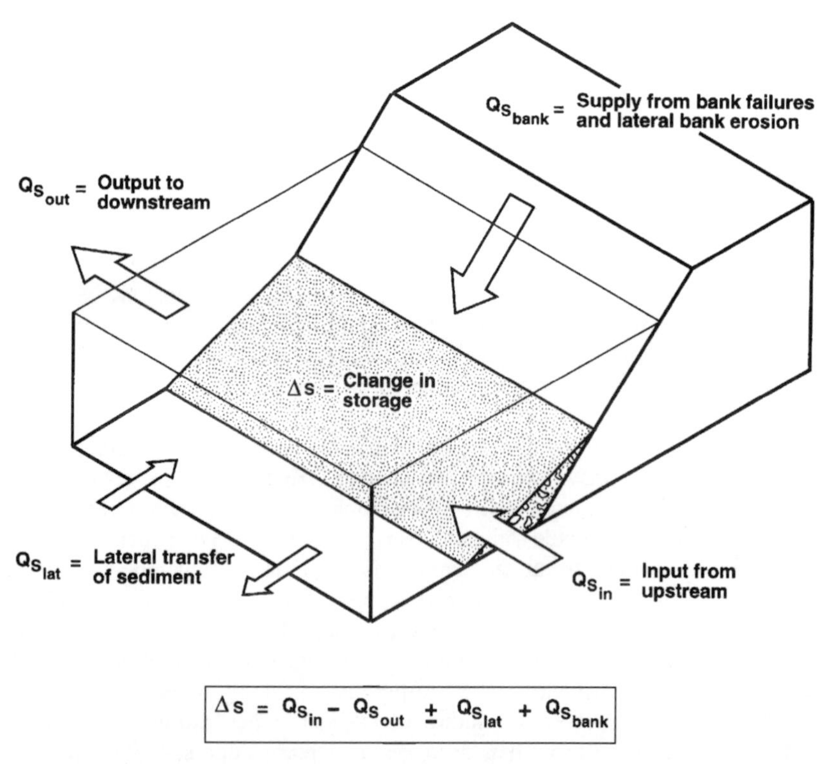

$$\Delta s = Qs_{in} - Qs_{out} \pm Qs_{lat} + Qs_{bank}$$

Figure 6.16 Schematic representation of sediment fluxes to and from river bank basal zones (from Thorne and Osman, 1988a)

entirely on its entrainment by the flow, possibly aided by subaerial weathering processes when emergent. The amount and duration of basal sediment storage depends on the balance between the rates of supply from bank processes and removal by fluvial entrainment. Three states of 'basal endpoint control' of the bank profile can be stated on this basis.

(i) *Equilibrium condition.* A balance exists between the amount of material delivered to the base and that removed by fluid entrainment. No change in basal elevation or bank angle occurs. The bank recedes by parallel retreat at a rate determined by the degree of fluvial activity at the base.
(ii) *Accumulation condition.* Bank processes, plus inputs from upstream and lateral flow, supply material to the base at a higher rate than it is removed. Basal accumulation results, decreasing average bank angle and height, and increasing bank stability. The rate of basal sediment supply then decreases, which tends to encourage equilibrium conditions.
(iii) *Scour condition.* Basal scour has excess capacity over the supply from bank erosion, bank failures and fluvial inputs. Basal lowering and undercutting occur, increasing the bank height and angle. The rate of sediment supply to the base thus increases, shifting the system towards equilibrium.

Hence, the retreat rate over the medium and long term is fluvially controlled, regardless of the nature of the bank material, the actual erosion and weathering processes, and the precise failure mechanisms responsible for bank retreat. For fine-grained bank materials which are eroded as individual grains, however, entrainment potential in the basal area is seldom a limiting condition. Therefore, given that most rivers can transport as wash load almost all fine particles supplied to them (see Carson, 1971), the concept of basal endpoint control is possibly more appropriate for (coarser) bank materials that contribute significantly to the bed material load or where retreat is achieved by block failure. Where bank-toe vegetation plays an important role in trapping fines and building berms or benches at the base of banks, the basal endpoint control model may have applications to fine-grained systems also.

6.9 OTHER BANK PROCESSES

An interesting sequence of progressive bank erosion has recently been reported by Davis and Gregory (1994) on a small stream in a woodland area in southern England. This involves partial damming of the stream and flow diversion by large organic debris, washout/sapping of gravels at the bank base to create a riparian cavity, and slow progressive subsidence and collapse of the cavity roof over a period of months. Wind-generated waves can also be important, especially on wide, exposed waterways, and their impact depends on their size and energy (a function of fetch and wind velocity), and angle of incidence. Boat-generated effects usually have been considered of minor importance in affecting bank stability, although they may be locally significant (e.g. Hey, 1991; Thorne et al., 1996). The erosive effect is concentrated at or near the water surface, and it is at this level that a notch may be developed. Some localised bank erosion may be caused by trampling processes associated with fishing, footpath usage and stock grazing. Generally, these activities contribute to erosion after an embayment or sloping bank has been formed by fluvial processes. Trampling on the edge of the bank may lead to transient surcharging and increased instability. Once mechanical damage destroys the riparian vegetation, a wide variety of otherwise ineffectual processes can produce significant erosion of the fallow surface. Hence, stock-watering places and anglers 'pegs' are significant areas of erosion on many rivers.

6.10 CONCLUSIONS

The following points emerge from this chapter.

1. The bank erosion problem cannot simply be interpreted in terms of boundary shear stress alone. High, in-channel flows produced by intense rainstorms can result in a variety of retreat mechanisms and process combinations such as pre-wetting weakening processes, direct fluid entrainment, and mass failure brought about by basal scour, undercutting or excess porewater pressures. Therefore significant correlations between, for example, discharge and erosion rate do not necessarily mean that fluid entrainment is the dominant process in a given bank erosion system.
2. System scale is an important control over which process or process group achieves dominance in any given bank erosion system. For example, although on the low banks of small rivers, mass failure mechanisms are very spatially and temporally restricted, at the larger sites which present management problems to river engineers, these processes need to be explicitly addressed by stability analyses which are as much geotechnical as hydraulic in nature.
3. Bank material characteristics are also significant in erosion rates, erosion processes and failure modes, and no single measure can summarise resistance to all processes. For example, fine-grained materials are highly resistant to fluid shear but tend to have low shear strength and are susceptible to mass failure. There are clear differences between the behaviour of cohesive and non-cohesive materials.
4. Vegetation can have both positive and negative effects on bank stability, although it generally inhibits fluid entrainment. Impacts on soil strength and bank hydrology are probably significant, but are at present unpredictable in general.
5. Monitoring techniques are still relatively crude and labour-intensive. It is only recently that techniques such as the PEEP system have emerged which allow monitoring of retreat rates automatically and quasi-continuously, to generate high-resolution time series that can be related to fluctuations in applied stresses.

6.11 ACKNOWLEDGEMENTS

The authors are grateful for the assistance of Richard Masterman and Heather Lawler in the production of this chapter.

6.12 REFERENCES

Amarasinghe, I.V. 1993. *Effects of Bank Vegetation in Waterways with Special Reference to Bank Erosion, Shear Strength, Root Density and Hydraulics*. Unpublished PhD thesis, The Open University, Milton Keynes, UK.

Arulanandan, K., Gillogley, E. and Tully, R. 1980. *Development of a Quantitative Method to Predict Critical Shear Stress and Rate of Erosion of Natural Undisturbed Cohesive Soils*. Report GL-80-5, US Army Engineers, Waterways Experiment Station, Vicksburg, Miss.

Bathurst, J.C., Thorne, C.R. and Hey, R.D. 1979. Secondary flow and shear stress at river bends, *Proceedings of the American Society of Civil Engineers, Journal of the Hydraulics Division*, **105**, 1277–1295.

Beatty, D.A. 1984. Discussion of 'Channel migration and incision on the Beatton River', *Proceedings of the American Society of Civil Engineers, Journal of Hydraulic Engineering*, **110**, 1681–1682.

Bello, A., Day, D., Douglas, J., Field, J., Lam, K. and Soh, Z.B.H.A. 1978. Field experiments to analyse runoff, sediment and solute production in the New England region of Australia, *Zeitschrift für Geomorphologie N.F. Suppl. Bd.*, **29**, 180–190.

Blacknell, C. 1981. River erosion in an upland catchment. *Area*, **13**, 39–44.

Bluck, B.J. 1971. Sedimentation in the meandering River Endrick, *Scottish Journal of Geology*, **7**(2), 93–138.

Branson, J., Lawler, D.M. and Glen, J.W. 1992. The laboratory simulation of needle ice. In: Maeno, N. and Hondoh, T. (Eds), *Physics and Chemistry of Ice*. Hokkaido University Press, Sapporo, Japan, 357–363.

Brice, J. 1977. *Lateral Migration of the Middle Sacramento River, California*. US Geological Survey Water Resources Investigations, 77/052, NTIS Report PB-271 662.

Carson, M.A. 1971. *Mechanics of Erosion*, Pion, London, 174 pp.

Carson, M.A. and Kirkby, M.J. 1972. *Hillslope Form and Process*. Cambridge University Press, Cambridge, 475 pp.

Chang, H.H. 1988. *Fluvial Processes in River Engineering*. Wiley-Interscience, New York, 432 pp.

Church, M. and Miles, M.J. 1982. Discussion of 'Processes and mechanisms of bank erosion'. In: Hey, R.D., Bathurst, J.C. and Thorne, C.R. (Eds), *Gravel Bed Rivers*. Wiley, Chichester, 259–268.

Collins, S.H. and Moon, G.C. 1979. Stereometric measurement of streambank erosion. *Photogrammetric Engineering and Remote Sensing*, **45**, 183–190.

Collison, A.J.C. and Anderson, M.G. 1996. Using a combined slope hydrology/stability model to identify suitable conditions for landslide prevention by vegetation in the humid tropics. *Earth Surface Processes and Landforms*, **21**(8), 737–748.

Darby, S.E. and Thorne C.R. 1992. Impact of channelisation on the Mimms Hall Brook, Hertfordshire, UK. *Regulated Rivers*, **7**, 193–204.

Darby, S.E. and Thorne, C.R. 1996. Development and testing of riverbank-stability analysis. *Journal of Hydraulic Engineering*, **122**(8), 443–454.

Davis, R.J. and Gregory, K.J. 1994. A new distinct mechanism of river bank erosion in a forested catchment. *Journal of Hydrology*, **157**, 1–11.

Downward, S.R., Gurnell, A.M. and Brookes, A. 1994. A methodology for quantifying river planform change using GIS, In: Olive, L. (Ed.) *Variability in Stream Erosion and Sediment Transport*. IAHS Publication.

Ergenzinger, P. 1992. Riverbed adjustments in a step-pool system: Lainbach, Upper Bavaria. In: Billi, P., Hey, R.D., Thorne, C.R. and Tacconi, P. (Eds), *Dynamics of Gravel-Bed Rivers*. Wiley, Chichester, 415–429.

Fredlund, D.G. 1987. Slope stability analysis incorporating the effect of soil suction. In, Anderson, M.G. and Richards, K.S. (Eds), *Slope Stability*. Wiley, Chichester, 113–144.

Gardiner, T. 1983. Some factors promoting channel bank erosion, River Lagan, County Down. *Journal of Earth Science, Royal Dublin Society*, **5**, 231–239.

Graf, W.L. 1982. Spatial variations of fluvial processes in semi-arid lands. In: Thorn, C.E. (Ed.), *Space and Time in Geomorphology*. Allen and Unwin, 193–217.

Gray, D.H. and Leiser, A.T. 1983. *Biotechnical Slope Protection and Erosion Control*. Van Norstrand Reinhold, New York.

Gray, D.H. and MacDonald, A. 1989. The role of vegetation in river bank erosion. In: Ports, M.A. (Ed.), *Hydraulic Engineering. Proceedings of the 1989 National Conference*, American Society of Civil Engineers, 218–223.

Gregory, K.J. and Walling, D.E. 1973. *Drainage Basin Form and Process*. Arnold, 456 pp.

Grissinger, E.H. 1982. Bank erosion of cohesive materials. In: Hey, R.D., Bathurst, J.C. and Thorne, C.R. (Eds), *Gravel-bed Rivers*, Wiley, Chichester, 273–287.

Hagerty, D.J. 1991. Piping/sapping erosion. I: Basic considerations. *Proceedings of the American Society of Civil Engineers, Journal of Hydraulic Engineering*, **117**, 991–1008.

Harvey, A.M. 1977. Event frequency in sediment production and channel change. In: Gregory, K.J. (Ed.), *River Channel Changes*. Wiley, Chichester, 301–315.

Hasan, A.A. 1991. *Erosion Rates and Processes on The River Arrow, Warwickshire, Midlands, U.K.* Unpublished PhD thesis, University of Birmingham, 373 pp.

Hasegawa, K. 1989. Studies on qualitative and quantitative prediction of meander channel shift. In: Ikeda, S. and Parker, G. (Eds), *River Meandering*. American Geophysical Union, Washington, DC, 215–235.

Hemphill, R.W. and Bramley, M.E. 1989. *Protection of River and Canal Banks*. CIRIA/Butterworths, London.

Hey, R.D., Heritage, G.L., Tovey, N.K., Boar, R.R., Grant, N. and Turner, R.K. 1991. *Streambank Protection in England and Wales* R&D Note 22, National Rivers Authority, London, 75 pp.

Hickin, E.J. 1983. River channel changes: retrospect and prospect. In: Collinson, J.D. and Lewin, J. (Eds), *Modern and Ancient Fluvial Systems*. Blackwell, Oxford, 61–83.

Hickin, E.J. and Nanson, G.C. 1975. The character of channel migration on the Beatton River, Northeast British Columbia, Canada. *Geological Society of America Bulletin*, **86**, 487–494.

Hickin, E.J. and Nanson, G.C. 1984. Lateral migration of river bends. *Proceedings of the American Society of Civil Engineers, Journal of Hydraulic Engineering*, **110**, 1557–1567.

Hill, A.R. 1973. Erosion of river banks composed of glacial till near Belfast, Northern Ireland. *Zeitschrift für Geomorphologie*, **17**, 428–442.

Hooke, J.M. 1977a. The distribution and nature of changes in river channel patterns: the example of Devon. In: Gregory, K.J. (Ed.), *River Channel Changes*. Wiley, Chichester, 265–280.

Hooke, J.M. 1977b. *An Analysis of Changes in River Channel Patterns: the Example of Streams in Devon*. Unpublished PhD thesis, University of Exeter, 452 pp.

Hooke, J.M. 1979. An analysis of the processes of river bank erosion. *Journal of Hydrology*, **42**, 39–62.

Hooke, J.M. 1980. Magnitude and distribution of rates of river bank erosion. *Earth Surface Processes*, **5**, 143–157.

Hooke, J. and Perry, R.A. 1976. The planimetric accuracy of tithe maps, *Cartographic Journal*, December, 177–183.

Hooke, J.M. and Redmond, C.E. 1989a. Use of cartographic sources for analysing river channel change with examples from Britain. In: Petts, G.E., Möller, H. and Roux, R.L. (Eds), *Historical Change of Large Alluvial Rivers: Western Europe*. Wiley, Chichester, 79–93.

Hooke, J.M. and Redmond, C.E. 1989b. River-channel changes in England and Wales. *Journal of the Institute of Water Engineers and Managers*, **3**, 328–335.

Hudson, H.R. 1982. A field technique to directly measure river bank erosion. *Canadian Journal of Earth Sciences*, **19**, 381–383.

Hughes, D.J. 1977. Rates of erosion on meander arcs. In: Gregory, K.J. (Ed.), *River Channel Changes*. Wiley, Chichester, 193–205.

Kamphuis, J.W. and Hall, K.R. 1983. Cohesive material erosion by unidirectional current. *Proceedings of the American Society of Civil Engineers, Journal of Hydraulic Engineering*, **109**, 49–61.

Kirkby, M.J. and Morgan, R.P.C. 1980. *Soil Erosion*, Wiley, Chichester.

Knighton, A.D. 1973. Riverbank erosion in relation to streamflow conditions, River Bollin-Dean, Cheshire. *East Midland Geographer*, **5**, 416–426.

Kouwen, N. 1987. Velocity distribution coefficients for grass-lined channels. Discussion of paper 20435 by D.M. Temple. *Proceedings of the American Society of Civil Engineers, Journal of Hydraulic Engineering*, **113**(9), 1221–1224.

Kouwen, N. and Li, R.M. 1979. Biomechanics of vegetated channel linings. *Proceedings of the American Society of Civil Engineers, Journal of the Hydraulics Division*, **106**(HY6), 1085–1103.

Lane, S.N., Richards, K.S. and Chandler, J.H. 1993. Developments in photogrammetry; the geomorphological potential. *Progress in Physical Geography*, **17**, 306–328

Lane, S.N., Chandler, J.H. and Richards, K.S. 1994. Developments in monitoring and modelling small-scale river bed topography. *Earth Surface Processes and Landforms*, **19**, 349–368.

Lawler, D.M. 1984. *Processes of River Bank Erosion: the River Ilston, South Wales*. Unpublished PhD thesis, University of Wales, 518 pp.

Lawler, D.M. 1986. River bank erosion and the influence of frost: a statistical examination. *Transactions of the Institute of British Geographers*, **11**, 227–242.

Lawler, D.M. 1987. Bank erosion and frost action: an example from South Wales. In: Gardiner, V. (Ed.), *International Geomorphology 1986, Part 1*. Wiley, Chichester, 575–590.

Lawler, D.M. 1988. Environmental limits of needle ice: a global survey. *Arctic and Alpine Research*, **20**, 137–159.

Lawler, D.M. 1989. *Some new developments in erosion monitoring: 2. The potential of terrestrial photogrammetric methods*. School of Geography, University of Birmingham Working Paper, 48, 21 pp.

Lawler, D.M. 1991. A new technique for the automatic monitoring of erosion and deposition rates. *Water Resources Research*, **27**, 2125–2128.

Lawler, D.M. 1992a. Process dominance in bank erosion systems. In: Carling, P. and Petts, G.E. (Eds), *Lowland Floodplain Rivers: Geomorphological Perspectives*. Wiley, Chichester, 117–143.

Lawler, D.M. 1992b. Design and installation of a novel automatic erosion monitoring system. *Earth Surface Processes and Landforms*, **17**, 455–463.

Lawler, D.M. 1993a. The measurement of river bank erosion and lateral channel change: a review. *Earth Surface Processes and Landforms*, **18**, 777–821.

Lawler, D.M. 1993b. Needle ice processes and sediment mobilization on river banks: the River Ilston, West Glamorgan, UK. *Journal of Hydrology*, **150**, 81–114.

Lawler, D.M. 1994. Temporal variability in streambank response to individual flow events: the River Arrow, Warwickshire, UK. In: Olive, L. et al. (Eds), *Variability in Stream Erosion and Sediment Transport*. IAHS Publication no. 224, 171–180.

Lawler, D.M. 1995. The impact of scale on the processes of channel-side sediment supply: a conceptual model. In: Osterkamp, W.R. (Ed.), *Effects of Scale on the Interpretation and Management of Sediment and Water Quality*. IAHS Publication no. 226, 175–184.

Lawler, D.M. and Leeks, G.J.L. 1992. River bank erosion events on the Upper Severn detected by the Photo-Electronic Erosion Pin (PEEP) system. In: Bogen, J., Walling, D.E. and Day, T.J. (Eds), *Erosion and Sediment Transport Monitoring Programmes in River Basins*. IAHS Publication no. 210, 95–105.

Lawler, D.M., Harris, N. and Leeks, G.J.L. 1997. Automated monitoring of bank erosion dynamics: applications of the novel Photo-Electronic Erosion Pin (PEEP) System in upland and lowland river basins. In: Wang, S.Y. et al., (Ed.) *Management of Landscapes Disturbed by Channel Incision, Oxford, Mississippi, May 1997*, 249–255.

Lawson, D.E. 1983. *Erosion of Perennially Frozen Streambanks*. CRREL Report, 83–29, 26 pp.

Leeks, G.J., Lewin, J. and Newson, M.D. 1988. Channel change, fluvial geomorphology and river engineering: the case of the Afon Trannon, mid-Wales. *Earth Surface Processes and Landforms*, **13**, 207–223.

Leopold, L.B. 1973. River channel change with time: an example. *Geological Society of America Bulletin*, **84**, 1845–1860.

Leopold, L.B. and Maddock, T. 1953. *The Hydraulic Geometry of Stream Channels and Some Physiographic Implications*. US Geological Survey Professional Paper, 252, 57 pp.

Leopold, L.B. and Skibitzke, H.E. 1967. Observations on unmeasured rivers. *Geografiska Annaler*, **49**, 247–285.

Lewin, J. 1987. Historical river channel changes. In: Gregory, K.J., Lewin, J. and Thornes, J.B. (Eds), *Palaeohydrology in Practice*. Wiley, Chichester, 161–175.

Lewin, J. 1990. River channels. In: Goudie, A. (Ed.), *Geomorphological Techniques*. Allen and Unwin, 196–221.

Lewin, J. and Hughes, D. 1976. Assessing channel change on Welsh rivers. *Cambria*, **3**, 1–10.

Lewin, J. and Manton, M.M.M. 1975. Welsh floodplain studies: the nature of floodplain geometry. *Journal of Hydrology*, **25**, 37–50.

Li, R. and Shen, H.W. 1973. Effect of tall vegetations on flow and sediment. *Proc. A.S.C.E.*, **99**, 793–814.

Little, W.C., Thorne, C.R. and Murphey, J.B. 1982. Mass bank failure of selected Yazoo Basin streams. *Transactions of the American Society of Agricultural Engineers*, **25**, 1321–1328.

Lohnes, R. and Handy, R.L. 1968. Slope angles in friable loess. *Journal of Geology*, **76**, 247–258.

Maling, D.H. 1989. *Measurement From Maps: Principles and Methods of Cartometry*. Pergamon, Oxford, 577 pp.

McGreal, W.S. and Gardiner, T. 1977. Short-term measurements of erosion from a marine and a fluvial environment in County Down, Northern Ireland. *Area*, **9**, 285–289.

Masterman, R.J.W. 1994. *Vegetation Effects on River Bank Stability*. Unpublished PhD Thesis, University of Nottingham, 200 pp.

Masterman, R. and Thorne, C.R. 1992. Predicting the influence of bank vegetation on channel capacity. *Proceedings of the American Society of Civil Engineers, Journal of Hydraulic Engineering*, **118**, 1052–1059.

Masterman, R. and Thorne, C.R. 1994. Analytical approach to predicting vegetation effects on flow resistance. In: Kirkby, M.J. (Ed.), *Process Models and Theoretical Geomorphology*. BGRG Special Publication Series, John Wiley, Chichester, 201–218.

May, J.R. 1982. *Engineering Geology and Geomorphology of Streambank Erosion. Report 3. Application of Waterborne Geophysical Techniques in Fluvial Environments*. US Army Engineer Waterways Experiment Station, Vicksburg, Mississippi, USA, Geotechnical Lab., Report WES/TR/GL-79-7/3, 243 pp.

Murgatroyd, A.L. and Ternan, J.L. 1983. The impact of afforestation on stream bank erosion and channel form. *Earth Surface Processes and Landforms*, **8**, 357–369.

Nanson, G.C. 1986. Episodes of vertical accretion and catastrophic stripping: a model of disequilibrium floodplain development. *Geological Society of America Bulletin*, **97**, 1467–1475.

Neill, C.R. and Yaremko, E.K. 1989. Identifying causes and predicting effects of bank erosion. In: Ports, M.A. (Ed.), *Hydraulic Engineering, Proceedings of a National Conference on Hydraulic Engineering*, 101–105.

Nickel, S.H. 1983. Erosion resistance of cohesive soils, *Proceedings of the American Society of Civil Engineers, Journal of Hydraulic Engineering*, **109**(1), 142–144.

Odgaard, A.J. 1991. Closure to 'River-meander model. I: Development'. *Proceedings of the American Society of Civil Engineers, Journal of Hydraulic Engineering*, **117**(8), 1091–1092.

Osman, A.M. 1986. *Channel Width Response to Changes in Flow Hydraulics and Sediment Load*. Unpublished PhD Thesis, Colorado State University, Fort Collins, Colorado, USA, 171 pp.

Osman, A.M. and Thorne, C.R. 1988. Riverbank stability analysis. I: Theory, *Proceedings of the American Society of Civil Engineers, Journal of Hydraulic Engineering*, **114**, 134–150.

Outcalt, S.I. 1971. An algorithm for needle ice growth. *Water Resources Research*, **7**, 394–400.

Oxley, N.C. 1974. Suspended sediment delivery rates and solute concentration of stream discharge in two Welsh catchments. In: Gregory, K.J. and Walling, D.E. (Eds), *Fluvial Processes in Instrumented Watersheds*. Institute of British Geographers, Special Publication 6, 141–154.

Painter, R.B., Blyth, K., Mosedale, J.C. and Kelly, M. 1974. The effect of afforestation on erosion processes and sediment yield. In: *Effects of Man on the Interface of the Hydrological Cycle with the Physical-Environment*. IAHS Publication no. 113, 62–68.

Park, C.C. 1975. Stream channel morphology in mid-Devon. *Report and Transactions of the Devonshire Association for the Advancement of Science*, **107**, 25–41.

Pizzuto, J.E. 1984. Bank erodibility of shallow sandbed streams. *Earth Surface Processes and Landforms*, **9**, 113–124.

Pizzuto, J.E. and Meckelnburg, T.S. 1989. Evaluation of a linear bank erosion equation. *Water Resources Research*, **25**, 1005–1013.

Richards, K.S. and Lorriman, N.R. 1987. Basal erosion and mass movement. In: Anderson, M.G. and Richards, K.S. (Eds), *Slope Stability*. Wiley, Chichester, 331–358.

Ritchie, J.C., Murphey, J.B., Grissinger, E.H. and Garbrecht, J.D. 1993. Monitoring streambank and gully erosion by airborne laser. In: Hadley, R.F. and Mizuyama, T. (Eds), *Sediment Problems: Strategies for Monitoring, Prediction and Control*. IAHS Publication no. 217, 161–166.

Salo, J., Kalliola, R., Hakkinen, I., Makinen, Y., Niemala, P., Puhakka, M. and Coley, P.D. 1986. River dynamics and the diversity of the Amazon lowland forest. *Nature*, **322**, 254–258.

Schofield, A.N. and Wroth, P. 1968. *Critical State Soil Mechanics*. McGraw-Hill.

Schumm, S.A. 1961. *Dimensions of some stable alluvial channels*., US Geological Survey Professional Paper 424-B.

Selby, M.J. 1982. *Hillslope Materials and Processes*, Oxford University Press, Oxford, 264 pp.

Sharp, J.M. 1977. Limitations of bank storage model assumptions. *Journal of Hydrology*, **35**, 31–47.

Smith, D.G. 1976. Effect of vegetation on lateral migration of anastomosed channels of a glacier meltwater river. *Geological Society of America Bulletin*, **87**, 857–860.

Stott, T.A., Ferguson, R.I., Johnson, R.C. and Newson, M.D. 1986. Sediment budgets in forested and unforested basins in upland Scotland. In: Hadley, R.F. (Ed.), *Drainage Basin Sediment Delivery*. IAHS Publication no. 159, 57–68.

Thomson, S. 1970. Riverbank stability at the University of Alberta. *Canadian Geotechnical Journal*, **7**, 157–168.

Thorne, C.R. 1978. *Processes of Bank Erosion in River Channels*, Unpublished PhD thesis, University of East Anglia, 447 pp.

Thorne, C.R. 1981. Field measurements of rates of bank erosion and bank material strength. In: *Erosion and Sediment Transport Measurement*. IAHS Publication no. 133, 503–512.

Thorne, C.R. 1982. Processes and mechanisms of river bank erosion. In:Hey, R.D., Bathurst, J.C. and Thorne, C.R., (Eds), *Gravel-bed Rivers*. Wiley, Chichester, 227–259.

Thorne, C.R. 1988. *Analysis of Bank Stability in the DEC Watersheds, Mississippi*. Final Technical Report to the US Army European Research Office, London, under Contract No. DAJA45-87-C-0021, Queen Mary College, University of London, 38 pp.

Thorne, C.R. 1990. Effects of vegetation on riverbank erosion and stability. I,: Thornes, J.B. (Ed.), *Vegetation and Erosion*. Wiley, Chichester, 125–144.

Thorne, C.R. 1991. *Field assessment techniques for bank erosion modeling*. Final Report for US Army European Research Office, Research Contract R&D 6560-EN-09, 75 pp.

Thorne, C.R. and Abt, S.R. 1993. Analysis of riverbank instability due to toe scour and lateral erosion. *Earth Surface Processes and Landforms*, **18**, 835–843.

Thorne, C.R. and Lewin, J. 1979. Bank processes, bed material movement and planform development in a meandering river. In: Rhodes, D.D. and Williams, G.P. (Eds), *Adjustments of the Fluvial System*. Kendall/Hunt, Dubuque, Iowa, 117–137.

Thorne, C.R. and Osman, A.M. 1988a. The influence of bank stability on regime geometry of natural channels. In: White, W.R. (Ed.), *International Conference on River Regime*. Hydraulics Research Limited, John Wiley, Chichester, 135–147.

Thorne, C.R. and Osman, A.M. 1988b. Riverbank stability analysis. II: Applications, *Proceedings of the American Society of Civil Engineers, Journal of Hydraulic Engineering*, **114**, 151–172.

Thorne, C.R. and Tovey, N.K. 1981. Stability of composite river banks. *Earth Surface Processes and Landforms*, **6**, 469–484.

Thorne, C.R., Murphey, J.B. and Little, W.C. 1981. *Bank Stability and Bank Material Properties in the Bluff Line Streams of Northwest Mississippi*. Report to the Vicksburg District, US Army Corps of Engineers, USDA Sedimentation Lab., Oxford, Mississippi, USA, 258 pp.

Thorne, C.R., Zevenbergen, L.W., Pitlick, J.C., Rais, B., Bradley, J.B. and Julien, P.Y. 1985. Direct measurements of secondary currents in a meandering sand-bed river. *Nature*, **316**, 746–747.

Thorne, C.R., Russell, A.P.G. and Alam, M.K. 1993. Planform pattern and channel evolution of the Brahmaputra River, Bangladesh. In: Best, J.L. and Bristow, C.S. (Eds), *Braided Rivers*. Geological Society Special Publication no. 75, 257–276.

Thorne, C.R., Reed, S. and Doornkamp, J.C. 1996. *A Procedure for Assessing River Bank Erosion Problems and Solutions*. R&D Report 28, National Rivers Authority, Bristol, 48 pp.

Twidale, C.R. 1964. Erosion of an alluvial bank at Birdwood, South Australia. *Zeitschrift für Geomorphologie*, **8**, 189–211.

US Army Corps of Engineers. 1981. *The Streambank Erosion Control Evaluation and Demonstrations Act of 1974 Section 32, Public Law 93-251*. Final report to Congress Waterways Experiment Station, Vicksburg, Miss.

Waldron, L.J. 1977. Shear resistance of root-permeated homogeneous and stratified soil. *Soil Science Society of America Journal*, **41**, 843–849.

Walker, H.J., Arnborg, L. and Peippo, J. 1987. Riverbank erosion in the Colville Delta, Alaska. *Geografiska Annaler*, **69A**, 61–70.

Wiberg, P.L. and Smith, J.D. 1987. Calculations of the critical shear stress for motion of uniform and heterogeneous sediments. *Water Resources Research*, **23**, 1471–1480.

Wolman, M.G. 1955. *The Natural Channel of Brandywine Creek, Pennsylvania*. US Geological Survey Professional Paper, 271.

Wolman, M.G. 1959. Factors influencing erosion of a cohesive river bank. *American Journal of Science*, **257**, 204–216.

Section IV

Channel Morphology and Dynamics

SECTION CO-ORDINATOR: K. S. RICHARDS

7 Channel Types and Morphological Classification

COLIN R. THORNE
Department of Geography, University of Nottingham, UK

7.1 INTRODUCTION

River channel patterns are characterised by a range of forms and geometries. For engineering and management purposes it is often useful to classify channels using a range of geomorphological channel types that minimise variability within them and maximise variability between them. Most classification systems centre on the planform pattern of the river, but others include consideration of the cross-sectional geometry, longitudinal profile and type of bed material (gravel, sand, or silt/clay). The objective of this chapter is to review the basis for the identification of channel type and the classification of rivers and to examine briefly the utility of channel classification to engineers and river managers.

7.2 DRAINAGE PATTERNS: THE ROLE OF REGIONAL GEOLOGY AND TERRAIN IN INFLUENCING THE PATTERN OF CHANNELS AT CATCHMENT SCALE

Morphological classification must start by considering the geology and physiography of the river basin as they affect and, in some cases, control river form and processes. Examination of maps showing the topography, solid geology and surficial deposits is essential and assistance may be sought from the geologists and physical geographers in determining the significance to the river of various terrain features, rock formations, tectonic movements and sedimentary units. The influence and impacts of these factors on the fluvial system can also be gauged to some extent by tracing-out and interpreting the *pattern* of drainage channels in the catchment. For this purpose a topographic map covering the drainage basin (watershed), such as US Geological Survey quad sheets or their equivalent, is usually ideal.

A great deal of work on the analysis and morphometric interpretation of drainage patterns has been undertaken, a substantial proportion of which is concerned with topological analyses of channel networks that centre on the concept of 'stream ordering' (Strahler, 1964; Shreve, 1966). However, this type of approach is prone to subjectivity in the way that data are extracted from maps, and different operators invariably produce

Applied Fluvial Geomorphology for River Engineering and Management, Edited by C. R. Thorne, R. D. Hey and M. D. Newson.

different statistical parameters such as drainage density and texture (see, for example, Chorley et al. (1984, p. 321)). In any case, such derived parameters have limited practical applications in river engineering and management. Direct examination of the overall pattern of drainage can be more useful. Howard (1967) grouped drainage patterns into eight categories which may be used to make useful inferences about the degree of influence of geology and terrain on the fluvial system (Figure 7.1). A summary of part of Howard's classification is given in Box 7.1.

Box 7.1 Drainage patterns and their geomorphic interpretation

A *dendritic* pattern is regarded as the simplest form of drainage system that results from the operation of fluvial processes in areas of homogeneous terrain with no strong geologic controls. Conversely, a *parallel* pattern develops where there is a steep regional dip (incline) to the terrain that imposes a preferred direction of drainage. A *trellis* pattern indicates both a regional dip and strong geologic control through the existence of folded sedimentary rock. A *rectangular* pattern is also associated with strong geologic control, this time through right-angled jointing and faulting. A *radial* pattern occurs around an eroded structural dome or volcano and is indicative of past or continuing tectonic and/or volcanic activity. Similarly, an *annular* pattern is associated with an eroded dome, the difference from the radial pattern being due to the channels forming where the fluvial system follows weaker strata in layered rocks. *Multi-basinal* drainage occurs in hummocky deposits such as those left by glacial deposition, and in areas of limestone solution. Finally, complex *contorted* drainage may be found where the terrain is heavily impacted by geology through structures produced by neotectonics and metamorphic activity.

7.3 THE CONTINUUM OF CHANNEL PATTERNS

7.3.1 Controls of Channel Form

The form, or morphology, of the channel (including its size, cross-sectional shape, longitudinal profile and planform pattern) is the result of processes of sediment erosion, transport and deposition operating within the constraints imposed by the geology and terrain of the drainage basin. Streams are constantly adjusting and evolving in response to the sequence of normal flow, flood flow and drought events which are associated with regional climate, local weather and catchment hydrology. In this respect, channel form can only be explained rationally if distinctions are made between those factors which drive the fluvial system (driving variables) in producing the channel, those which characterise the physical boundaries within which the channel is found (boundary conditions), and those which respond to the driving and boundary conditions to define the three-dimensional geometry of the channel (channel form) (Figure 7.2).

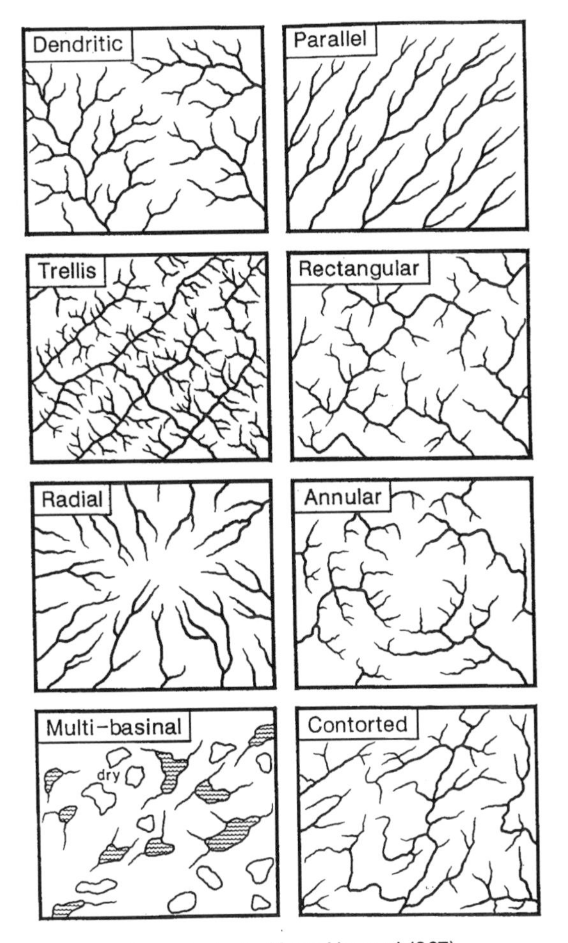

Figure 7.1 Basic drainage patterns (adapted from Howard, 1967)

Driving Variables

The inputs of water and sediment to the channel are not constant through time but vary widely. The input of water from drainage basin runoff drives the flow in the river, while the input of sediment from landscape erosion supplies some proportion of the sediment transported by the river. The balance between water and sediment inputs in turn controls the aggradation or degradational tendencies of the channel. Both the instantaneous values and time distribution of water and sediment are controlled by the climatic, terrain, geological and vegetational characteristics of the hydrological basin. These characteristics are themselves dynamic and they change in response to long-term climatic, geomorphological and biogeographical trends. However, such changes are not usually significant over human and engineering timescales. Fluvial processes may alter runoff characteristics over long timespans, as is clearly demonstrated in Chapters 2 and 3, but for the purpose of channel classification the inputs of water and sediment may be considered as *driving variables* which are effectively independent of channel morphology.

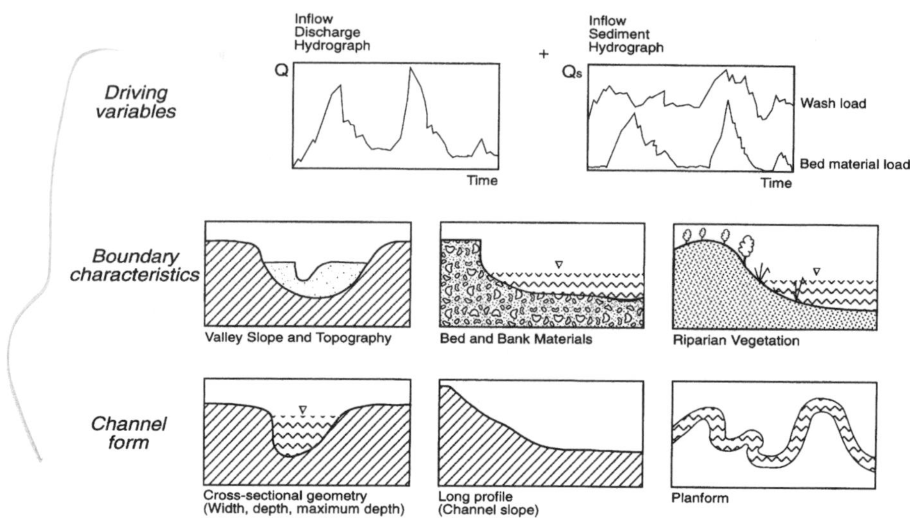

Figure 7.2 Independent and dependent controls of channel form

Boundary Conditions: Confined and Unconfined Channels; Bedrock vs. Alluvial Channels

The water and sediment inputs, or hydrographs, illustrated in Figure 7.2, interact with the landscape to form the channel. 'Landscape' in this respect can be defined in terms of the characteristics of the terrain and materials through which the river flows, and in which the channel is formed. These comprise the valley topography, and particularly the valley slope, together with the bed and bank materials and riparian vegetation.

Valley slope determines the overall rate of conversion of potential to kinetic energy and losses in the fluvial system and, hence, it controls the maximum stream power of a given water discharge, which is a function of the discharge–slope product. Stream power is a measure of the erosivity and sediment transport capacity of the flow for a given bed sediment size and input sediment load from upstream. The fact that it is a good parameter to represent the forces applied to the channel by the flow may explain why stream power is often used to classify channel type and to predict channel form, as discussed later in this chapter.

The bed and bank materials control the erosive resistance, or erodibility, of the channel boundaries. Here, important distinctions can be drawn between channels formed in bedrock and those formed in alluvium or sediment and, hence, between confined and unconfined channels. Channels formed in sediment that can be eroded, transported and deposited by the flow can be classified as 'self-formed', or *alluvial.* The nature and form of these channels is constantly being adjusted by the flow, and their dimensions obey the laws of hydraulic geometry or regime theory which are somewhat transferable between fluvial systems of various scales and geographical locations. Conversely, channels formed in bedrock only occasionally obey these laws because their forms and dimensions are governed directly by geological and structural influences.

A further distinction can be made between confined and unconfined channels. A channel flowing through a narrow valley interacts frequently with the valley sides.

Geomorphologically, fluvial and hillslope systems are closely coupled together. Hence, slope processes, such as soil creep and mass failure, may be driven directly by fluvial undercutting of the valley side by the river. Under these circumstances there may be a substantial supply of debris directly from valley-side processes into the stream channel. As a result, the morphologic development of the channel may well be confined by the valley sides in two ways. Firstly, if the valley sides are formed in consolidated, lithified materials such as rock, then outcrops of erosion-resistant materials in the channel bed and banks may restrict the development of a hydraulically 'self-formed' channel. Secondly, if the valley sides are formed in unconsolidated materials such as loose rock (talus) or soil, then mass failures may deliver such large volumes of sediment that the channel is unable to transport all of the debris away. Hence, the course and planform pattern of the river will be at least partly controlled by the spatial distribution of major sediment sources along the valley. Such streams, where the lateral development of the planform is restricted by interaction with the valley sides, are said to be confined.

Conversely, channels flowing through broad valleys with floodplains on either side rarely interact directly with the valley sides. The products of hillslope processes are stored as colluvium at the foot of the valley side and these are only attacked by the river infrequently during high out-of-bank floods or where in its lateral wanderings the channel encounters the edge of the floodplain. For the most part the channel is formed in erodible sediments and the river is said to be unconfined.

Floodplain vegetation, and most importantly bank vegetation, also plays a role in controlling the erodibility and stability of the channel boundaries. It is the balance between the erosivity of the flow and the erodibility of the boundary materials which controls the rate and direction of channel changes and the ultimate, stable form of the channel. Significant relationships between riparian vegetation and channel-forming processes have been demonstrated in hydraulic (Masterman and Thorne, 1992), geotechnical (Gray and Leiser, 1983) and geomorphological studies (Simon and Hupp, 1986) of channel flow and morphology. For example, research on the stable hydraulic geometry of gravel-bed rivers both in the USA by Andrews (1980) and in the UK by Hey and Thorne (1986) concluded that streams with heavily vegetated banks are narrower than those with thinly vegetated banks, for similar formative discharges. As the planform pattern of an alluvial channel is scaled closely on the width, the influence of vegetation on width will also, indirectly, affect the planform morphology and geometry of the channel.

7.3.2 Channel Morphology

The action of the driving variables of water and sediment inputs on the boundary conditions presented by the floodplain topography, bed sediments, bank materials and riparian vegetation produces the characteristic channel morphology of an unconfined, alluvial stream. Geomorphological classifications of channel type have established qualitative links between channel process, form and stability. In an important paper, Leopold and Wolman (1957) undertook a detailed examination of river form and concluded that natural channels form a continuous spectrum of patterns from straight, single-thread channels through to multithread, braided systems. The title of the 1957 paper by Leopold and Wolman, *River Channel Patterns – Braided, Meandering and Straight*, has been taken to infer that there are actually distinct types of pattern with clearly defined breaks between them, although the text of the paper actually stresses the continuity of channel planform geometries. In the paper an attempt was made to discriminate between meandering and braiding on the basis of formative discharge and channel slope.

The theory that there is a simple geomorphic threshold between meandering and braided planforms has been perpetuated through the quest for a numerical equation that can define this threshold quantitatively in terms of just two or three parameters representing the complex range of driving variables and boundary conditions responsible for controlling channel form. This quest is understandable from the point of view of the river engineer wishing to gauge the sensitivity of channel planform to engineering or river training, but it can obscure the fact that a distinct threshold does not actually exist. A more useful approach is to accept that there is a continuum of planform patterns and use an examination of the geomorphological features displayed by the channel to classify stream type. It is then possible to infer sensitivity from geomorphic classification. For example, Figure 7.3 shows a general relationship between sediment load, channel stability and channel form first proposed by Schumm (1977) that grades from straight, through meandering to braided channels with no abrupt breaks in between.

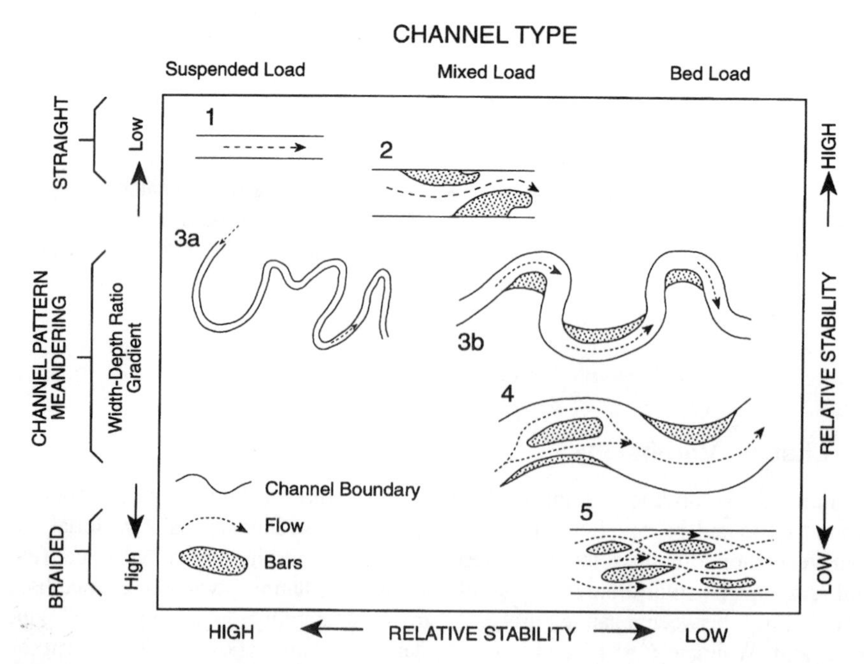

Figure 7.3 Classification of channel pattern based on sediment load and system stability (adapted from Schumm, 1977)

7.4 CHANNEL PLANFORM CLASSIFICATIONS AND CHARACTERISTICS

7.4.1 Channel Form and Processes

Channels with fine sediment moving in suspension and highly erosion-resistant boundary materials are relatively the most stable and follow straight and slightly sinuous courses (Types 1 and 2 in Figure 7.3). Such channels are often effectively confined by their bank materials and display rates of lateral shifting and planform evolution that are slow, or imperceptible. Leopold and Wolman (1957) classified a stream as straight if its sinuosity (ratio of channel length to valley length) was less than 1.1, sinuous if it was between 1.1 and 1.5 and meandering if it exceeded 1.5. Although these limiting values are somewhat arbitrary, they have become entrenched in the literature and remain widely accepted as the critical limits of sinuosity for a stream to be classed as straight, sinuous or meandering.

Mixed-load streams, with more mobile bed materials, greater sediment supply and resistant but somewhat erodible banks, adopt dynamic, meandering courses (Types 3 and 4 in Figure 7.3). These channels migrate freely across their floodplains through a combination of cut-bank erosion and point bar growth interspersed with neck and chute cutoffs of tight bends.

Rivers with sufficiently high energy to transport abundant, relatively coarse sediment moving as significant bedload, and with weak bank materials (which erode and thereby also contribute to the sediment load), tend to have very wide channels that feature multithreaded, braided patterns (Type 5 in Figure 7.3). Braided channels are made up of subchannels called *anabranches* which are separated by braid bars that are inundated at bankfull stage. Such channels are of low stability and they wander across their floodplains unpredictably through a combination of rapid, localised bank erosion and frequent anabranch avulsions. In some rivers the braid bars grow to the extent that they are not inundated even at bankfull stage, allowing them to vegetate and stabilise as semi-permanent islands. In this case the channel pattern is conventionally classified as *anastomosed*.

Anastomosing was for many years viewed as a particularly intense form of braiding (Leopold and Wolman, 1957). However, more recent research on anastomosed channels suggests that they may often in fact be geomorphologically distinct from braided systems. Work in the 1970s by Miall (1977) and by Smith and Smith (1980) showed that anastomosed rivers, with highly sinuous anabranches separated by large, vegetated areas of land at about the same elevation as the floodplain, are actually associated with low-energy fluvial systems. This led Rust (1978) to propose another qualitative diagram for the continuum of patterns, using sinuosity and degree of channel division as its axes and allowing subdivision of divided rivers based on their sinuosity (Figure 7.4). The forms and features of low-energy, anastomosing channels are now accepted as sufficiently different from conventional high-energy braided systems to merit a separate classification (Nanson and Croke, 1992). As a result there are now four generally accepted operational classes of channel, rather than Leopold and Wolman's original three.

Having recognised the fact that channel patterns form a continuum, when undertaking a closer examination of the geomorphological forms and features of alluvial channels it is still convenient to consider channels separately according to whether the channel at

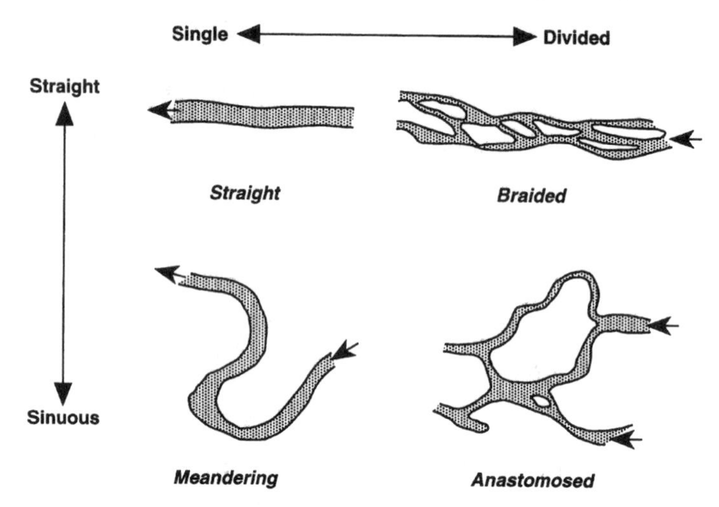

Figure 7.4 Classification of channel pattern based on sinuosity and degree of channel division (adapted from Rust, 1978)

formative flow is straight, meandering, braided or anastomosed. This convention is therefore adopted here.

7.4.2 Straight Channels

The relative rarity of straight alluvial channels has been much commented on by geomorphologists. While this rarity may partly be attributed to variability in local floodplain topography, bank material properties and riparian vegetation that drive random bank collapses, there remains the fact that the vast majority of unconfined, single-thread streams follow a sinuous or meandering course. Even where a channel does follow a straight course for a significant distance, it is usually found that the paths of both the filament of maximum velocity and the line of the deepest point, or *thalweg*, oscillate across the width to describe a sinuous pattern within the straight alignment of the banks. The tendency to produce a sinuous thalweg is closely related to vertical oscillations in the bed elevation termed *pools* (deeps) and *riffles* (shallows) which are clearly defined in gravel-bed rivers but can also be detected in sand-bed streams.

The pool–riffle couplet represents the basic geomorphic unit of the straight river and the overall form and features of the stream can be explained in terms of pool–riffle combinations and their impacts on the channel geometry. Hence, a morphological description of the features of straight alluvial channels must still account for the presence of three-dimensional features and must explain the link between planform and cross-sectional geometries.

Relation Between Channel Pattern and Cross-sectional Geometry in Straight Rivers: Pool–Riffle Sequences, Channel Asymmetry and the Distribution of Bank Erosion

The bed topography in straight alluvial channels is non-uniform, especially where the bed material is sufficiently widely graded that selective entrainment, transport and deposition produces systematic sorting of grain sizes between scour pools and riffle bars. Riffles are the topographic high points in the undulating long profile, while pools are the intervening low points (Figure 7.5). In gravel- and cobble-bed streams it is generally found that bed materials on riffles are coarser than those in pools, at least at low flows when sampling usually takes place. Working on straight and meandering gravel-bed rivers, Hey and Thorne (1986) found that:

$$RD_{50} = 1.19\,D_{50} \qquad (r^2 = 0.95) \tag{7.1}$$

where, RD_{50} = riffle bed material median size (mm) and D_{50} = channel average bed material median size (mm). The occurrence of coarser bed materials with open structures and voids between them in riffles is not only important morphologically, but is also crucial in providing spawning habitat for fish.

At low and intermediate flows riffles act as natural, in-channel weirs that pond water in the pool upstream. The head of water in the pool upstream of a riffle drives flow through the bar that keeps the voids between coarse particles clear of silt. This, again, is not only important morphologically, but is also vital to oxygenate fish eggs buried in redds in the riffle. Flow in the pools is deeper and slower than would be expected in a channel of uniform cross-section, while flow over the riffle is shallow, rapid and tumbling. Pools not only tend to trap fine sediment during low-flow periods, but they also provide refuges for fish to rest and to hide from predators.

The details of flow behind, through and over riffles and the existence of local low-flow variability produced by pool–riffle bed topography are vital to providing a diverse habitat in the stream and in this respect the importance of these natural geomorphic bed controls in supporting valuable ecosystems cannot be over-emphasised.

Riffles are usually spaced fairly evenly along the channel at a distance scaled on the top-width. Leopold et al. (1964) noted that the riffle spacing was five to seven times the channel width. Thirty years of further observations and measurements has not altered this assertion. For example, Hey and Thorne (1986), working in British gravel-bed rivers with single-thread channels and a mixture of straight, sinuous and meandering planforms, found a strong correlation between width and riffle spacing (Figure 7.6a). They found that riffle spacing (measured along the line of the channel centreline) could be defined by:

$$z = 6.31w \tag{7.2}$$

where z = riffle spacing (m) and w = bankfull width (m). The coefficient of determination was 0.88 and the range on riffle spacing was between four and 10 times the width for the

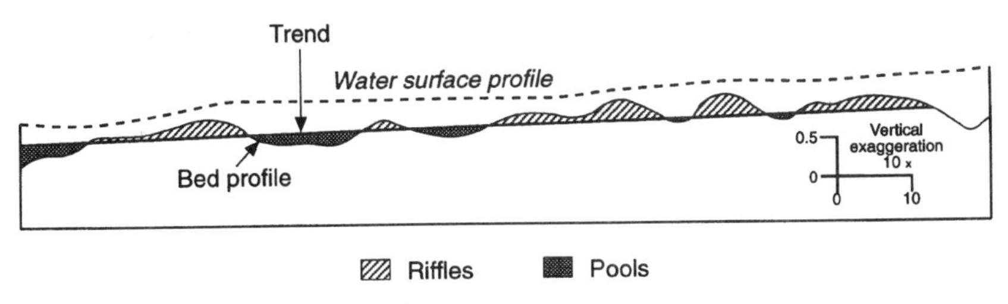

Figure 7.5 Pool–riffle sequence in a straight, gravel-bed channel: the River Fowey, England (modified from Richards, 1982)

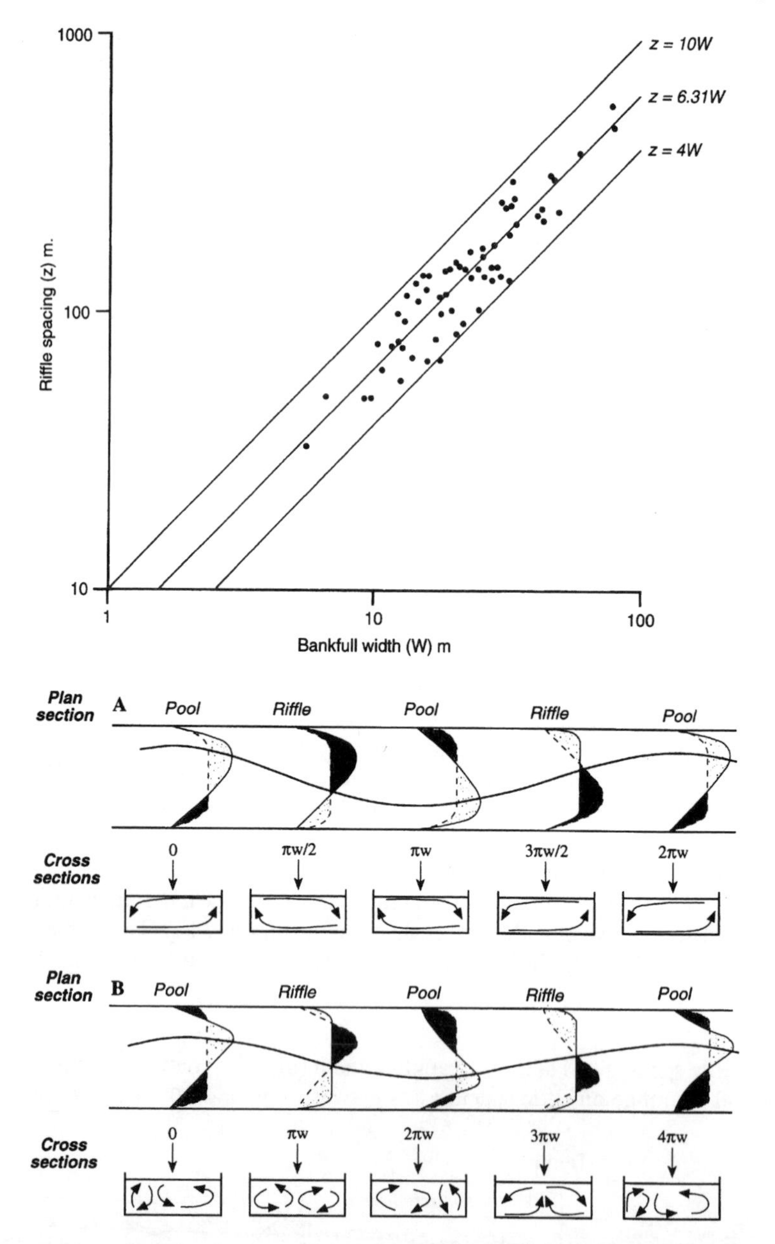

Figure 7.6 (a) Riffle spacing as a function of bankfull width (modified from Hey and Thorne, 1986). (b) Rational explanation of riffle spacing by Hey (1976) based on the theoretical work of Yalin (1972)

great majority of sites surveyed. However, the geometric regularity of pool and riffle spacing does not itself explain their formation.

It is generally accepted that pools and riffles are a dynamic response in the form of the channel to large-scale non-uniformity in the distributions of velocity, boundary shear stress and sediment transport. Theoretical work on the geometry and spacing of macro-turbulence and large-scale flow structures by Yalin (1972) suggested that riffle spacing should be π times the width, but this is half the observed spacing (Eqn 8.2). Hey (1976) re-examined Yalin's theory. He noted that while Yalin had assumed that secondary flow in straight channels was dominated by a single large cell extending across the whole width, field and flume observations showed that secondary flow in straight channels actually features twin cells of secondary circulation, which alternately dominate the pattern (Figure 7.6b). Based on Hey's reanalysis it would be expected that riffle spacing should be 2π times the width. The coefficient in Eqn 7.2 is in fact practically identical to 2π (6.28). This very strong periodicity in riffle spacing indicates a close analogy to meander arc length in sinuous streams (which is also about 2π times the width) and suggests that the processes responsible for meandering also operate in straight streams.

The local variability associated with pool–riffle bed topography in straight and meandering streams was also characterised by Hey and Thorne (1986) in the form of modifications to the equations defining the stable, or regime, hydraulic geometry of the channel as a whole:

$$R_w = 1.034w \qquad (r^2 = 0.97) \tag{7.3}$$

$$R_d = 0.951d \qquad (r^2 = 0.97) \tag{7.4}$$

$$R_{dm} = 0.912d_m \qquad (r^2 = 0.96) \tag{7.5}$$

$$R_v = 1.033v \qquad (r^2 = 0.92) \tag{7.6}$$

where R_w = riffle bankfull width (m), w = channel bankfull width (m), R_d = riffle bankfull mean depth (m), d = channel bankfull mean depth (m), R_{dm} = riffle bankfull maximum depth (m), d_m = channel bankfull maximum depth (m), R_v = riffle bankfull mean velocity (m/s) and v = channel bankfull mean velocity (m/s). These relationships show that, morphologically, riffles are a little shallower and wider than the average dimensions of the channel, even at bankfull stage. This variability is much greater at lower flows and accounts for many of the aesthetic features provided by natural alluvial channels that are often lacking in engineered channels. Differences between pools and riffles decrease as flow stage increases and probably disappear at about bankfull flow.

The pool–riffle sequence in the bed is generated by a combination of turbulent velocity fluctuations and large-scale, coherent flow structures which drive sediment pulsing and produce alternating areas of scour and fill along the axis of the flow (see Section III and the other chapters in this Section). These flow structures are three-dimensional and they generate lateral as well as vertical non-uniformity. The morphological result of this lateral non-uniformity of the flow is for pools to develop asymmetrically, with deep scour adjacent to one bank and a shoaling bar at the opposite bank, the sense of asymmetry alternating from one side of the channel to the other between consecutive pools (Figure 7.7). Riffles, between pools which are on opposite sides of the channel centreline, then become locations where the thalweg and maximum velocity filament cross the channel from one pool to the next.

Deep scour and high-velocity flow close to one bank in the asymmetrical pools and impinging flow attacking one bank just downstream of the riffles often generates bank instability and retreat. In such cases, the stream will not remain straight since retreat of alternate banks in pools along its length leads directly to the development of a sinuous

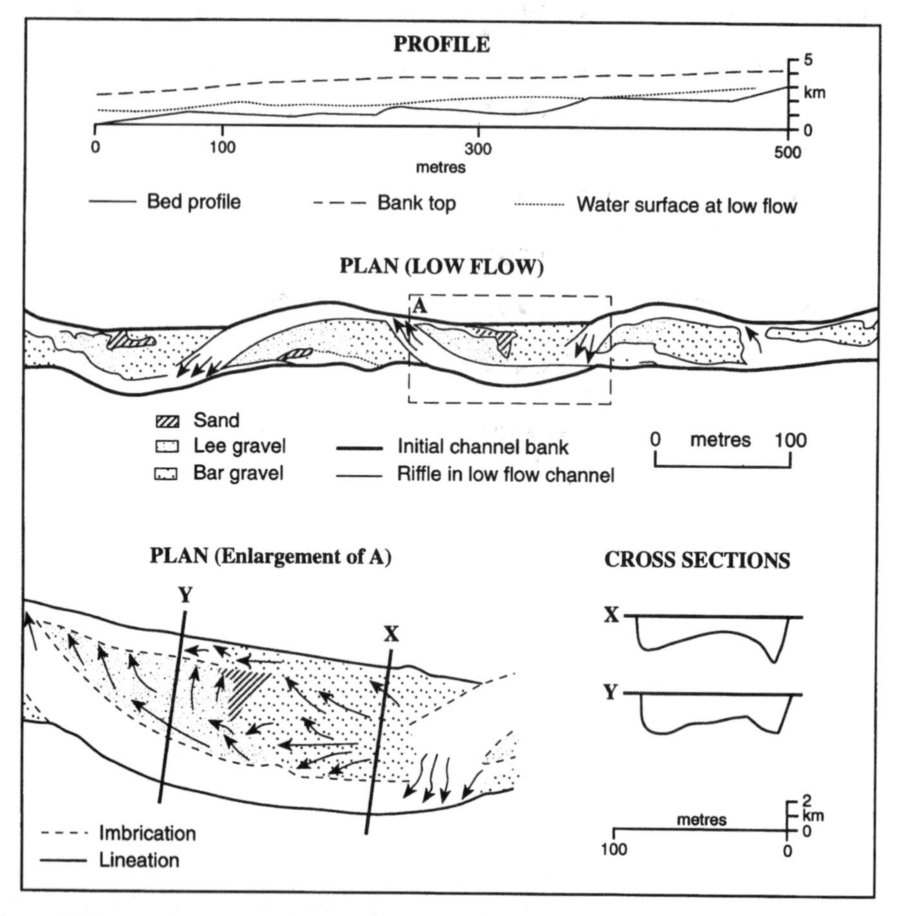

Figure 7.7 Formation of asymmetrical pools, alternate bars and riffle crossings in a straight alluvial channel (adapted from Richards, 1982)

planform. In most models of channel planform evolution the riffles become points of inflection in the sinuous pattern, with cut banks persisting at the outside of bends developing in the pools and the alternate bars growing at the inside of the bends becoming point bars (Figure 7.8). Since this development is the consequence on flow structures and bed asymmetry that definitely pre-existed in the straight channel, it is apparent that meandering is a natural progression of tendencies found even in entirely straight streams. This makes it hard to argue that an abrupt 'geomorphic threshold' exists between the straight and meandering forms, other than that the ability to erode the banks is essential if the planform as defined by the banklines is to be made sinuous.

The topics of bank erosion and retreat are dealt with in detail in Chapter 6, but it is relevant to point out here that, since bank stability and retreat are closely linked to processes operating at the bed through the concept of 'basal endpoint control', bank retreat adjacent to deeply scoured, asymmetrical pools is almost a certainty if the bank materials are alluvial. Only in confined channels can meandering tendencies due to active

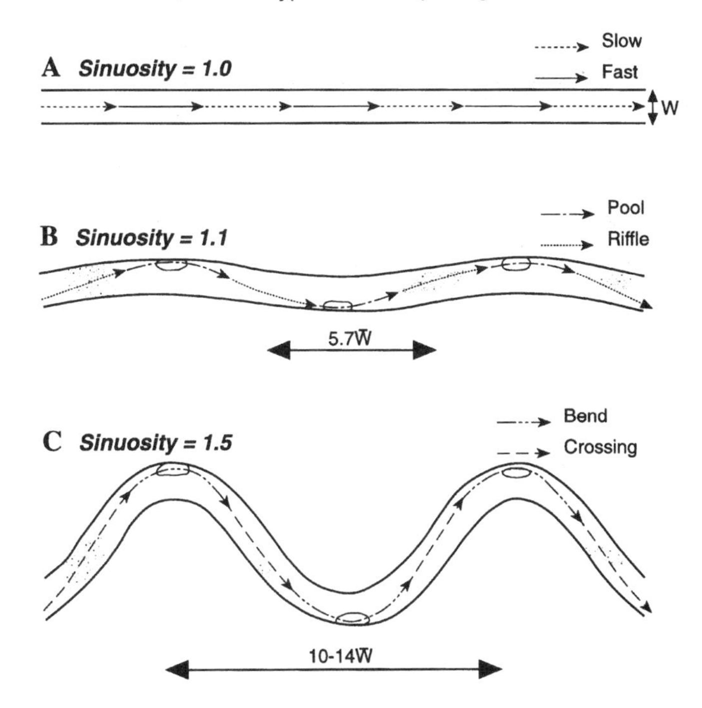

Figure 7.8 Transition from a straight to a meandering course through bank erosion and point-bar growth (adapted from Chorley et al. (1984)).

development of asymmetrical bed topography be frustrated by bank resistance to erosion and mass instability. However, the power of the flow to erode the channel boundaries must never be underestimated, and meanders incised into solid bedrock bear witness to the fact that flow scour and mass-wasting will usually prevail over bank resistance, given sufficient time. Viewed in this light, the rarity of straight, natural channels is no longer surprising.

7.4.3 Meandering Channels

Meander Planform Geometry

Meanders are usually defined geometrically in terms of their shape, bend radius of curvature and wavelength (Figure 7.9). The channel width at the dominant discharge or 'channel-forming flow' is used to scale the geometric relationships.

The vast majority of streams follow a winding, more or less sinuous course and are usually morphologically classified as meandering. However, it is important to recognise at the outset that not all sinuous channels with bends are necessarily actively meandering through cut-bank erosion and point-bar growth. Unless a further distinction is made by classifying sinuous channels as exhibiting either *active* or *passive* meandering, then correct interpretation of the morphological forms and sensitivity of the channel will be difficult. The remainder of this discussion of meandering channels is relevant to active meandering.

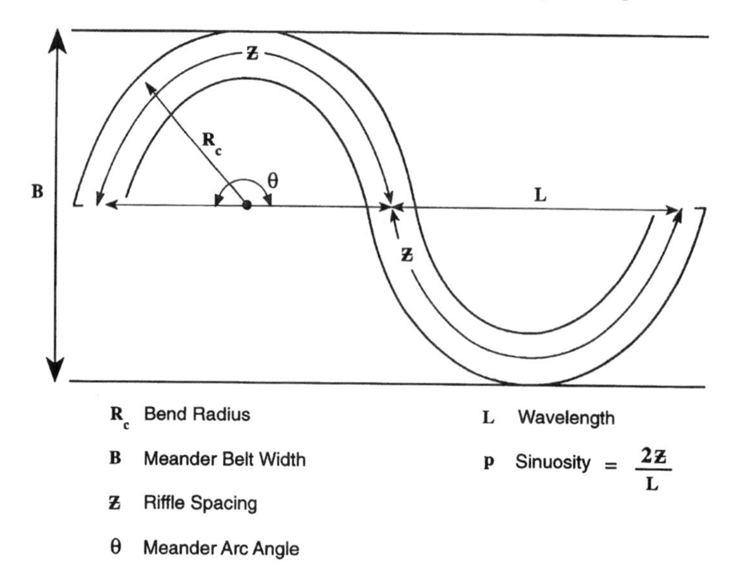

Figure 7.9 Definition diagram for meander planform

A brief outline of active and passive meandering, and how to tell the difference in the field, is given in Box 7.2.

Studies of meander shape were initiated by Leopold and Langbein (1966) who attempted to characterise the planform of meanders in terms of a generalised geometric shape. Figure 7.11 shows the four types of curve proposed by Leopold and Langbein.

Box 7.2 Active versus passive meandering

Active meandering is the result of on-going bed and bank deformation by the flow in a self-formed alluvial channel. The topography of pools and riffles in the bed is matched to the pattern of bends and crossings in the planform, with pools being located at bends and riffles being found at crossings. The riffle spacing (five to seven times the width) is very close to half the meander wavelength (10 to 14 times the width), so that there is in general only one deep pool in each bendway and only one distinct riffle in each crossing reach.

Streams with sinuous courses which do not meet these criteria should be classified as having *passive meandering*. For example, Richards (1982) used the Afon Elan in Wales to show how an apparently meandering stream may actually be following a sinuous course only because of planform patterns imposed by the local terrain. The Elan (Figure 7.10) is an underfit stream (see Chapters 2 and 3) which no longer has the stream power necessary to deform its channel boundaries through active bed scour and bank erosion. The channel follows a sinuous course, but meander wavelength is much greater than 10–14 times the width and there are several pool–riffle units in each bendway. Bends occur because the bluffs confining the stream deflect it back and forth across the comparatively narrow floodplain. Morphologically, passive streams of this type are distinct from freely meandering systems which are more actively forming the landscape. They have more in common with straight streams and are better classified as either confined, or geomorphologically straight.

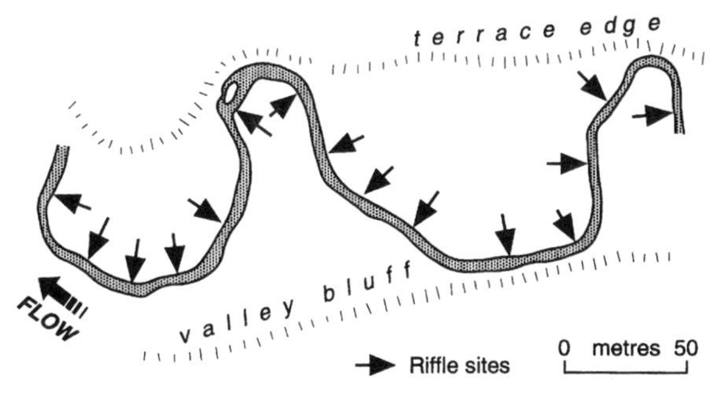

Figure 7.10 Planform of the Afon Elan, Wales. The sweeping bends appear to be classic alluvial features but are in fact the result of diversions of the stream by valley side bluffs. The channel is more properly classified as confined with passive meandering (adapted from Richards, 1982)

They found that a sine-generated curve resembled an idealised meandering river. This curve closely approximates the curve of least work in turning around the bend and they put this forward as an explanation of the form of natural meanders. The path of the river following a sine-generated curve is defined by:

$$\phi = \phi_{\max} \sin[(x/T)2\pi] \tag{7.7}$$

Leopold and Langbein (1966) noted at the time that real bendways are asymmetrical and deviate significantly from the idealised, perfect symmetry of the sine-generated curve. This asymmetry is associated with the fact that the points of deepest bed scour and of maximum attack on the outer bank in bends are usually located downstream of the geometric apex of the bend, so that through time the bends migrate downstream, becoming skewed in the downvalley direction as they shift. Several researchers, including notably Ferguson (1973)

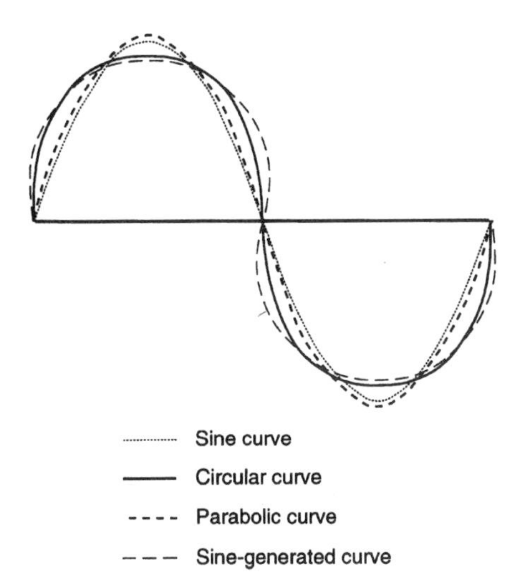

Figure 7.11 Geometric curves investigated by Leopold and Langbein to define meander shape in terms of minimisation of work (adapted from Leopold, 1994)

and Carson and Lapointe (1983), have examined many models of bend shape and concluded that symmetrical models cannot correctly reproduce the downvalley asymmetry that is an essential feature of real meanders.

In nature, every meandering river has a pattern made up of a complicated and unique series of bends connected by short, more or less straight, intervening reaches. If valley terrain and sedimentary variability were the primary controls on meander form then it would be expected that meander patterns would produce random planform attributes. However, while irregular planform paths do occur, in general this is not the case. Leopold and Wolman (1957, 1960) produced graphs linking meander wavelength to channel width over several orders of scale of flow (Figure 7.12) and in a variety of natural environments. They found that power law relationships described the range of wavelengths observed and these were defined by:

$$L = 7.32w^{1.1} \tag{7.8}$$

to

$$L = 12.13w^{1.09} \tag{7.9}$$

where L = meander wavelength measured along the axis of the channel (m), and w = channel top width at the dominant discharge (m). It is important to note the range in the multiplier of width, which indicates that there is real variability in the wavelength to width relationship of natural meanders.

Subsequent reanalysis of Leopold and Wolman's data has shown that because the exponents in Eqns 7.8 and 7.9 are not significantly different from one another, a linear function fitted through the data is acceptable. This has been defined by Richards (1982) as:

$$L = 12.34w \tag{7.10}$$

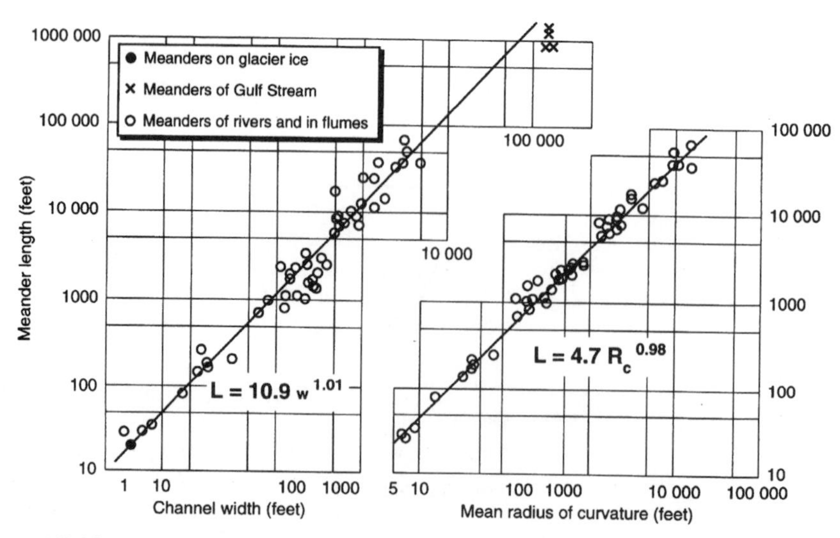

Figure 7.12 Relationship between meander wavelength and channel width (adapted from Leopold, 1994)

In this equation the coefficient is numerically very close to 4π (12.57), which is twice the riffle spacing in a straight channel. Although, strictly, the riffle spacing in a meandering channel should be measured along the channel rather than along the axis of the meanders, this matching of waveforms in the bed topography and planform is almost certainly related to turbulent flow structures and secondary currents in the flowing water that are responsible for the genesis of non-uniformity in the channel in both straight and meandering channels.

Because the channel top width of an alluvial channel is closely related to discharge through hydraulic geometry relationships, it follows that there should be a relationship between discharge and wavelength. Allen (1970) found such a link and suggested the equation:

$$L = 168Q_{a}^{0.46} \tag{7.11}$$

where Q_a = the mean annual discharge (m^3/s). In fact, mean annual discharge has little or no geomorphic significance and so a relationship based on bankfull discharge (often taken as the channel-forming flow) is more meaningful, morphologically. Dury (1956) suggested:

$$L = 54.3Q_{b}^{0.5} \tag{7.12}$$

where Q_b = bankfull discharge (m^3/s). However, it is known that sediment load and boundary materials have real impacts on channel geometry as well as discharge (see Chapter 8). While the use of channel width as a scaling factor for meander wavelength to some extent incorporates these impacts implicitly, the use of discharge alone ignores them completely. This may explain why the relationships based on discharge are less general and much less popular than those based on width.

Schumm (1968) attempted to take account of the effect of boundary materials on meander wavelength explicitly by using a weighted silt-clay index of the bed and bank sediments. He analysed large empirical data sets for sand-bed rivers and streams to produce:

$$L = 1935Q_{m}^{0.34}M^{-0.74} \tag{7.13}$$

$$L = 618Q_{b}^{0.43}M^{-0.74} \tag{7.14}$$

$$L = 395Q_{ma}^{0.74}M^{-0.74} \tag{7.15}$$

where Q_m = mean annual discharge (m^3/s), Q_b = bankfull discharge (m^3/s), Q_{ma} = mean annual flood (m^3/s), and M = weighted silt-clay index. As expected, each equation shows that as the proportion of fine material in the bed and banks increases, the meander wavelength for a given discharge decreases. This is taken to indicate that the greater erosion resistance of silt-clay banks allows a narrow cross-section with steeper banks and tighter, shorter wavelength bends to develop than is the case for friable, easily eroded banks in sand.

Schumm (1963) had already demonstrated that channel sinuosity was related to the weighted silt-clay index and the form ratio (width/depth) using the relations:

$$p = 0.94M^{0.25} \tag{7.16}$$

$$p = 3.50F^{-0.27} \tag{7.17}$$

where p = planform sinuosity and F = width/depth ratio. These relationships form a rational and logical set of empirical equations linking the characteristic wavelength of meandering channels to the formative flow in the channel, its width and the nature of the boundary sediments.

Meander wavelength and bend radius of curvature are closely related, since as the wavelength shortens, bends, necessarily, tend to tighten. Leopold and Wolman (1960) derived an equation describing this relationship:

$$L = 4.59R_c^{0.98} \tag{7.18}$$

where R_c = bend radius of curvature (m). Combining the wavelength relations with width and with bend radius, it follows that:

$$R_c \approx 2 - 3w \tag{7.19}$$

This morphological relationship, arrived at empirically by Leopold and Wolman, was shown at the same time by Bagnold (1960) to have a basis in the theory of physics. Bagnold's work on flow hydraulics and energy losses at bends indicated that at a bend radius-to-width ratio of 2 to 3, energy losses due to the curving of flow in the bend were minimised. Tighter bends produced extensive areas of flow separation at both the outer bank at the bend entrance, and the inner bank at the bend exit. Separation produced large energy losses due to flow constriction, large-scale eddying and distortion of the free surface (Bagnold termed this spill-resistance) so that bends tighter than an R_c/w of 2 exhibited a disrupted flow pattern and high flow resistance. Plots of both meander migration rate and bend scour depth as a function of bend tightness also peak sharply at an R_c/w of between 2 and 3, indicating that such bends are the most effective at eroding their bed and banks (see below and Chapter 9). The fact that in nature many bends develop to an R_c/w value of 2 to 3 and then retain that form while migrating across the floodplain may, therefore, be consistent with their conforming to the most efficient hydraulic shape, which also maximises their geomorphic effectiveness. However, Hey (1976) pointed out that the relationship between width and bend radius also depends on the arc length of the bend. He plotted a graph which generalises Leopold and Wolman's relationship for bends with various arc angles and in various intermediate stages of evolution from meander genesis to loop cutoff and abandonment (Figure 7.13). Data from the Rivers Tweed and Wye were used to validate the geometric relationship between these bend parameters. The data also illustrate the fact that an R_c/w of 2.4 combined with an arc angle of 150° forms a boundary to bend evolution for these particular rivers. This geometry is typical of many natural, alluvial streams with freely meandering planforms.

Figure 7.13, taken together with the various equations and rules-of-thumb for meander morphology quoted here, could be used to assess the form of existing meander bends in engineering-geomorphic studies, or could form the basis for restoring a straightened

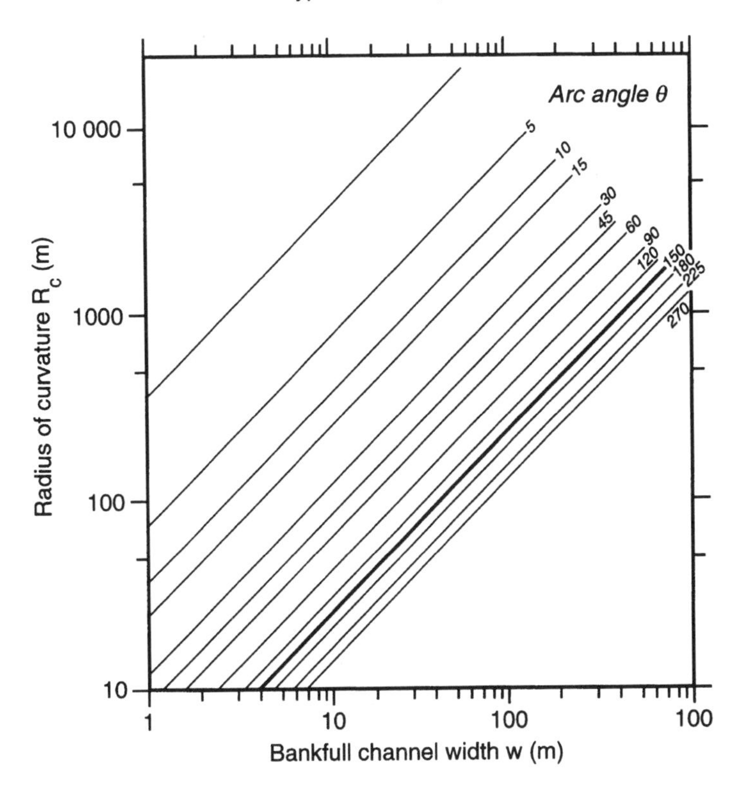

Figure 7.13 Relationship between bankfull width, meander bend radius and bend length (represented by meander arc angle) (after Hey 1976)

stream to a sinuous pattern that mimics the planform of a natural single-thread channel. But it is important to remember that despite such generalities of alluvial meander geometry, in real rivers perfectly formed meanders are, in fact, the exception rather than the rule.

Fisk (1944,1947), working on the Lower Mississippi, identified that the form of most meanders was influenced by variations in the erodibility of the materials encountered in the outer bank. He concluded that outcrops of erosion-resistant clays in the bank have the strongest influence and that such outcrops slowed bank erosion locally, distorting the curve of the outer bank, changing the flow direction and inducing a decrease in the bend radius of curvature. 'Clay plugs' are frequently encountered by rivers meandering across alluvial floodplains. They are produced by infilling of old meander bend scars and abandoned channels by overbank and backswamp deposition of fine sediment. Fisk (1944) set out examples of the effect of clay plugs on meander form and Schumm and Thorne (1989) suggested how these can be used to identify the presence of a resistant hard point in the bank, in the field, from channel maps or from aerial photographs (Figure 7.14b). Thorne (1992) described specific examples of bend deformation by clay plugs and other resistant outcrops. Salient points from these papers are given in Box 7.3.

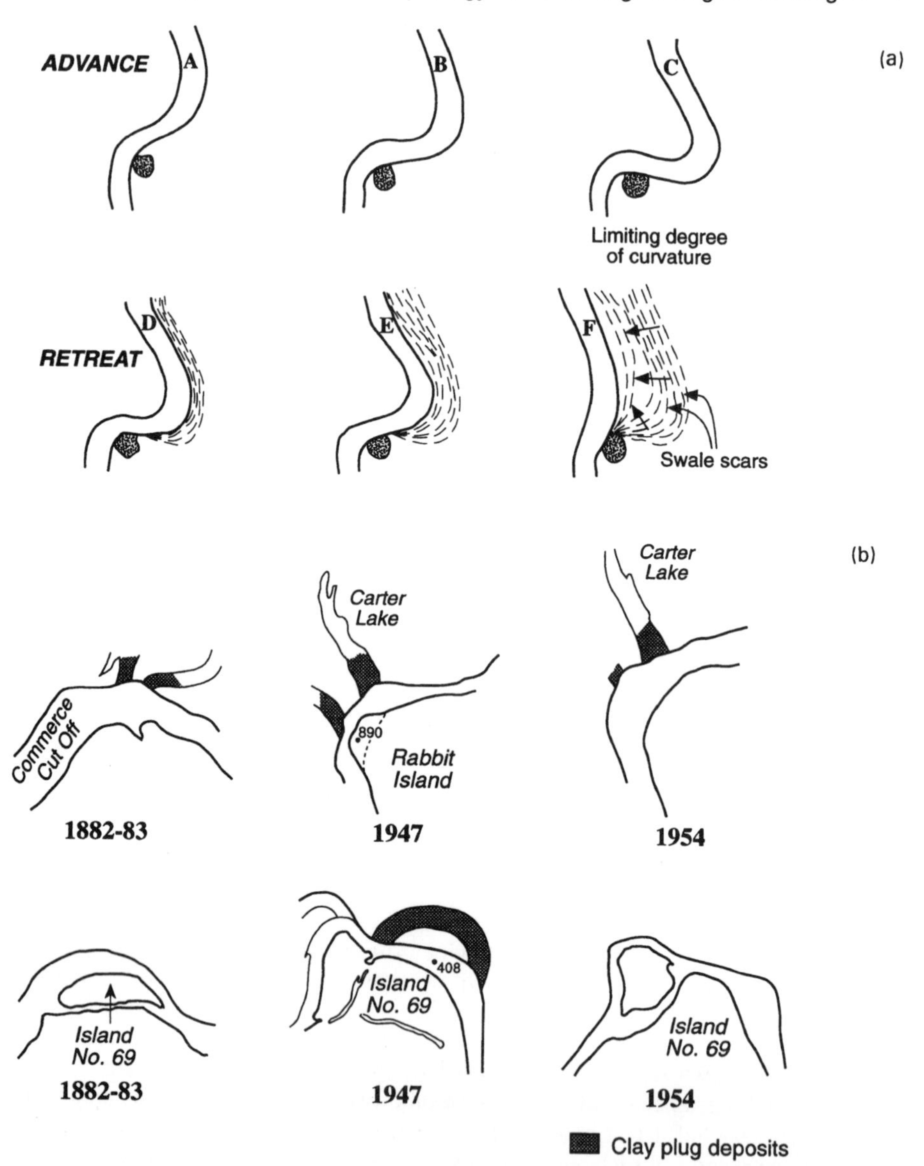

Figure 7.14 (a) Theoretical impact of a hard point on meander morphology and evolution (after Reid, 1984). (b) Empirical impact of hard points on meander morphology and evolution: examples from the Lower Mississippi (after Schumm and Thorne, 1989)

Relationship Between Channel Pattern and Cross-sectional Geometry in Meandering Rivers

The hydraulics and morphology of meandering rivers have received close attention from fluvial geomorphologists over many years. Although much is now known about the flow processes, sediment dynamics and morphological features of actively meandering

Box 7.3 Deformation of meander bends due to clay plugs and hard points

When the river encounters erosion-resistant material in the outer, retreating bank, there are morphological responses both locally and throughout the bend. Irregularities in the planform of the bend can, therefore, be used to detect the influence of resistant materials and their presence can then be taken into account when analysing and predicting channel evolution and sensitivity to river engineering and management.

The immediate effect of resistant material is to slow the local rate of bank retreat. If the longstream extent of the resistant material is short compared to the length of the bend then the outcrop constitutes a *hard point*. As the surrounding bank continues to retreat, the hard point develops into a local bank promontory. This deflects the flow, inducing local acceleration of the primary flow, intense turbulence and strong secondary currents which cause deep bed scour and increased erosion of the surrounding, weaker bank materials. This usually leads to flanking of the hard point. However, in cases where the hard point cannot be flanked easily, the bend may become so deflected that flow adjacent to the outer bank stalls and separates, leading to bar deposition of an outer bank bench and flow attack of the point bar opposite. This in turn leads either to the active channel progressively 'backing out' of the bend, as described by Reid (1984) (Figure 7.14a), or to a chute cutoff across the point bar at the inner bank.

If the resistant material is more extensive, as in the case of most clay plugs, this may be identified as a convexity in the otherwise concave curve of the outer bankline. This deforms the planform of the bend, with particular impacts that depend on the location of the clay plug in the bend. Fisk (1944) described two basic patterns of deformation: when a clay plug is encountered at the bend apex, the apex is flattened and in many cases a compound or double-headed bend develops; if the clay plug is encountered downstream of the bend apex, the downstream limb is fixed in position while the upstream limb continues to shift downvalley, compressing the bend and leading to a neck cutoff. If the presence of a clay plug directs the flow into highly erodible adjacent sediments, a bend of abnormally high amplitude develops that will eventually cutoff. Numerous clay plugs flanking a channel can inhibit meander development and a relatively straight channel will be confined to a narrow zone of the floodplain.

If the resistant material is very extensive, as is the case where a migrating bend comes up against rock or consolidated materials in the valley side, deep scour may develop all along the bank, effectively locking the channel against the valley side for a considerable period until the bend is overtaken by a more mobile bend from upstream.

channels, there remain severe limitations to the applicability of this knowledge in predicting channel cross-sectional parameters such as scour depth for practical river engineering and management (see, for example, publications by Ikeda and Parker (1989) and by Markham and Thorne (1992)).

The topography of the bed and pattern of the planform are closely related, at least for rivers which are freely meandering. It is well known that pools usually occur in bendways and riffles occur in the intervening straight reaches or crossings. It is further known that the depth of pool scour is in some way related to the geometry of the bend. Data assembled from hydrographic surveys of the meandering Red River in Louisiana and Arkansas are typical (Figure 7.15). Scour depth is a function of river size as well as bend geometry and in Figure 8.15 the scour depth is made non-dimensional by dividing the maximum scour depth (BD_m) and the mean scour depth in the bend (BD_b) by the mean depth at the crossing

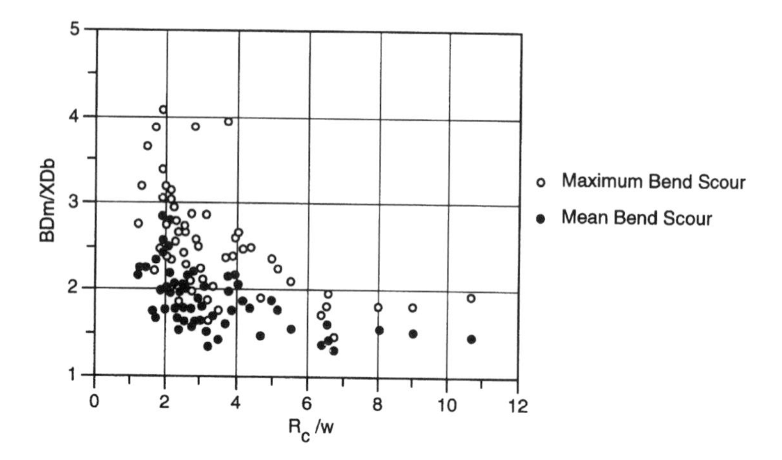

Figure 7.15 Bend pool depths from the Red River, USA (after Thorne, 1989)

upstream (XD_b). The geometry of the bend is represented by the ratio of meander bend radius (R_c) to the channel top width (w), measured at the inflection point upstream. It is important to use the crossing width to non-dimensionalise the bend radius, rather than the width at the bend, because the wide expanse of the point bar at the inner bank often makes it difficult to identify the top bank width in a bend. Also, a reference discharge and associated water level must be used to define the channel dimensions. Ideally, this should be the geomorphologically important 'formative flow', that is the discharge responsible for forming most of the features of the channel. This may be taken as the dominant flow, bankfull discharge or two-year flow. In the case of the Red River data, the two-year flow was used (Biedenharn et al., 1987).

In a study of the Red River, Thorne (1988, 1992) examined the distribution of bend scour with bend geometry and found that in very long radius bends ($R_c/w > 10$) mean scour pool depth is about 1.5 times the mean riffle (crossing) depth and the maximum scour depth is between 1.7 and 2 times the mean crossing depth. This geometry is probably representative of local variability in the parabolic, 'regime' cross-section of the alluvial channel when bend effects are small.

The data from free alluvial meanders show how both the mean and maximum scour depths in the bendway pools increase as a long radius bend becomes tighter and more pronounced. The relationship is non-linear, with scour depths increasing markedly once the R_c/w value decreases to a value below about 5.

For bends with R_c/w values between 2 and 4, scour depths may be anywhere between two and four times the mean crossing depth, with the deepest scour being associated with an R_c/w of about 2. For extremely tight bends with R_c/w less than 2, there is evidence that maximum scour depths decrease with decreasing bend radius. This is consistent with the theoretical and empirical work of Bagnold (1960) which showed that at an R_c/w a distinct change in bend flow hydraulics took place. He found that energy losses at a bend were minimised and the flow efficiency of the bend maximised. For tighter bends the flow pattern broke down to produce large-scale separation at both the outer bank near the entrance and the inner bank at the exit, leading to gross changes in the pattern of erosion, transport and deposition of sediment. Leopold and Wolman (1960) found that most natural bends tend towards R_c/w values in the range 2 to 3, which is consistent with Bagnold's findings. These results also demonstrate the significance of an R_c/w value of about 2 to bend morphology, and suggest that bends with R_c/w values less than 2 must be treated separately when analysing or predicting scour depth. In his study of the Red River, Thorne (1989, 1992) fitted a semi-logarithmic function to the data for maximum scour depth in bends with $R_c/w > 2$. The resulting equation is defined by:

$$(BD_m/XD_b) = 2.07 - 0.19\ \log_e((R_c/w) - 2) \tag{7.20}$$

where BD_m = maximum scour depth in bendway pool (m), XD_b = mean depth at crossing (m), R_c = bend radius of curvature (m), and w = channel width at the crossing (m). This curve fitted the Red River data, from which it was derived, with a statistically significant r^2 of 0.66, but a more stringent test is required if the relationship is to be applicable to any other rivers. In a subsequent study, Thorne and Abt (1993) compiled data from 256 bends on a wide variety of rivers, streams and flume channels. They then used the analytical bend-flow models of Bridge (1982) and Odgaard (1989) and the empirical equation of Thorne (1988) to predict the expected scour depth and compare the results to observed scour depths. The results are plotted in Figure 7.16a. The empirical relationship clearly performs more reliably than the analytical methods, with the great majority of the predictions being within $\pm 30\%$ of the observed value. Figure 7.16b shows the errors plotted as a function of R_c/w. This diagram shows that Odgaard's model actually does quite well for very tight bends with $R_c/w < 2$, although errors of up to 80% may still occur. The empirical equation is inapplicable to these bends. Bridge's model should not be used for such tight bends as it is liable to produce errors of as much as 300%. For longer radius bends Odgaard's model systematically under-predicts scour depth, while Bridge's model is prone to over-prediction. The empirical equation tends to over-predict somewhat, which puts it on the safe side in engineering terms.

It is perhaps disappointing that in 1997 a relatively crude empirical equation can outperform more process-based analytical models. Hopefully, as our ability to model bend flow and sediment interactions improves, this situation will change. At the moment, however, the extremely stringent data requirements of sophisticated and conceptually strong models of bend flow, such as the model of Smith and McLean (1984), make them impractical for day-to-day use as bend scour predictors.

7.4.4 Braided Rivers

Braiding Forms and Processes

Compared to single-thread, meandering channels, much less is known about the morphology of braided rivers. This is partly due to the fact that they have, until recently, received less attention from fluvial geomorphologists, but mostly because their morphology is much more complicated and, therefore, more difficult to define and classify.

The origins of contemporary morphological descriptions of braided rivers may be traced back to Leopold and Wolman's paper *River Channel Patterns – Braided, Meandering and Straight* of 1957. Their paper reported the results of a laboratory flume experiment to

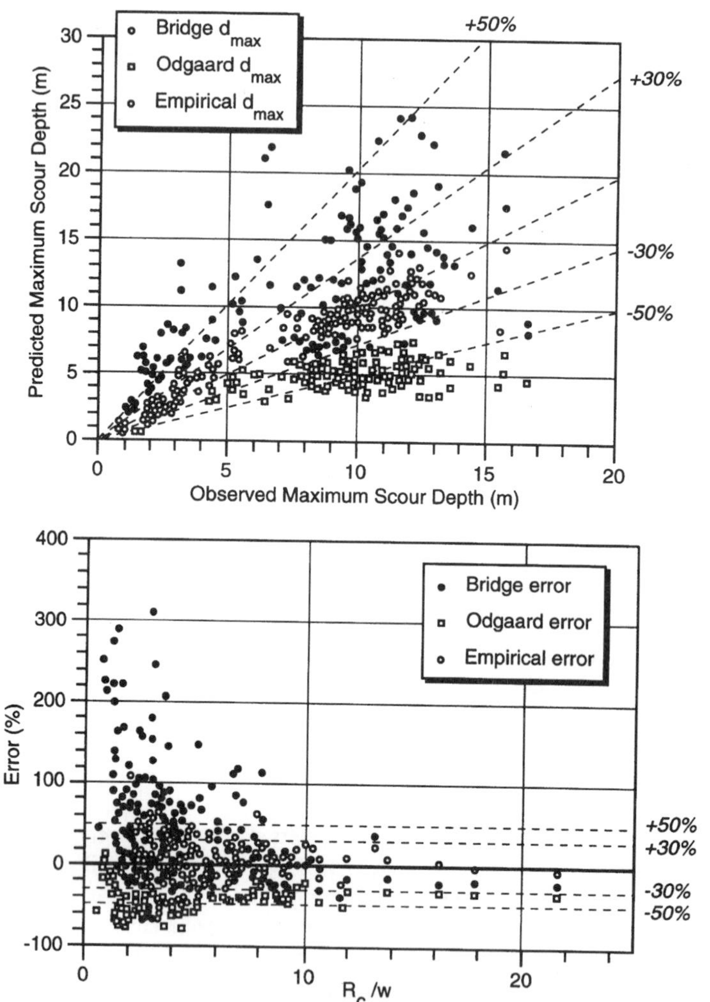

Figure 7.16 (a) Observed and predicted maximum bendway scour depths, and (b) errors as a function of bend geometry (after Thorne and Abt, 1993)

simulate the processes by which a single-thread channel evolved into a multithreaded, braided channel. Figure 7.17 shows the sequence of observed channel changes.

In a channel with abundant bedload, deposition of a mid-channel bar deflects the flow first to one side and then the other, to attack and erode the banks (Figure 7.17, A). The resulting bank retreat feeds sediment to the channel, supporting further bar growth. It also produces a lenticular planform shape to the channel that creates space for lateral expansion of the mid-channel bar (B). As the bar grows and the banks retreat, the subchannels on either side of the bar become increasingly curved, inducing strong secondary currents. The curved flow scours the bed and further erodes the banks at the outer margins of the channel

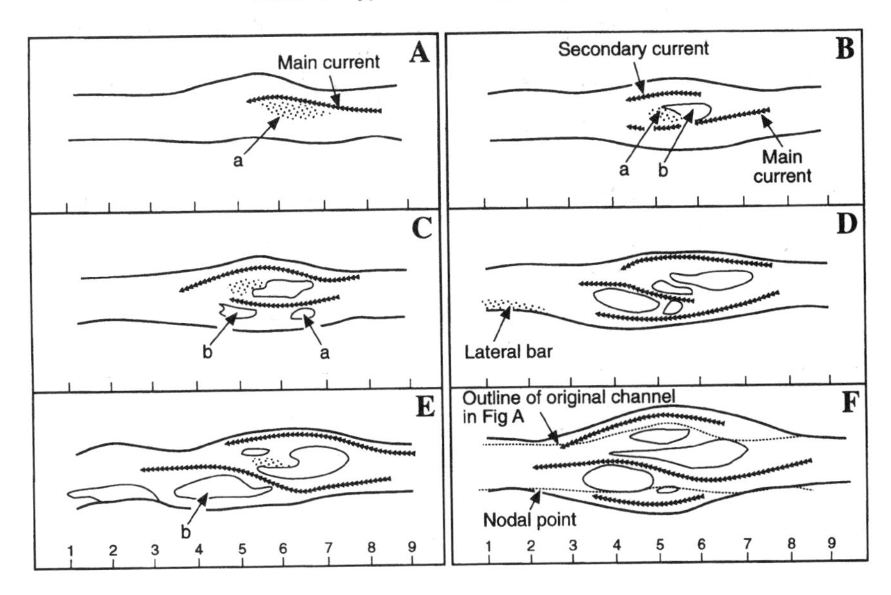

Figure 7.17 Progress in the development of a braided channel (after Leopold and Wolman, 1957)

while driving bed deposition along the inner margins of the anabranches (C). Bed scour in the divided reach lowers the water surface elevation so that the top of the mid-channel bar emerges as an island. Through time, a bar–island complex develops, with multiple flow divisions and subchannels (D and E). Eventually, as the width increases, the bar–island complex may coalesce to form a much larger, semi-permanent island. Mid-channel bar formation in each of the anabranch channels on either side of the island may then lead to further braiding through division of the flow following the same sequence of events. The resulting planform morphology resembles a string of beads, with relatively long, wide, multithreaded island reaches interspersed with shorter, narrower, single-thread nodes (F).

Differentiation of Islands and Bars

Brice (1964) built on Leopold and Wolman's identification of the difference between bars and islands to define these two features of braided river morphology. Bars are defined as dynamic features which are unvegetated and submerged at bankfull stage. Islands are more stable features, emergent at bankfull stage and vegetated. In practice, it is usually possible to differentiate between islands and bars although, as pointed out by Bridge (1993), the terms used to define them are purely qualitative and should be replaced by quantitative terms based on their rates of creation, migration and destruction.

The Braided Pattern: Nodes and Island Reaches

The idea that braided rivers display a node–island pattern was taken up by Coleman (1969) in an important paper describing the morphology of the Brahmaputra River in

Bangladesh. He generalised Leopold and Wolman's findings to produce a generic diagram for the planform/cross-section associations in braided channels that draws parallels with the geometry and wavelength of meandering channels (Figure 7.18).

Coleman's diagram shows how braid bars, asymmetrical cross-sectional geometries in the flanking anabranches, and deep scour holes at confluences combine to link planform geometry to cross-sectional shape in braided rivers. At a node (a–a′ in Figure 7.18), the single-thread channel is narrow and relatively deep owing to confluence scour, although often there may be a pronounced medial bar. At an island reach, the multithread channel is very wide, with deep scour in some anabranches due to flow curvature and shallow channels running across the intervening islands.

The spacing of nodes along the length of the river appears to be scaled on the channel width, although this relationship is not nearly so well established as that for meander wavelength in single-thread channels. For example, in a morphological study of the Brahmaputra River in Bangladesh, Thorne et al. (1993) identified seven islands and eight nodes somewhat evenly spaced along a 220 km reach (Figure 7.19). The average node spacing was about 30 km, which approximates to about six times the 5 km average width of the braided channel. This finding is consistent with the theory of Yalin as modified by Hey (1976), which predicts that nodes should be spaced at about 2π times the width.

Braiding Intensity

Leopold and Wolman (1957) noted how division and subdivision of the channel into increasing numbers of anabranches continued until the flow in the outer, flanking channels was no longer able to erode the banks, input sediment for bar building or increase the braid

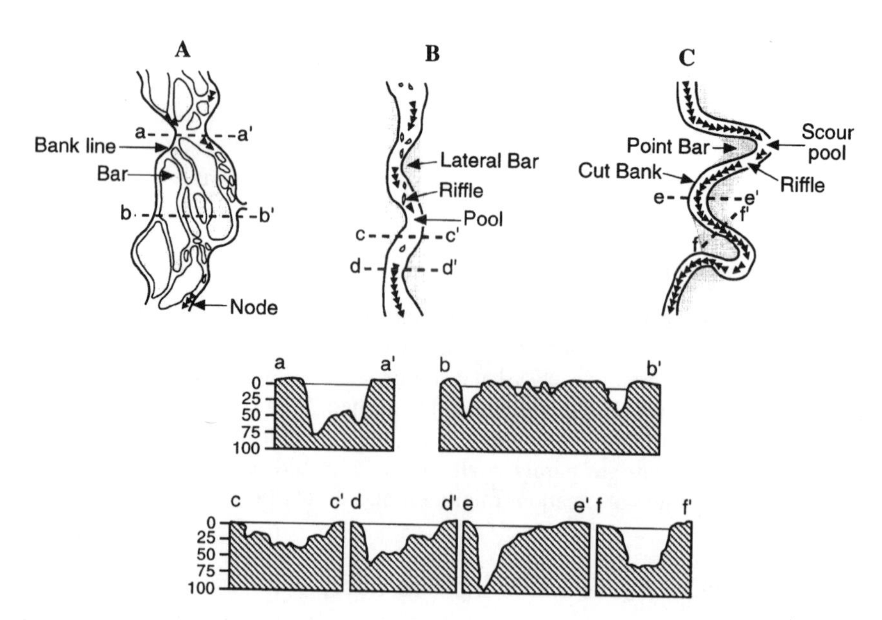

Figure 7.18 Planform/cross-section associations in braided and meandering channels (after Coleman, 1969)

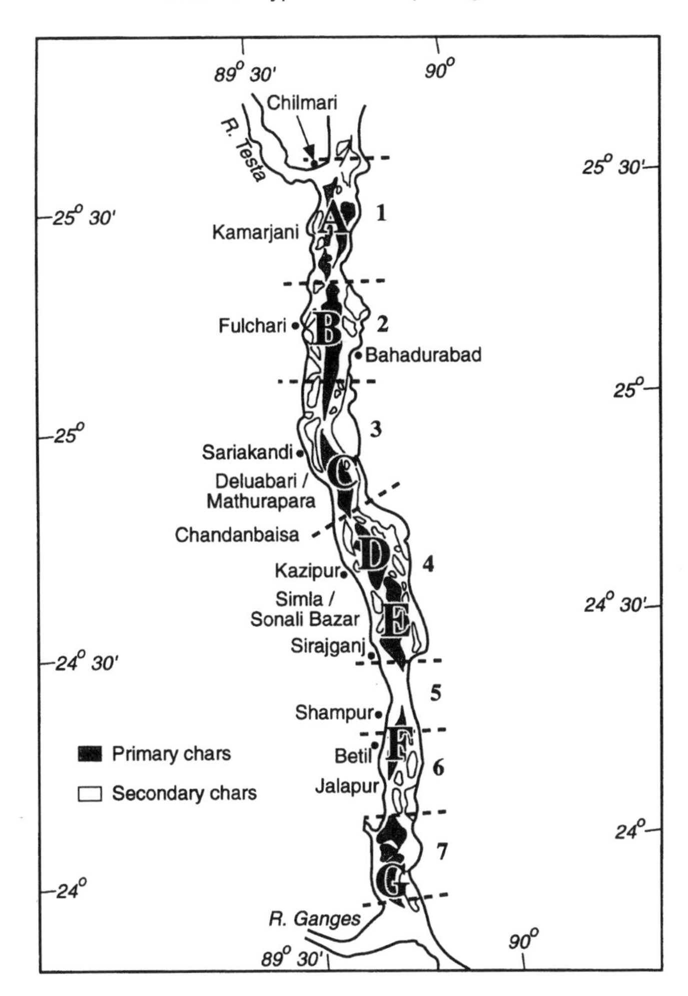

Figure 7.19 Islands, nodes and morphological reaches in the braided pattern of the Brahmaputra River, Bangladesh (after Thorne et al., 1993)

belt width. This infers a relationship between the competence of the stream to erode and transport sediment and the degree of braiding. On this basis researchers have subsequently attempted to develop a quantitative index of braiding intensity to characterise the degree of braiding. Bridge (1993) presented a useful summary of some of the more commonly used indices (Table 7.1).

These indices generally fall into two categories: those based on the number of active subchannels or braid bars at a section across the braid belt; and those based on the ratio of the sum of the channel lengths within a reach to a measure of the reach length. These latter types are actually measures of total sinuosity (as noted by Richards (1982)). In fact, these two types of index are measuring different aspects of braiding, both of which are informative in their own way.

Table 7.1 Braiding indices (modified from Bridge, 1993)

Author	Braiding index
Brice (1960, 1964)	$\text{Braid index} = \dfrac{2(\text{sum of lengths of all bars + islands in the reach})}{\text{centreline length of the reach}}$
Howard et a. (1970)	Braid index = (Av. no. of anabranches per cross-section) – 1
Engelund and Skovgaard (1973), Parker (1976), Fujita (1989)	Mode = number of rows of alternate bars (and sinuous flow paths) = 2 × the number of braid and side bars per cross-section
Rust (1978)	Mode = number of braids per meander wavelength
Hong and Davies (1979)	$\text{Total sinuosity} = \dfrac{\text{length of channel segments}}{\text{channel belt length}}$
Mosley (1981)	$\text{Braiding index} = \dfrac{\text{total length of bankfull channels}}{\text{distance along main channel}}$
Richards (1982)	$\text{Total sinuosity} = \dfrac{\text{total active channel length}}{\text{valley length}}$
Ashmore (1991)	Mean number of active channels per transect, or Mean number of active channel links in braided network
Friend and Sinha (1993)	$\text{Braid channel ratio} = \dfrac{\text{sum of mid-channel lengths of all channels}}{\text{length of mid-line of widest channel}}$

Generally, the first type of braiding index is preferable because it is a measure of the intensity of flow division that is the essence of braiding. This type of index can be used to characterise and compare the intensity of braiding in adjacent reaches and to identify time trends in braiding intensity of particular reaches. For example, Figure 7.20 shows the results of an engineering-geomorphic study of the Brahmaputra River in Bangladesh which were used to establish spatial variations in Howard et al.'s (1970) braiding intensity upstream and downstream of Sirajganj and to identify contrasting time trends in braiding intensity within morphologically defined subreaches (see Figure 7.19 for the locations of the reaches).

The second type of index (based on total sinuosity) combines the intensity of splitting of the flow with a measure of the sinuosity of various channels and subchannels which is, in fact, an entirely different morphological characteristic. Such indices are indeterminate morphologically because (as Bridge (1993) points out) it is possible for a braided river with a large number of relatively straight subchannels to have the same total sinuosity as one with a few, highly sinuous subchannels. Ideally, both a measure of flow division and a measure of total sinuosity should be used to define the planform morphology of a braided reach.

Braiding as an Equilibrium Channel Form

The shifting, changing nature of braided channels and the fact that they are often generated by sediment deposition and bed aggradation has led many engineers and river scientists to associate them almost exclusively with disequilibrium in the fluvial system. Yet Leopold

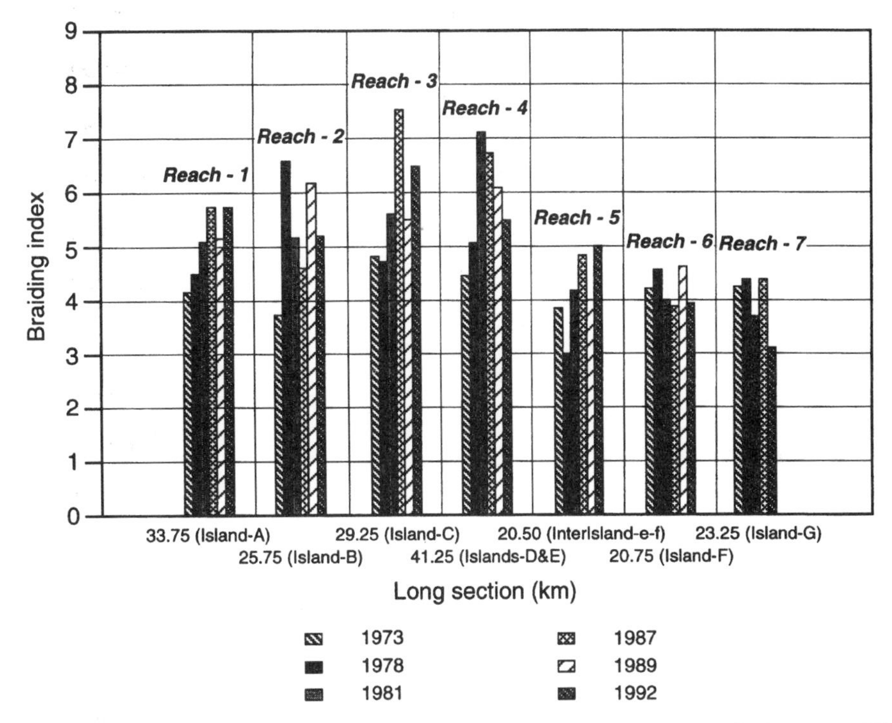

Figure 7.20 Spatial and temporal changes in the braiding intensity of the Brahmaputra River, Bangladesh, measured using the index developed by Howard et al. (1970). Locations of study reaches are marked in Figure 7.19 (after Sir William Halcrow and Partners, 1992)

and Wolman were at pains to point out as long ago as 1957 that braided rivers are a distinct and viable category of dynamically stable planform, along with straight and meandering configurations. The fact is that it is difficult to recognise this stability in systems which exhibit rapid and unpredictable channel changes owing to high mobility of bed and bank sediments and frequent adjustments of the positions and patterns of bars and anabranches. For example, specific gauge analysis of the records for Bahadurabad, on the braided Brahmaputra River in Bangladesh, indicated no significant change in stage levels over a 30 year period during which around 15 billion tonnes of sediment was transported through the section (Sir William Halcrow and Partners, 1992). This demonstrates that it is possible for a braided pattern to be associated with a graded profile, at least over engineering timescales.

Similarly, if a global view is taken of channel pattern, then the state of adjustment of channel form can be revealed. Analysis of satellite images of the Brahmaputra River, using LANDSAT images covering the period 1973–1992, has allowed insights into the overall adjustment of the system that were previously impossible using ground-based observations. In the study, the area of the braid plain was categorised from false-colour images as being water, sand, vegetation or cultivation. Taken together, water and sand represent the active channels and bars of the river, while the areas covered by vegetation and cultivation represent islands. A plot of the areas covered by active channels and bars and by islands reveals organisation and progressive change where formerly there was thought to be disorganisation and disequilibrium (Figure 7.21). The data show how, as the overall area of the braid plain has increased progressively due to widening, the area of active channels and bars has been maintained at between 48% and 52% of the total area. That is, it has been constant to within $\pm 2\%$. This is certainly a form of dynamic equilibrium quite different to that found in single-thread channels, but nevertheless it displays a degree of mutual adjustment not usually recognised in braided channels.

There is a great deal of fundamental research that must be performed before the fluvial forms and processes of braided channels will be properly understood. Until this work has been completed, morphological classifications and characterisations will remain sketchy at best.

7.4.5 Anastomosed Rivers

Differences Between Anastomosed and Braided Rivers

Anastomosed rivers are the fourth and most recently recognised type of channel pattern. The term 'anastomosing' comes from medicine and is used to describe a distributary system of arteries in the body at locations such as the back of the hand. The term seems first to have been applied to rivers by Lane (1957), but it only came into wide usage following work by Miall (1977), Rust (1978) and Smith and Smith (1980). Like braided

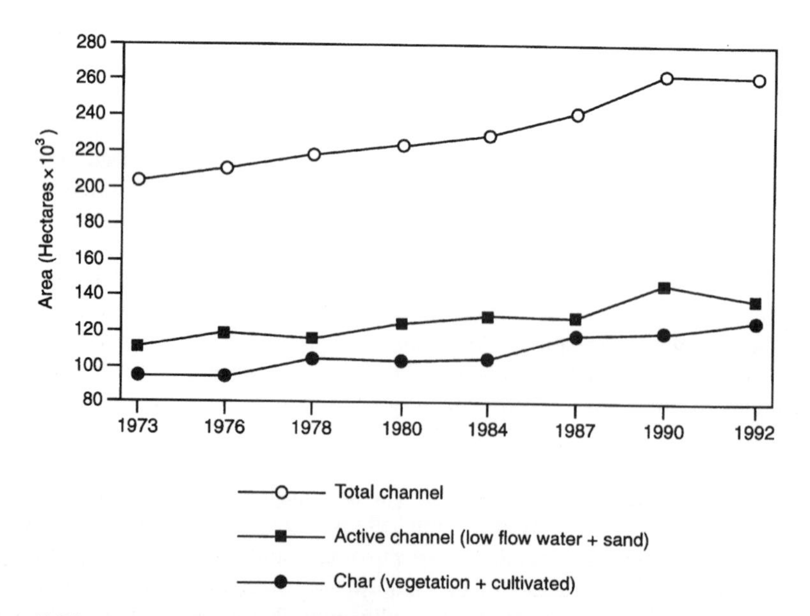

Figure 7.21 Time trends in the areas of water and sand (active channels and bars) and vegetation and cultivation (islands) for the Brahmaputra River, Bangladesh, between 1973 and 1992 (data from ISPAN, FAP-19, Dhaka, courtesy of Mr Tim Martin)

rivers, anastomosed rivers are multithreaded, but they differ fundamentally in at least two important respects.

The first distinctive difference relates to channel morphology (Ferguson, 1987). Braided rivers have a straight or slightly sinuous channel in which the flow diverges, divides and converges around relatively small, mobile, unvegetated sand or gravel bars. The highest elevations of the bars are, on average, a little less than that of the surrounding floodplain so that the bars are submerged and the anabranched flows combined at stages approaching bankfull. Anastomosed rivers have two or more channels, each of high sinuosity, separated by large, semi-permanent, vegetated islands capped by fine-grained sediments such as silts and clays. The highest elevations of the islands are about equal to those of the surrounding floodplain, so that the pattern remains multithreaded even at bankfull stage.

The second distinctive difference relates to channel slope (Ferguson, 1987). Braided rivers generally have relatively steep slopes compared to rivers with similar discharges and single-thread, meandering channels. The steep channel slope is closely associated with the abundant, relatively coarse sediment load which is another characteristic of braided rivers. Anastomosed rivers have slopes which are as low or lower than those associated with equivalent meandering rivers. This fact was first noted by Lane (1957), although the significance of the fact that there were two suites of multithread channels in his plots was not fully realised at the time.

Morphology of Anastomosed Rivers

It is now recognised that anastomosed rivers represent a channel type that is genetically different from braiding. Anastomosed systems comprise of a number of sinuous, low-energy anabranches following paths across the floodplain that, although they cross occasionally, operate independently over considerable distances. In an anastomosed system the lengths of anabranches between junctions is much longer than the characteristic bar length, so that individual anabranch reaches contain their own bars, scaled on the width of that particular anabranch. Anabranches usually meander, often with the highly sinuous, even tortuous, planform associated with low-energy streams flowing through cohesive sediments. On this basis, use of braiding indices based both on flow division and a measure of total sinuosity should allow quantitative differentiation of braided and anastomosed rivers. Braided rivers will display a high degree of flow division and a low total sinuosity, while anastomosed rivers will characteristically have a low degree of flow division and a high total sinuosity. In practice, however, many large alluvial rivers display elements of both braiding and anastomosing at the same time and in the same reaches (Coleman, 1969; Bristow, 1987; Bridge, 1993; Thorne et al., 1993).

Anabranch meanders share the same geometric relationships between channel width, meander wavelength and bend radius as those for single-thread meandering channels. Compared to both single-thread meandering channels and the subchannels in a braided system, anabranches are relatively stable. Rates of bank erosion, bend migration and planform evolution are characteristically small. The channel dynamics, floodplain environment and sedimentary deposits associated with anastomosed rivers are sufficiently distinguishable for them to be classified separately to those of other systems (Nanson and Croke, 1992).

Floodplain Classification of Anastomosed Rivers

In their genetic classification, Nanson and Croke (1992) define anastomosed rivers as producing low-energy, cohesive floodplains. Braided rivers form high-energy, non-cohesive floodplains and meandering systems form medium-energy, non-cohesive floodplains. Consequently, floodplain deposition is dominated by vertical accretion during overbank flows, setting anastomosed systems apart from meandering and braided systems which both tend to build floodplains by lateral point-bar accretion, mid-channel bar accretion and infilling of abandoned channels, as well as vertical accretion.

7.4.6 Prediction of Channel Planform Morphology

Given the marked contrasts of geometry, sedimentology and stability between rivers with different channel types and morphologies, it is not surprising that engineers and scientists need to predict channel pattern and channel pattern changes that might occur in response to changes in the river regime, engineering or management practice. This chapter has emphasised that, notwithstanding the usefulness of considering channel patterns under the headings straight, meandering, braided and anastomosed, there is actually a continuum of planform morphologies. In practice, it is probably the intermediate and transitional forms that occur most frequently, with easily classified forms being the exception rather than the rule. Hence, it is not meaningful to search for sharp dividing lines or geomorphological thresholds between different patterns because, in reality, a range of transitional patterns exists. This is not really a problem; in fact it actually makes life easier for professional engineers. As pointed out by Ferguson (1987), if a river is actually susceptible to pattern transformation from, say, meandering to braided (with serious implications for bankside and floodplain structures and human activities on the floodplain), this should be apparent through the prior existence of a range of channel forms and features that are recognisably transitional between meandering and braiding. If the channel displays only the features of an archetypal meandering stream then geomorphologically it is probably safely remote from the braided threshold in any case.

It is, however, unlikely that such qualitative arguments will convince team leaders, planners and managers and, usually, recourse to a quantitative analysis will be unavoidable. Several criteria exist and Bridge (1993) summarised many of them in a table which is reproduced, in modified form and with some additions, in Table 7.2.

Predictors such as those listed in Table 7.2 are currently out of fashion with geomorphological thinking and are subject to heavy criticism in learned journals and texts. Despite this, they can be used to add a quantitative dimension to qualitative arguments concerning planform evolution and the potential for climate change, sea level rise or engineering intervention in the fluvial system to trigger abrupt changes in channel planform type and morphology. Accepting this, the problem which remains is that of selecting the appropriate predictor for a given situation. A number of studies have been performed to evaluate these predictive models (Julien, 1986, 1987).

Ahmed (1986) used a hydrodynamic stability analysis to predict whether an initially straight channel would remain straight or would tend to either meander or braid. Re-examining the stability approach of Fredsøe (1978), he found that a channel will remain straight if its width/depth ratio is less than 8 and will always braid if its width/depth ratio is greater than 60. Diagrams based on Ahmed's analyses are reproduced in Figure 7.22. His

Table 7.2 Predictors of Channel Pattern (modified from Bridge, 1993)

Author	Function*	Explanation
Lane (1957)	$S < 0.007Q_m^{-0.25}$	Meandering, sand-bed channels
	$0.0041Q_m^{-0.25} > S < 0.007Q_m^{-0.25}$	Meandering–braiding transition
	$S > 0.0041Q_m^{-0.25}$	Braided, sand-bed channels
Leopold and Wolman (1957)	$S = 0.013Q_b^{-0.44}$	Meandering–braiding threshold
Henderson (1961)	$S = 0.000196D^{1.14}Q_b^{-0.44}$	Meandering–braiding threshold
Antropovsky (1972)	$S = 1.4Q_{ma}^{-1}$	Meandering–braiding threshold
Parker (1976)	$S/Fr \approx d/w$	Meandering–braiding threshold
Fredsøe (1978)	$\theta = (\tau/(s-1))D_{50}$	Straight–meandering–braided thresholds (see Figure 7.22)
Begin et al. (1981)	$S = 0.0016Q_m^{-0.3}$	Meandering–braiding threshold for a standard channel with $\tau = \tau_{ave}$
	$S = 0.0016(\tau/\tau_{ave})Q_m^{-0.3}$	Relations for non-standard channels (braided channels: $\tau > \tau_{ave}$; meandering: $\tau < \tau_{ave}$)
Ackers (1982)	$S = 0.0008Q^{-0.21}$	Meandering–braiding threshold for sand-bed flumes and rivers
Bray (1982)	$S = 0.07Q_{2f}^{-0.44}$	Meandering–braiding threshold for gravel-bed rivers
Ferguson (1984)	$S = 0.042Q^{-0.49}D_{50}^{0.09}$	Meandering–braiding threshold for gravel-bed rivers
	$S = 0.056Q^{-0.5}$	Meandering–braiding threshold for any river
	$S = 0.0049Q^{-0.21}D_{50}^{0.52}$	Meandering–braiding threshold based on Parker's theory and hydraulic geometry
Chang (1985)	$S \approx aQ^{-0.5}D^{0.5}$	Meandering–braiding threshold
Robertson-Rintoul and Richards (1993)	$\Sigma P = 1 + 5.52(QS_v)^{0.38}D_{84}^{-0.44}$	Meandering–braiding threshold for gravel-bed rivers (Figure 7.23)
	$\Sigma P = 1 + 2.64(QS_v)^{0.4}D_{84}^{-0.14}$	Meandering–braiding threshold for sand-bed rivers (see Figure 7.23)

*SI units

analyses did not, however, establish whether these findings were a cause or an effect of planform development.

Stubblefield (1986) tested the methods of Lane (1957), Leopold and Wolman (1957) and Parker (1976) using information for 56 streams extracted from a database published by Church and Rood (1983). His overall finding was that each of the methods was of limited accuracy and suggested that in practice all three should be used and the results combined to increase confidence in the predictions. Lane's (1957) method was found to give the best results for sand-bed streams.

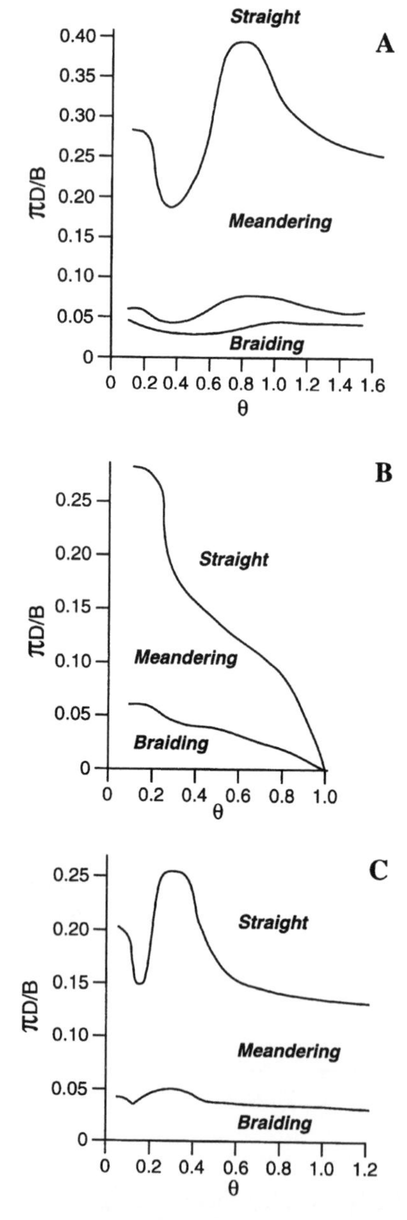

Figure 7.22 Fredsøe's stability diagrams for planform prediction in: (A) a sand-bed river with dunes ($s=2.65$, $d/D=1000$, $Cd=7$); (B) a flat-bed channel; and (C) a dune-bed with suspended load neglected (modified from Ahmed, 1986)

Smith (1987) used a more extensive data set from 101 channels to investigate the accuracy of nine methods in defining the meandering/braiding threshold in alluvial rivers. The methods tested were the empirical relations. of Lane (1957), Leopold and Wolman

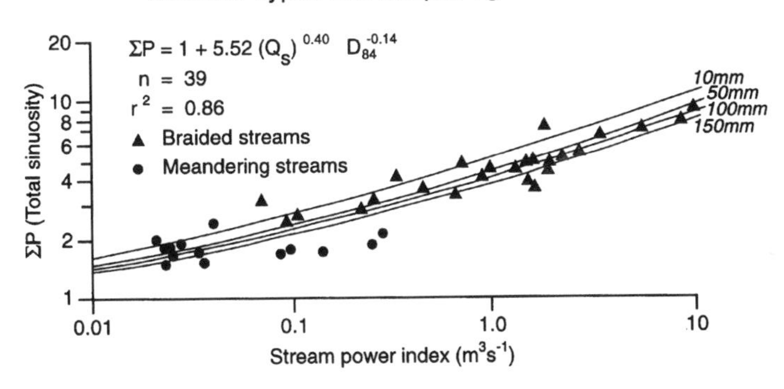

Figure 7.23 Relationship between total sinuosity and stream power for single-thread and multi-thread channels (adapted from Robertson-Rintoul and Richards, 1993)

(1957), Henderson (1961), Osterkamp (1978), Begin et al. (1981), Bray (1982) and Ferguson (1984), and the theoretical relations of Parker (1976) and Fredsøe (1978).

Smith's results emphasised the importance of considering the size of the bed sediment when attempting to predict channel planform. In practice this may be achieved either by selecting a method which was developed for river conditions similar to those being studied, or by using a method which explicitly accounts for bed material size. Smith confirmed Stubblefield's finding that Lane's (1957) method gave the best results for sand-bed rivers, but also demonstrated that it must *not* be applied to gravel-bed rivers. Ferguson's (1984) method was the most reliable for gravel-bed rivers. Fredsøe's (1978) method was, overall, the best predictor for streams of all types, although the requirement that width and depth be specified as input parameters limits its applicability compared to that of both Lane's and Ferguson's methods, which do not require the user to specify a cross-sectional geometry. This is potentially a great advantage, as the cross-sectional geometry may well be unknown when predictions are being made of channel planform response to changes in the driving variables or to the impacts of engineering intervention.

Most recently, van den Berg (1995) has re-examined the prediction of alluvial planforms and presents a new method which uses as input variables the bed material median grain size and the potential specific stream power based on bankfull discharge and valley slope. A data set of 228 streams was used to develop a discriminant function between meandering rivers with sinuosities greater than 1.5 and less sinuous, braided rivers. It is defined by:

$$\omega_{\mathrm{vt}} = 843 D_{50}^{0.41} \tag{7.21}$$

where ω_{vt} = specific stream power at the transition between meandering and braiding (W/m²). Specific stream power (stream power per unit bed area) is defined by:

$$\omega_{\mathrm{v}} = 2.1 S_{\mathrm{v}} Q_{\mathrm{b}}^{0.5} \quad \text{for sand-bed rivers} \tag{7.22}$$

$$\omega_{\mathrm{v}} = 3.3 S_{\mathrm{v}} Q_{\mathrm{b}}^{0.5} \quad \text{for gravel-bed rivers} \tag{7.23}$$

This reflects the different cross-sectional geometry for the same discharge in sand-bed and gravel-bed rivers. Streams with potential specific stream powers greater than the threshold value will braid and those with values less than the threshold value will meander. The limits to the applicability of the function are $Q > 10 m^3/s$ and $0.1\ mm < D_{50} < 100\ mm$. The data and threshold line are shown in Figure 7.24.

van den Berg's approach takes into account bed material size and, by using stream power in place of discharge, it better accounts for the competence of the river to entrain and transport bed sediment. A discriminant function of this type may well represent the logical endpoint of the line of investigation into the meander/braiding threshold begun by Lane and by Leopold and Wolman nearly 40 years ago.

7.5 STREAM CLASSIFICATION FOR ANALYSIS, ENGINEERING AND MANAGEMENT: THE FUTURE?

In terms of channel pattern classification, the diagram produced by Brice (1975) covers the entire range of planforms identified in this paper and is recommended for use in engineering geomorphic studies (Figure 7.25). However, planform is only one aspect of channel form and the cross-sectional and longitudinal dimensions should also be considered for completeness.

Perhaps the most comprehensive system for classification yet devised is that of Rosgen (1994). This divides streams into seven major types on the basis of degree of entrenchment, gradient, width/depth ratio, and sinuosity. Within each major category there are six subcategories depending on the dominant type of bed/bank materials.

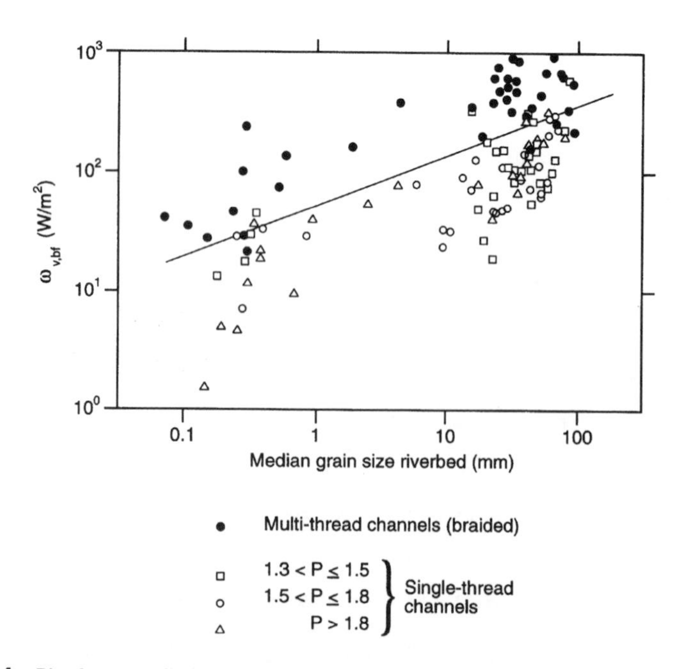

Figure 7.24 Planform prediction diagram developed by van den Berg (1995)

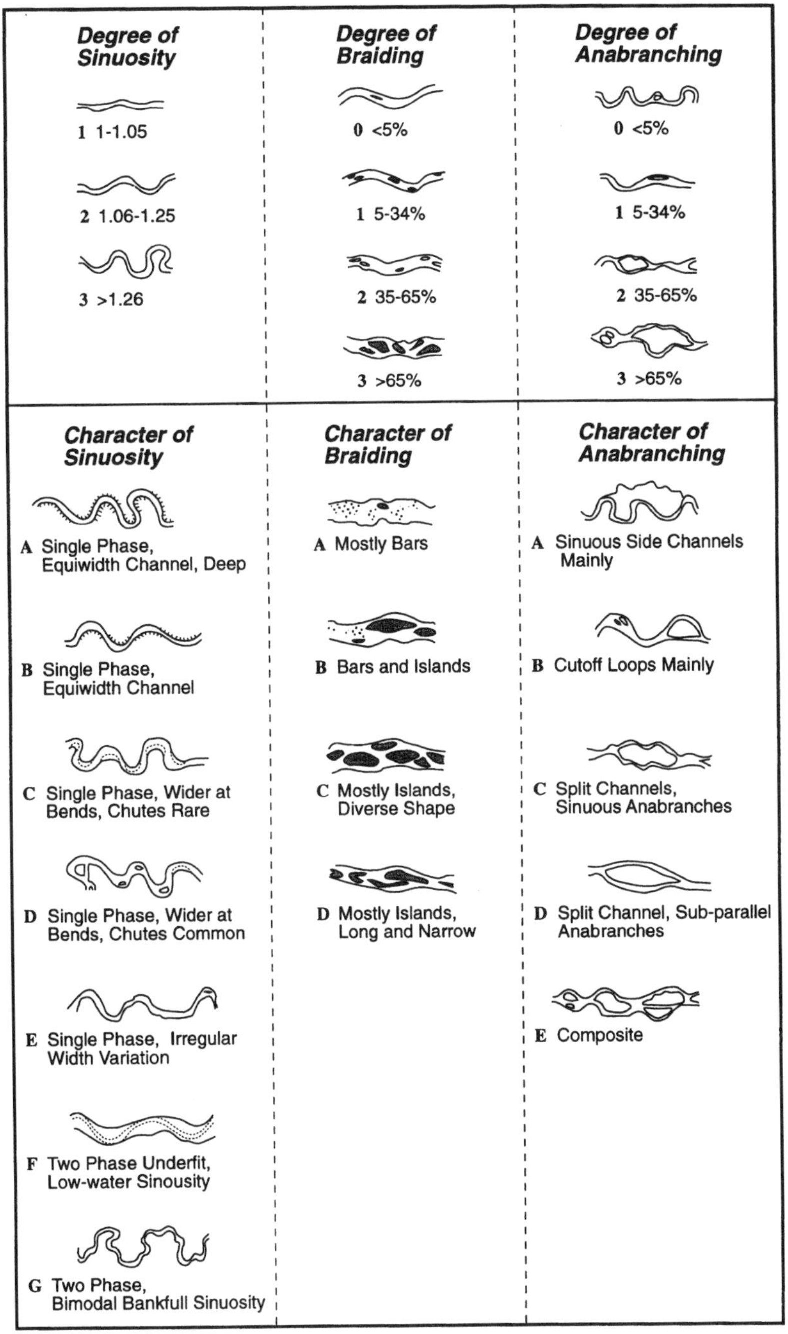

Degree of Sinuosity

1 1-1.05

2 1.06-1.25

3 >1.26

Degree of Braiding

0 <5%

1 5-34%

2 35-65%

3 >65%

Degree of Anabranching

0 <5%

1 5-34%

2 35-65%

3 >65%

Character of Sinuosity

A Single Phase, Equiwidth Channel, Deep

B Single Phase, Equiwidth Channel

C Single Phase, Wider at Bends, Chutes Rare

D Single Phase, Wider at Bends, Chutes Common

E Single Phase, Irregular Width Variation

F Two Phase Underfit, Low-water Sinousity

G Two Phase, Bimodal Bankfull Sinuosity

Character of Braiding

A Mostly Bars

B Bars and Islands

C Mostly Islands, Diverse Shape

D Mostly Islands, Long and Narrow

Character of Anabranching

A Sinuous Side Channels Mainly

B Cutoff Loops Mainly

C Split Channels, Sinuous Anabranches

D Split Channel, Sub-parallel Anabranches

E Composite

Figure 7.25 Channel pattern classification devised by Brice (after Brice, 1975)

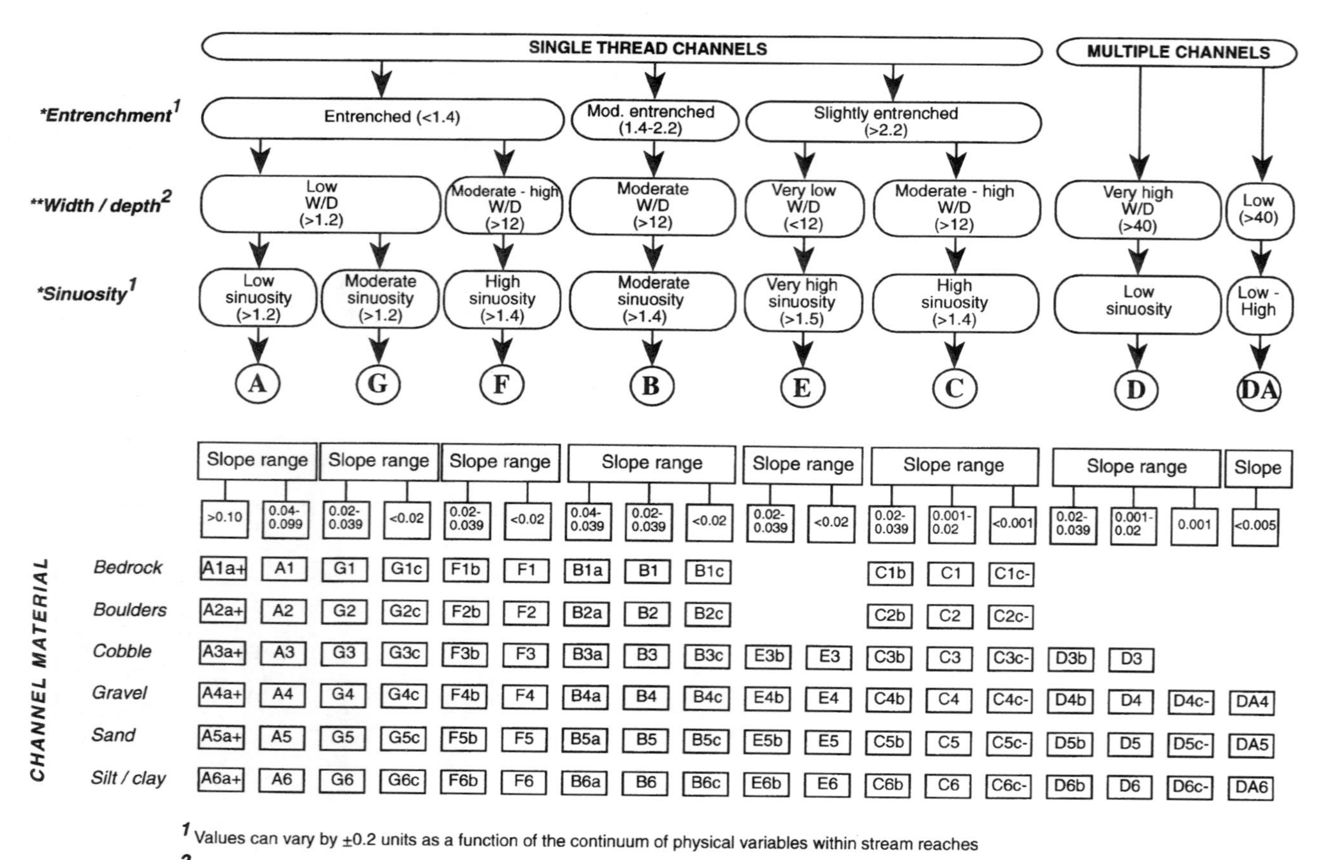

[1] Values can vary by ±0.2 units as a function of the continuum of physical variables within stream reaches

[2] Values can vary by ±0.2 units as a function of the continuum of physical variables within stream reaches

Figure 7.26 Key to classification of rivers in Rosgen's method (modified from Rosgen, 1994)

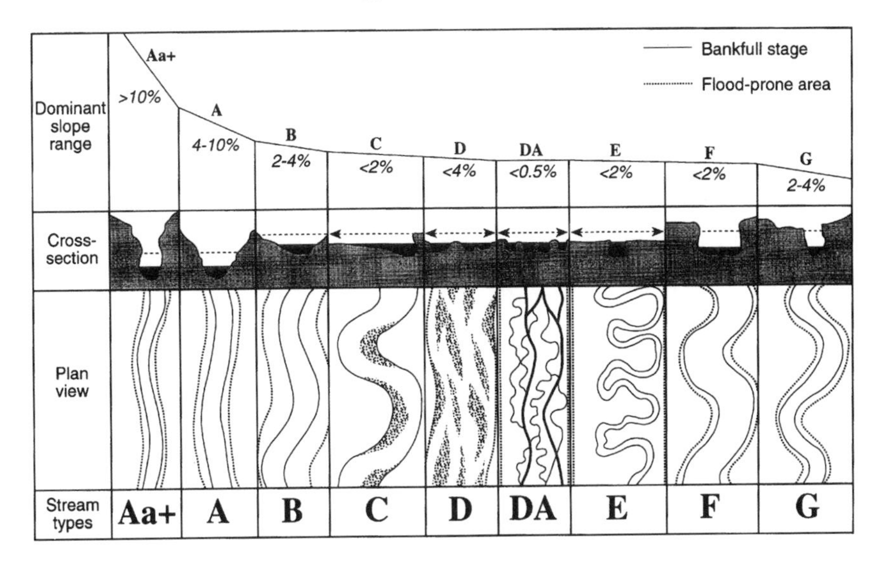

Figure 7.27 Longitudinal, cross-sectional and planform views of major stream types in Rosgen's method (modified from Rosgen, 1994)

The basic framework of Rosgen's method is set out in Figures 7.26 and 7.27. Criteria for the classification system and descriptions of the salient forms and features of each type are listed in Table 7.3. Examination of the criteria, forms and features listed in Table 7.3 illustrates that Rosgen has synthesised much of the material covered in this chapter. The result is a classification which is comprehensive in its scope, but which requires a strong geomorphological insight and understanding to apply consistently and usefully. It is at present too early to judge the usefulness and reliability of Rosgen's method when applied by engineers and managers with only a limited background in fluvial geomorphology, although indications are that users can gain the knowledge required through intensive, short-course training.

A more serious problem with all classifications based on existing channel morphology is that they fail to account for dynamic adjustment or evolution of the fluvial system. Increasing recognition of the fact that rivers are seldom in dynamic equilibrium has driven a desire on the part of engineers and managers to be able to predict channel changes in the short and medium term. In response, geomorphologists have begun to develop new schemes of river classification based on adjustment processes and trends of channel change rather than existing channel morphology and sediment features. The relatively simple adjustment classification of Brice (1981) identified channels as degrading, aggrading, widening, shifting at both banks, or shifting laterally at points of flow impingement. Brookes (1988) accentuated instream adjustments with adjustment classes that accounted for bed degradation, armouring, thalweg sinuosity, bar development and bank erosion. Downs (1995) developed a comprehensive system that incorporates the classifications of Brice and Brookes but builds on their earlier work by linking observed trends and patterns of adjustment to the fluvial and sediment processes responsible for driving channel change (Figure 7.28).

Table 7.3 Summary of criteria used for broad level classification in the Rosgen method (redrafted from Rosgen (1994))

Stream type	General description	Entrenchment ratio	*W/D* ratio	Sinuosity	Slope	Landform/soils/features
Aa +	Very steep, deeply entrenched, debris transport streams.	< 1.4	< 12	1.0 to 1.1	> 0.10	Very high relief. Erosional, bedrock or deposition features; debris flow potential. Deeply entrenched streams. Vertical steps with deep scour pools; waterfalls.
A	Steep, entrenched, cascading, step/pool streams. High energy/debris transport associated with depositional soils. Very stable if bedrock- or boulder-dominated channel.	< 1.4	< 12	1.0 to 1.2	0.04 to 0.10	High relief. Erosional or depositional and bedrock forms. Entrenched and confined streams with cascading reaches. Frequently spaced, deep pools, associated step-pool bed morphology.
B	Moderately entrenched, moderate gradient, riffle-dominated channel, with infrequently spaced pools. Very stable plan and profile. Stable banks.	1.4 to 2.2	> 12	> 1.2	0.02 to 0.039	Moderate relief, colluvial deposition and/or residual soils. Moderate entrenchment and *W/D* ratio. Narrow, gently sloping valleys. Rapids predominate with occasional pools.
C	Low gradient, meandering, point-bar, riffle/pool, alluvial channels with broad, well defined floodplains	> 2.2	> 12	> 1.4	< 0.02	Broad valleys with terraces, in association with floodplains, alluvial soils. Slightly entrenched with well defined meandering channel. Riffle–pool bed morphology.
D	Braided channel with longitudinal and transverse bars. Very wide channel with eroding banks.	n/a	> 40	n/a	< 0.04	Broad valleys with alluvial and colluvial fans. Glacial debris and depositional features. Active lateral adjustment, with abundance of sediment supply.

DA	Anastomosing (multiple channels) narrow and deep with expansive well vegetated floodplain and associated wetlands. Very gentle relief with highly variable sinuosities. Stable streambanks.	>4.0	<40	variable	<0.005	Broad, low-gradient valleys with fine alluvium and/or lacustrine soils. Anastomosed (multiple channel geologic control creating fine deposition with well vegetated bars that are laterally stable with broad wetland floodplains.
E	Low gradient, meandering riffle/pool stream with low width/depth ratio and little deposition. Very efficient and stable. High meander width ratio.	>2.2	<12	>1.5	<0.02	Broad valley/meadows. Alluvial materials with floodplain. Highly sinuous with stable, well vegetated banks. Riffle–pool morphology with very low width/depth ratio.
F	Entrenched meandering riffle/pool channel on low gradients with high width/depth ratio.	<1.4	>12	>1.4	<0.02	Entrenched in highly weathered material. Gentle gradients, with a high *W/D* ratio. Meandering, laterally unstable with high bank-erosion rates. Riffle–pool morphology.
G	Entrenched 'gulley' step/pool and low width/depth ratio on moderate gradients.	<1.4	<12	>1.2	0.02 to 0.039	Gulley, step–pool morphology with moderate slope and low *W/D* ratio. Narrow valleys, or deeply incised in alluvial or colluvial materials, i.e. fans, deltas. Unstable, with grade control problems and high bank erosion rates.

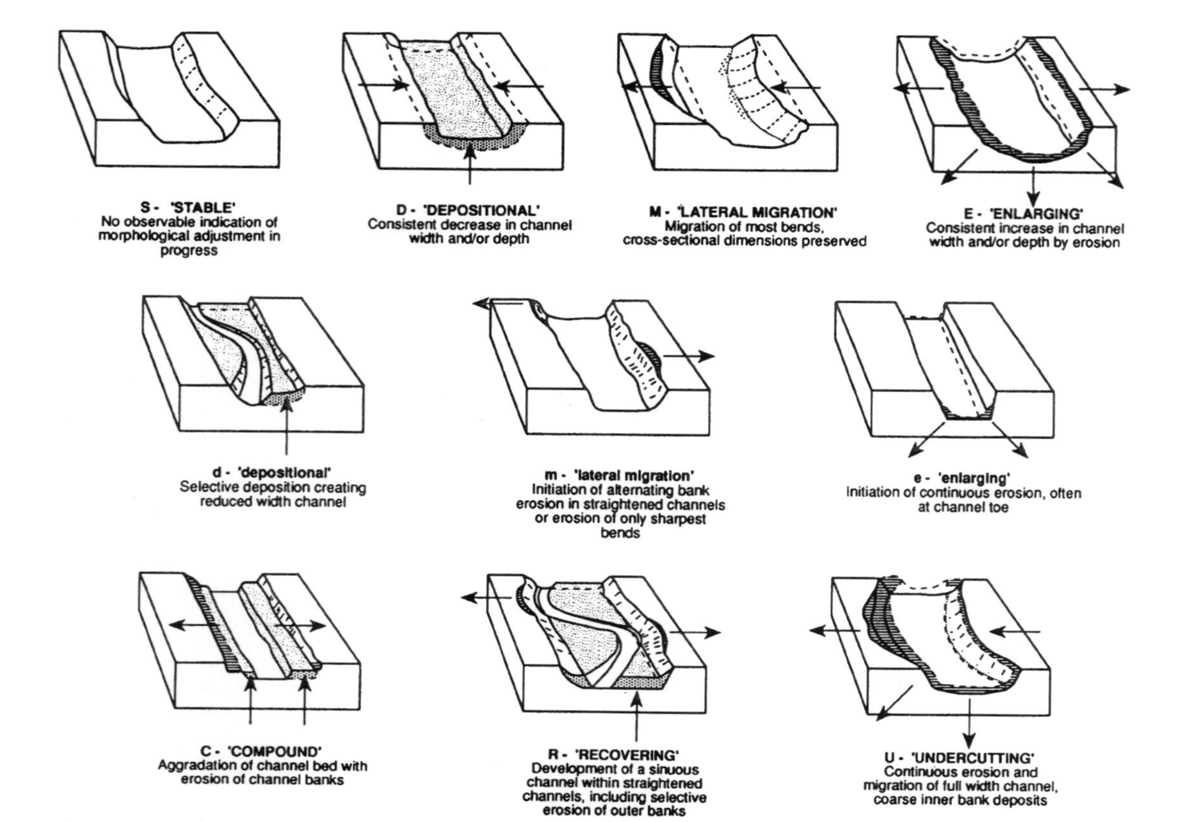

Figure 7.28 Downs' channel classification, based on trends and types of morphological change (modified from Downs, 1995)

Adjustment-based classifications differ fundamentally from morphology-based schemes in that they require the individual performing the classification to determine the current nature of channel adjustment processes. While historical records of types, trends and rates of channel change are very useful as the basis for determination of the current situation, such information is not always available and, even if it is, ongoing changes in catchment characteristics, alterations to channel management, or complexity in the response of the fluvial system often mean that past changes are not representative of current or future adjustments (Downs and Thorne, 1996). For these reasons, classification of channel adjustment requires judgement on the part of the engineer or scientist, who must infer adjustment processes from channel form. Evaluations of this demand careful observation coupled to insight into process–form linkages that support qualitative interpretation and evaluation. This places additional emphasis on the need for reliable and repeatable methods of stream reconnaissance to support rapid acquisition of the observational data necessary to support classification (Thorne, 1993; Simon and Downs, 1995). Also, training in applied fluvial geomorphology is essential to equip those responsible for stream classification with the skills necessary to allow sound interpretation of morphological data (Downs, 1995).

7.6 CONCLUSIONS

Scientists, engineers and water resource managers in the mid-1990s are expected to take a broad, environmentally oriented view of the river that recognises the need to work with, rather than against, nature. Environmental considerations do not, however, absolve the engineer of the obligation to account for flood defence, land drainage, channel stability and navigation interests. The need to balance the needs of different interests, sometimes with conflicting aims, makes it essential to take a multifunctional approach. Engineers seek to solve river-related problems while retaining those natural forms and features that allow the river to transmit the inputs of water and sediment, support diverse habitats and provide a pleasing landscape for river-centred recreation. A comprehensive and reliable morphological analysis and classification system forms the essential basis to sound engineering geomorphology.

The identification of channel type and the classification of channel morphology are fairly new additions to the methodologies routinely used by river engineers and managers. While care must be exercised by users unfamiliar with the limitations of geomorphic parameters and discriminants, even a rudimentary classification of channel morphology puts significant flesh on the bones provided by a standard channel survey consisting of cross-sections, plan maps and a long profile.

The ambitious method recently developed by Rosgen (1994) attempts to produce a comprehensive, semi-quantitative, holistic morphological classification system that incorporates all three dimensions of channel form while also accounting for differences in channel-forming materials. This approach, combining qualitative description and quantitative parameters in the definition of channel type, no doubt represents the way forward, although Rosgen's method does not represent the final product in terms of classification systems.

This chapter has shown that at present the key to identifying and classifying channels correctly lies in selecting techniques appropriate to the fluvial system in question and in using a variety of approaches in order to increase confidence in decision-making.

7.7 REFERENCES

Ackers, P. 1982. Meandering channels and the influence of bed material. In: Hey, R.D., Bathurst, J.C. and Thorne, C.R. (Eds), *Gravel-bed Rivers*. Wiley, Chichester, 389–414.

Ahmed, M.M. 1986. Meandering and braiding of rivers. In: Julien, P.Y. (Ed.), *Essays on River Mechanics*, Colorado State University, Fort Collins, Colorado, 1–18.

Allen, J.R.L. 1970. A quantitative model of grain size and sedimentary structure in lateral deposits. *Geological Journal*, **7**, 129–46.

Andrews, E.D. 1980. Effective and bankfull discharges of streams in the Yampa River Basin, Colorado and Wyoming. *Journal of Hydrology*, **46**, 311–330.

Antropovsky, V.I. 1972. Criterial relations of types of channel processes. *Soviet Hydrology*, **11**, 371–381.

Ashmore, P.E. 1991. How do gravel rivers braid?, *Canadian Journal of Earth Science*, **28**, 326–341.

Bagnold, R.A. 1960. Some Aspects of the Shape of River Meanders. United States Geological Survey. *Professional Paper 282E.*

Begin, Z.B., Meyer, D.F. and Schumm, S.A. 1981. Development of longitudinal profiles of alluvial channels in response to base-level lowering, *Earth Surface Processes*, **6**, 49–68.

Biedenharn, D.S., Little, C.D. and Thorne, C.R. 1987. Magnitude and frequency analysis in large rivers. In: *Hydraulic Engineering, Proceedings of the Williamsburg Symposium*, August 1987, ASCE, 782–787.

Bray, D.I. 1980. Regime equations for gravel-bed rivers. In: Hey, R.D., Bathurst, J.C. and Thorne, C.R. (Eds), *Gravel-bed Rivers*. Wiley, Chichester, 517–552.

Brice, J.C. 1960. Index for description of channel braiding. *Bulletin of the Geological Society of America*, **71**, 1833.

Brice, J.C. 1964. *Channel patterns and terraces of the Loup Rivers, Nebraska*. US Geological Survey Professional Paper, 422-D.

Brice, J.C. 1973. Meandering pattern of the White River in Indiana – an analysis. In: Morisawa, M. (Ed.), *Fluvial Geomorphology*. Publications in Geomorphology, SUNY Binghamton. 179–200.

Brice, J.C. 1975. Air photo interpretation of the form and behavior of alluvial rivers. Final report to the US Army Research Office.

Brice, J.C. 1981. Stability of relocated stream channels. Report to Federal Highways Administration National Technical Information Service. Washington D.C. Report Number FHWA/RD–80/158, 177 pp.

Bridge, J.S. 1993. The interaction between channel geometry, water flow, sediment transport and deposition in braided rivers. In: Best, J.L. and Bristow, C.S. (Eds), *Braided Rivers*. Geological Society of London Special Publication no. 75, 13–63.

Bridge, J.S. 1982. A revised mathematical model and FORTRAN IV Program to predict flow, bed topography and grain size in open channel bends. *Computers and Geosciences*, **8**, 91–95.

Bristow, C.S. 1987. Brahmaputra River: Channel migration and deposition. In: Ethridge, F.G., Flores, R.M. and Harvey, M.D. (Eds), *Recent Developments in Fluvial Sedimentology*. Society of Economic Palaeontologists and Mineralogists, Special Publication no. 39, 63–74.

Brookes, 1988. *Channelized Rivers*. John Wiley, Chichester, UK, 336 pp.

Carson, M.A. and Lapointe, M.F. 1983. The inherent asymmetry of river meander planform. *Journal of Geology*, **91**, 41–55.

Chang, H.H. 1985. River morphology and thresholds. *Journal of Hydraulic Engineering*, **111**, 503–519.

Coleman, J.M. 1969. Brahmaputra River: Channel processes and sedimentation. *Sedimentary Geology*, **3**, 129–239.
Chorley, R.J., Schumm, S.A. and Sugden, D.E. 1984. *Geomorphology*. Methuen, London and New York, 605 pp.
Church and Rood. 1983. Catalogue of alluvial river channel regime data. Department of Geography, University of British Columbia, Vancouver, Canada, 99 pp.
Downs. 1995. Estimating the probability of river channel adjustment. *Earth Surface Processes and Landforms*, **20**, 687–705.
Downs and Thorne. 1996. A geomorphological justification of river channel reconnaissance surveys. *Transactions of the Institute of British Geographers*, New Series, **21**, 455–468.
Dury, G.H. 1955. Bed width and wave-length in meandering valleys. *Nature*, **176**, 31.
Engelund, F. and Skovgaard, O. 1973. On the origin of meandering and braiding in alluvial streams. *Journal of Fluid Mechanics*, **57**, 289–302.
Ferguson, R.I. 1973. Regular meander path models. *Water Resources Research*, **9**, 1079–1086.
Ferguson, R.I. 1984. The threshold between meandering and braiding. In: Smith, K.V.A. (Ed.), *Channels and Channel Control Structures, Proceedings of the First International Conference on Hydraulic Design in Water Resources Engineering*. Springer-Verlag, Berlin, 6.15–6.29.
Ferguson, R.I. 1987. Hydraulic and sedimentary controls of channel pattern. In: Richards, K.S. (Ed.), *Rivers – Environment and Process*. Blackwell, Oxford, 129–158.
Fisk, H.N. 1944. *Geological Investigation of the Alluvial Valley of the Lower Mississippi River*. US Army Corps Eng., Mississippi River Commission, Vicksburg, Mississippi.
Fisk, H.N. 1947. *Fine Grained Alluvial Deposits and their Effects on the Mississippi River Activity*. US Waterways Experimental Station, Vicksburg, Mississippi.
Fredsøe, J. 1978. Meandering and braiding of rivers. *Journal of Fluid Mechanics*, **84**, 609–624.
Friend, P.F. and Sinha, R. 1993. Braiding and meandering parameters. In: *Braided Rivers*. Best, J.L. and Bristow, C.S. (Eds), Geological Society of London Special Publication no. 75, 105–111.
Fujita, Y. 1989. Bar and channel formation in braided streams. In: Ikeda, S. and Parker, G. (Eds), *River Meandering*. American Geophysical Union, Water Resources Monographs 12, 417–462.
Gray, D.H. and Leiser, A.T. 1983. *Biotechnical Slope Protection and Erosion Control*. Van Nostrand Reinhold, New York.
Henderson, F.M. 1961. Stability of alluvial channels. *American Society of Civil Engineers, Journal of the Hydraulics Division*, **87**, 109–138.
Hey, R.D. 1976. Impact prediction in the physical environment. In: O'Riordan, T. and Hey, R.D. (Eds), *Environmental Impact Assessment*. Saxon House, Farnborough, 71–81.
Hey, R.D. and Thorne, C.R. 1986. Stable channels with mobile gravel beds. *Journal of Hydraulic Engineering*, **112**, 671–689.
Hong, L.B. and Davies, T.R.H. 1979. A study of stream braiding. *Geological Society of America Bulletin*, **79**, 391–394.
Howard, A.D. 1967. Drainage analysis in geologic interpretation: a summation. *Bulletin of the Geological Society of America*, **56**, 275–370.
Howard, A.D., Keetch, M.E. and Vincent, C.L. 1970. Topological and geomorphic properties of braided streams. *Water Resources Research*, **6**, 1647–1688.
Ikeda, S. and Parker, G. (Eds), 1989. *River Meandering*. American Geophysical Union, Water Resources Monographs 12.
Julien, P.Y. 1986. *Essays on River Mechanics*, Colorado State University, Fort Collins, Colorado, Spring 1986.
Julien, P.Y. 1987. *Essays on River Mechanics*. Colorado State University, Fort Collins, Colorado, Spring 1987.
Lane, 1957. A study of the shape of channels formed by natural streams flowing in an erodible material, *MRD Sediment Series 9*, US Army Engineer Division, Missouri River.
Leopold. 1994. *A View of the River*. Harvard University Press, Cambridge, Massachusetts, 298 pp.
Leopold, L.B. and Langbein, W.B. 1966. River meanders. *Scientific American*, **214**(6), 60–69.
Leopold, L.B. and Wolman, M.G. 1957. *River Channel Patterns – Braided, Meandering and Straight*. United States Geological Survey, Professional Paper 282B.

Leopold, L.B. and Wolman, M.G. 1960. River meanders. *Bulletin of the Geological Society of America*, **71**, 769–794.

Leopold, L.B., Wolman, M.G. and Miller, J.P. 1964. *Fluvial Processes in Geomorphology*. Freeman, San Francisco, 522 pp.

Markham, A.J. and Thorne, C.R. 1992. Geomorphology of gravel-bed river bends. In: Billi, P., Hey, R.D., Thorne, C.R. and Tacconi, P. (Eds), *Dynamics of Gravel-bed Rivers*. Wiley, Chichester, 433–456.

Masterman, R.J. and Thorne, C.R. 1992. Predicting the influence of bank vegetation on channel capacity. *American Society of Civil Engineers, Journal of Hydraulic Engineering*, **118**, 1052–1059.

Miall, A.D. 1977. A review of the braided-river depositional environment. *Earth-Science Reviews*, **13**, 1–62.

Mosley, M.P. 1981. Semi-determinate hydraulic geometry of river channels, South Island, New Zealand. *Earth Surface Processes and Landforms*, **6**, 127–137.

Nanson, G.C. and Croke, J.C. 1992. A genetic classification of floodplains. *Geomorphology*, **4**, 459–486.

Odgaard, A.J. 1989. River meander model. I: Development, and II: Applications. *American Society of Civil Engineers, Journal of Hydraulic Engineering*, **112**, 1117–1150.

Osterkamp, W.R. 1978. Gradient, discharge and particle-size relations of alluvial channels in Kansas, with observations on braiding. *American Journal of Science*, **278**, 1253–1268.

Parker, G. 1976. On the cause and characteristic scales of meandering and braiding in rivers. *Journal of Fluid Mechanics*, **76**, 457–480.

Reid, J.B. 1984. Artificially induced concave bank deposition as a means of floodplain erosion control. In: *River Meandering, Proceedings of the Conference on Rivers '83*, New Orleans, ASCE, New York, 295–305.

Richards, K.S. 1982. *Rivers: Form and Process in Alluvial Channels*. Methuen, London.

Robertson-Rintoul, M.S.E. and Richards, K.S. 1993. Braided channel pattern and paleohydrology using an index of total sinuosity. In: Best, J.L. and Bristow, C.S. (Eds), *Braided Rivers*. Geological Society of London Special Publication no. 75, 113–118.

Rosgen D.L. 1994. A classification of natural rivers. *Catena*, **22**, 169–199.

Rust, B.R. 1978. A classification of alluvial channel systems. In: Miall, A.D. (Ed.), *Fluvial Sedimentology*. Canadian Society of Petroleum Geologists, Calgary, Memoir no. 5, 187–198.

Schumm, S.A. 1963. Sinuosity of alluvial rivers in the Great Plains. *Geological Society of America Bulletin*, **74**, 1089–1100.

Schumm, S.A. 1968. River adjustment to altered hydrologic regimen – Murrumbidgee River and palaeochannels, Australia. US Geological Survey Professional Paper no. 596, 65 pp.

Schumm, S.A. 1977. *The Fluvial System*. Wiley, New York, 338 pp.

Schumm, S.A. and Thorne, C.R. 1989. Geologic and geomorphic controls of bank erosion. In: Ports, M.A. (Ed.), *Hydraulic Engineering, Proceedings of the National Conference on Hydraulic Engineering*. ASCE, New Orleans, 106–111.

Shreve, R.L. 1966. Statistical law of stream numbers. *Journal of Geology*, **75**, 178–186.

Simon and Downs. 1995. An interdisciplinary approach to evaluation of potential instability in alluvial channels. *Geomorphology*, **12**(3), 215–232.

Simon, A. and Hupp, C.R. 1986. Channel evolution in modified Tennessee channels. *Proceedings of the 4th Interagency Sedimentation Conference*, Las Vegas, Nevada, vol. 2, 5-71–5-82.

Sir William Halcrow and Partners. 1992. *River Training Studies of the Brahmaputra River*. Report to the Peoples' Government of Bangladesh, Sir William Halcrow and Partners, Banani, Dhaka.

Smith, S.A. 1987. Gravel counterpoint bars: examples from the River Tywi, south Wales. In: Ethridge, F.G., Flores, R.M. and Harvey, M.D. (Eds), *Recent Developments in Fluvial Sedimentology*, Society of Economic Palaeontologists and Mineralogists, Special Publication no. 39, 75–81.

Smith and McLean. 1984. A model for meandering rivers. *Water Resources Research*, **20**, 1301–1315.

Smith, S.A. and Smith, N.D. 1980. Sedimentation in anastomosed river systems: examples from alluvial valleys near Banff, Alberta. *Journal of Sedimentary Petrology*, **50**, 157–164.

Stubblefield, W.G. 1986. An evaluation of criteria for predicting morphological patterns in alluvial rivers. In: Julien, P.Y. (Ed.), *Essays on River Mechanics*. Colorado State University, Fort Collins, Colorado.

Strahler, A.N. 1964. Quantitative geomorphology of drainage basins and channel networks. In: Chow, V.T. (Ed.), *Handbook of Applied Hydrology*. McGraw-Hill, New York, Section 4-11.

Thorne, C.R. 1988. *Bank Processes on the Red River between Index, Arkansas and Shreveport, Louisiana*. Final Report US Army Research Office under contract DAJA45-88-C-0018, Queen Mary College, University of London, December 1988, 50 pp.

Thorne, C.R. 1992. Bend scour and bank erosion on the meandering Red River, Louisiana. In: Carling, P.A. and Petts, G.E. (Eds), *Lowland Floodplain Rivers – Geomorphological Perspectives*. Wiley, Chichester, 95–115.

Thorne, C.R. 1993. Guildelines for the use of stream reconnaissance sheets in the field. Contract Report HL-93-2, US Army Engineer Waterways Experiment Station, Vicksburg, Mississippi 39180, 91 pp.

Thorne, C.R. and Abt. 1993. Velocity and Scour Prediction in river bends. Contract Report HL-93-1, March 1993, US Army Corps of Engineers, Waters Experiment Station, Vicksburg, Mississippi, 39180, 2 parts, 93 pp and 66 pp, respectively.

Thorne, C.R., Russell, A.P.G. and Alam, M.K. 1993. Planform pattern and channel evolution of the Brahmaputra River, Bangladesh. In: Best, J.L. and Bristow, C.S. (Eds), *Braided Rivers*. Geological Society of London Special Publication no. 75, 257–275.

van den Berg. 1995. Prediction of alluvial channel pattern of perennial rivers. *Geomorphology*, **12**(4), 259–280.

Yalin, M.S. 1972. *Mechanics of Sediment Transport*. Pergamon, Oxford.

APPENDIX: NOTATION

a	a constant in threshold equation of Chang (1985)
B	channel width or meander belt width (m)
BD_b	mean scour depth in bend (m)
BD_m	maximum scour depth in bendway pool (m)
d	channel depth (m) or bed material size
d_m	channel bankfull maximum depth (m)
D	average channel depth
D_{50}	bed material median size (mm or m)
D_{84}	bed material size for which 84% of the sediment is finer (mm or m)
F	width/depth ratio
Fr	Froude number
L	meander wavelength measured along the axis of the channel (m)
M	weighted silt-clay index
p	planform sinuosity = channel length/valley length
Q_a	mean annual discharge (m^3/s)
Q_b	bankfull discharge (m^3/s)
Q_m	mean annual discharge (m^3/s)
Q_{ma}	mean annual flood (m^3/s)
Q_{2f}	two year flood (m^3/s)
R_c	bend radius of curvature (m)
R_d	riffle bankfull mean depth (m)

R_{dm}	riffle bankfull maximum depth (m)
R_v	riffle bankfull mean velocity (m/s)
R_w	riffle bankfull width (m)
RD_{50}	riffle bed material median size (mm)
s	specific gravity of sediment (usually 2.65)
S	channel slope
S_v	valley slope
T	meander wavelength multiplied by sinuosity (m)
v	channel bankfull mean velocity (m/s)
w	bankfull width (m)
$\omega_{v,bf}$	specific stream power at bankfull discharge (W/m^2)
ω_v	specific stream power (W/m^2)
ω_{vt}	specific stream power at the braiding/meandering transition (W/m^2)
x	distance along channel centreline
XD_b	mean depth at crossing (m)
z	riffle spacing (m)
θ	meander arc angle and parameter in Fredsoe's method (Fig. 7.22)
Σp	total sinuosity
τ	boundary shear stress (N/m^2)
τ_{ave}	cross-section average boundary shear stress (N/m^2)
ω	stream power (W/m)
ϕ	local angle of channel planform curvature
ϕ_{max}	maximum angle of channel planfrom curvature

8 Stable River Morphology

RICHARD D. HEY
School of Environmental Sciences, University of East Anglia, UK

8.1 INTRODUCTION

Natural alluvial channels are noted for their morphological diversity which both reflects and influences hydraulic conditions. Such variety provides a range of habitats which are essential for preserving fisheries and instream and bankside macrophytes and for maintaining the amenity value of the riverscape. Stable single-thread rivers characteristically have a well defined channel and a sinuous course. Deeper sections are associated with meander bends and shallower ones with inflection points between bends. This produces the classic pool–riffle longitudinal profile which is typical of alluvial channels. Rivers have a similar range of features, irrespective of their absolute size, which implies that their equilibrium shape and dimensions are governed by the same physical processes, which differ only in scale, and by the same controlling factors.

In the course of carrying out river engineering works for flood alleviation and channel stabilisation, many rivers have been considerably modified. These changes have adversely affected the stability of the engineered and adjacent reaches and, in the process, destroyed the conservation and amenity value of riverine areas (Brookes, 1988; Purseglove, 1988). Consequently there is an urgent need to develop more sympathetic engineering design procedures which will preserve both the natural stability of the river and, by maintaining habitat diversity, its conservation and amenity value. By designing with nature rather than attempting to impose on nature, such approaches are likely to be more cost-effective than traditional heavy engineering solutions, to require less maintenance and, above all, to minimise environmental impact.

Increasingly there are demands to restore and rehabilitate heavily engineered reaches and this also requires the adoption of natural solutions in order to recreate channel features which are enduring and in harmony with local flow processes. Meandering channels with pools, riffles, glides, dead zones, vertical alluvial banks and point bars need to be recreated in order to restore the habitat features which were destroyed by previous engineering works. These features cannot be installed at random, and badly designed schemes will quickly be destroyed as the river reacts to the unnatural imposed conditions. This emphasises the need for the development of sympathetic design procedures which are in harmony with local river conditions.

In order to achieve these objectives it is necessary to identify the methods available for predicting the three-dimensional shape of alluvial channels as these are the basis for any natural engineering design or restoration work.

Applied Fluvial Geomorphology for River Engineering and Management, Edited by C. R. Thorne, R. D. Hey and M. D. Newson.

8.2 VARIABLES DEFINING AND CONTROLLING STABLE CHANNEL GEOMETRY

Alluvial rivers possess nine degrees of freedom since they can adjust their average bankfull width (W), depth (d), maximum depth (d_m), height (Δ) and wavelength (λ) of bedforms, slope (S), velocity (V), sinuosity (p), and meander arc length (z) through erosion and deposition. For river reaches that are in regime, implying that they do not systematically change their average shape and dimensions over a period of a few years, these can be regarded as dependent variables. In these circumstances the sediment load supplied from upstream can be transmitted without net erosion or deposition.

The variables controlling stable river dimensions are the discharge (Q), sediment load (Q_s), calibre of the bed material (D), bank material, bank vegetation, and valley slope (S_v). Change in any one of these independent variables will eventually result in the development of a new regime channel geometry which is in equilibrium with the changed conditions. When stable, the channel morphology will be uniquely defined by the new values of the controlling variables.

Under regime conditions most of the controlling variables are effectively constant. The two exceptions are discharge and bed material transport as they can vary considerably through time. To overcome this difficulty, Inglis (1946) suggested that a constant flow, at or about bankfull, produces the same gross shapes and dimensions as the natural sequence of events and could be regarded as the dominant or channel-forming flow. This does not preclude dramatic change during extreme flood events, but rather suggests that the channel morphology is modified by subsequent events such that over a period of years its long-term average channel dimensions are relatively invariant.

There is a considerable body of evidence to suggest that bankfull flow is the dominant or channel-forming flow. First, flume and field data indicate that flows at or about bankfull stage may be the dominant ones for channel development because of the strong empirical relations between channel dimensions and bankfull flow (Ackers and Charlton, 1970; Wolman and Leopold, 1959; Hey, 1975). Second, in terms of erosional and depositional processes affecting morphological change, bankfull stage is a major discontinuity. During overbank flows, a proportion of the total flow occurs at some distance from the river and, therefore, cannot affect bankfull channel dimensions. Equally, within-channel flow velocities are significantly higher than those on the floodplain and this ensures that inbank flows are independent of overbank flows. Consequently bankfull flow represents the highest flow that has morphological significance for the river. This is confirmed by Ackers and Charlton's (1970) experiments on small meandering channels since bankfull flow produced a comparable meander geometry to the variable flow regime.

The reason why bankfull flow is the channel-forming flow can be explained by considering the magnitude and frequency of sediment transport processes. Field studies indicate that the frequency of occurrence of the flow that transports most sediment in both sand- and gravel-bed rivers (Andrews, 1980; Wolman and Miller, 1960) corresponds to the frequency of bankfull flow (Hey, 1975; Wolman and Leopold, 1957). With regard to bed material load, it follows that its critical value is the transport rate associated with bankfull flow (Inglis, 1949).

For gravel-bed rivers the 1.5-year flood, based on the annual maximum series, is often regarded as the return period of bankfull flow, although there is considerable scatter around its value (Williams, 1978). As this approach disregards all lesser flood events above bed material transport thresholds, it is more appropriate to determine return periods using the partial duration series with a threshold discharge set at the initiation of bed material movement (Carling, 1988; Hey and Heritage, 1988). For UK gravel-bed rivers this gives a mean return period of once every 0.9 years for bankfull flow (Hey and Heritage, 1988).

8.3 RATIONAL EQUATIONS

The dependent variables describing alluvial channel morphology and the independent variables controlling them (Section 8.2) should be linked by nine governing or process equations, one for each degree of freedom, which define the adjustment mechanisms. If the relationship between the variables could be established (Table 8.1), it would be possible to predict the multidimensional morphology of alluvial channels by the simultaneous solution of the process equations. This is referred to as the rational approach to channel design.

If it can be assumed that certain of the dependent variables are invariant, by virtue of being artificially constrained, fewer degrees of freedom exist and fewer equations are required to provide a unique solution. For example, for straight fixed-bed canals, it is possible to determine the design width, depth and velocity given information on the bankfull discharge, sediment characteristics and channel gradient using tractive force design procedures (Lane, 1937). Essentially this uses the continuity, flow resistance and critical shear stress equations to ensure that the material lining the channel boundary is below the threshold of sediment transport.

For more natural mobile-bed channels it is only possible to specify three equations – continuity, flow resistance and sediment transport – and these are insufficient to define overall channel morphology. Until process-based equations are formulated to predict channel width and planform it precludes the use of this method for defining natural channel dimensions.

8.4 APPLICATION OF RATIONAL EQUATIONS

For rivers the method can be used to predict, for example, the width, depth and velocity of a channel with revetted banks given the slope, discharge, sediment load, bed material size and bank roughness. For straight channels the slope would be prescribed by the valley slope, but if a meandering channel was required, then a lower slope could be designated. In this case a flow resistance equation might be required which explicitly allowed for additional energy loss due to meandering. Other possible combinations of variables are listed in Table 8.2, but all would require some form of bank revetment to maintain channel width.

In canal design there would appear to be greater flexibility in the choice of the controlling variables, as discharge and sediment load could be dependent on prescribed channel dimensions. However, in practice this could not be achieved as it would require discharge and sediment load to be maintained at design values. While the former would be possible, the latter would be impracticable.

Inherent in this approach to channel design is the assumption that the chosen flow resistance and sediment transport functions apply to the conditions in question and that the

Table 8.1 Degrees of freedom and governing process equations* (after Hey 1982a)

Degrees of freedom	Dependent variables	Fixed variables	Independent variables	Type of flow	Governing equations
1	V	$d, S, W, d_m, \lambda, \Delta, p, z$	Q	Fixed bed	1 Continuity
2	V, d	$S, W, d_m, \lambda, \Delta, p, z$	Q, D, D_r, D_l	Fixed bed	+ 2 Flow resistance
3	V, d, S	$W, d_m, \lambda, \Delta, p, z$	Q, Q_s, D, D_r, D_l	Mobile bed	+ 3 Sediment transport
5	V, d, S, W, d_m	λ, Δ, p, z	Q, Q_s, D, D_r, D_l	Mobile bed	+ 4 Bank erosion + 5 Bar deposition
7	$V, d, S, W, d_m, \lambda, \Delta$	p, z	Q, Q_s, D, D_r, D_l	Mobile bed	+ 6 Bedforms + 7 Bedforms
9	$V, d, S, W, d_m, \lambda, \Delta, p, z$	–	Q, Q_s, D, D_r, D_l, S_v	Mobile bed	+ 8 Sinuosity + 9 Riffle spacing/meander arc length

*Average flow velocity = V; mean depth = d; channel slope = S; width = W; maximum flow depth = d_m; bedform wavelength = λ; bedform amplitude = Δ; sinuosity = p; meander arc length = z; discharge = Q; sediment discharge = Q_s; characteristic size of bed, right and left bank sediment = D, D_r, D_l; valley slope = S_v.

Table 8.2 Practical combinations of dependent and independent variables based on continuity, flow resistance and sediment transport equations

Dependent variables	Independent variables	
V, W, d	S, Q, Q_s, D	Revetted banks, fixed side slope
V, S, W	d, Q, Q_s, D	Revetted banks, fixed side slope
V, d, S	W, Q, Q_s, D	Revetted banks, fixed side slope

equations are accurate. Ideally each equation should be theoretically based and have general application. Unfortunately most have an empirical component that can restrict their range of application. Sediment transport equations can be particularly inaccurate, especially in gravel-bed rivers where transport can be supply-limited and bed armouring occurs, which can lead to significant design errors. Before applying process-based equations for design purposes, local checks should be made to verify their validity. This can be achieved either individually by checking each process equation or collectively by predicting the dimensions of stable channels in the locality.

8.5 EXTREMAL HYPOTHESES

Several extremal hypotheses have been formulated to provide the fourth equation to enable the width, depth, slope and velocity of alluvial channels to be determined. Effectively they rely on a variational argument in which the maximum or minimum of some parameter is sought; these include the following.

1. Minimum stream power (Chang, 1979, 1980). Given the discharge and sediment load, the river adjusts its width, depth and slope (+velocity) such that γQS is a minimum subject to given constraints, where γ is the specific weight of water. This implies that slope (S) is a minimum as discharge (Q) is prescribed.
2. Minimum unit stream power (Yang, 1976). The river adjusts its width, depth, slope, velocity and roughness to minimise the unit stream power required to transport a given sediment and water discharge. Unit stream power, or stream power per unit weight of water, is given by:

$$\frac{Q\gamma LS}{\rho g W d L} = VS$$

   where L is the length of the reach, W the width, d the mean depth, ρ the density of water, g the acceleration due to gravity, and V the mean velocity.
3. Minimum energy dissipation rate (Brebner and Wilson, 1967; Yang et al., 1981). The river at equilibrium adjusts to the condition where its rate of energy dissipation is a minimum. For a channel reach of length L, the latter is given by:

$$(Q\gamma + Q_s\gamma_s)LS$$

   where Q and Q_s are the discharge and sediment load and γ and γ_s the specific weights of water and sediment respectively.

4. Maximum friction factor (Davies and Sutherland, 1983). A channel will stabilise when its shape and dimensions correspond to a local maximum of the friction factor.
5. Maximum sediment transport rate (Kirkby, 1977; White et al., 1982). This hypothesis suggests that for a given discharge and slope, channel width adjusts to maximise the sediment transport rate.

Bettess and White (1987), in a review of the various methods, showed that maximum sediment transport rate and minimum energy dissipation rate were effectively equivalent to minimum stream power. This leaves three independent hypotheses: minimum stream power, minimum unit stream power, and maximum friction factor.

In order to predict width, depth, slope and velocity, one of the extremal hypotheses has to be used in conjunction with continuity, flow resistance and sediment transport equations.

Bettess and White compared the performance of these three extremal hypotheses when coupled with (i) the White et al. (1980) friction equation and the Ackers and White (1973) transport equation, and (ii) the Chang–Parker transport function and the Keulegan resistance equation (Griffiths, 1984). For each combination, widths were calculated for a bed material size (D_{35}) of 0.01 m and a sediment concentration of 10 ppm for a range of discharges 1–1000 m s^{-3} for comparison against empirical width equations.

The results showed that the Griffiths approach grossly overestimated width, which possibly reflects the inappropriateness of these particular resistance and transport equations for gravel-bed rivers (Figure 8.1). Predicted widths using the White et al. equations and minimum stream power and minimum unit stream power relations corresponded quite well with Hey and Thorne's (1986) full regime width equation,

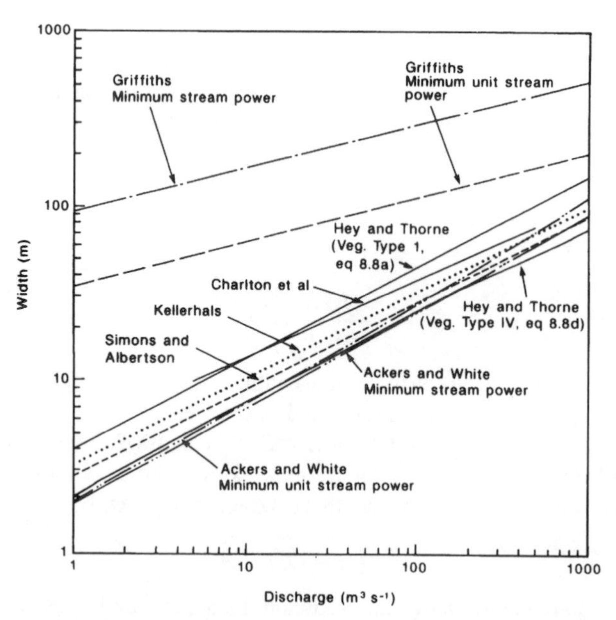

Figure 8.1 Regime widths for gravel-bed rivers (after Bettess and White 1987)

which includes sediment load, for rivers with over 50% tree and shrub cover on the banks, but significantly underestimated width for channels with less than 1% tree and shrub cover (Hey, 1988) (Figure 8.1).

It is apparent that there are several potential problems with the application of extremal methods for predicting channel width, even if the friction and transport functions used are applicable to the particular site. First, minimum (or maximum) values of stream power, slope or load may at best be difficult to identify because the turning points are very flat, leading to considerable range of near-optimal widths, depths and slopes, or at worst may not even reveal a minimum (or maximum) value at all. For example, Bettess and White were unable to identify a turning point for maximum friction factor. Second, the optimisation models assume that bed and banks are equally erodible. For rivers with more resistant banks, White et al. (1982) argue that the width will be narrower and slopes greater than for a maximum efficiency solution and vice versa with banks which are more susceptible to erosion than the bed. This is not borne out by empirical evidence since slope is not influenced by bank vegetation (Section 8.6). Third, each extremal hypothesis requires that dimensionless shear stresses at bankfull are the same on all rivers whereas, in reality, they vary considerably (Griffiths, 1984). Finally, the optimum dimensionless shear stress values implied by most hypotheses are unnaturally high (Ferguson, 1986). For example, Kirkby's (1977) optimum bankfull shear stress would be 10 times the threshold value, or even higher with the narrower of Chang's (1979) sand-bed solutions, when observed values for gravel-bed rivers are generally less than twice the threshold value. It is inconceivable that such values would not cause bank erosion and create a wider than optimum channel.

On the basis of this evidence, extremal hypotheses do not have any physical basis and fail to explain bank erosion and bar deposition which control width adjustment. Their predictive capability is also questionable, particularly as there is considerable scope for subjective interpretation.

8.6 REGIME EQUATIONS

A range of empirical regime equations has been developed to predict the geometry of stable alluvial channels. As with all empirical equations, care has to be exercised in their use, particularly with regard to their range of application. This should be restricted to channels which are similar to those used to derive the equations. Ideally, all variables controlling channel shape and dimensions should be used in the derivation of each set of equations and, to ensure their general application, the variability within and between the variables should be maximised. Equally, as the equations are not generally dimensionless, careful consideration needs to be given to the choice of units.

Regime equations can, for convenience, be classified on the basis of bed material size into cohesive-bed ($D < 0.0625$ mm), sand-bed ($0.0625 < D < 2$ mm) and gravel-bed equations ($D > 2$ mm). Equally, they can be divided into mobile- and quasi-fixed-bed channels, with regard to the presence or near-absence of bed material transport, or subdivided on the basis of bank material or vegetation type which affect the erosional resistance of the banks. For rivers, mobile bed conditions generally prevail, otherwise they would not be self-forming, yet relatively few equations have been developed which explicitly allow for bed

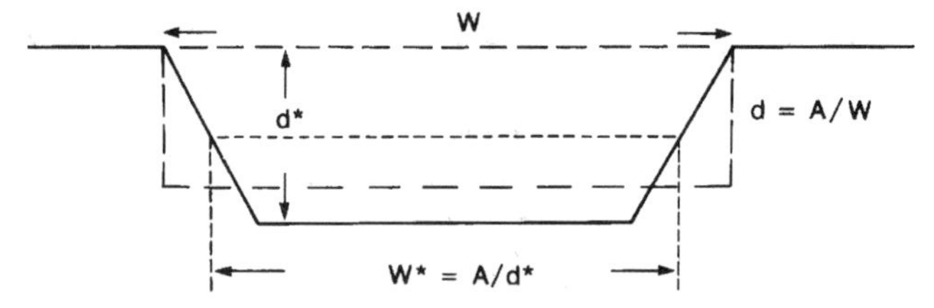

Figure 8.2 Definitions of width and depth used in regime equations (A = cross-sectional area)

material transport. Blench (1966) developed equations to predict the width, depth and slope of mobile sand-bed channels, while Hey and Thorne (1986) have presented equations which define the three-dimensional geometry of mobile gravel-bed rivers.

These are outlined below. For each set of equations the independent variables and their units are listed, together with their range of application. Figure 8.2 illustrates the various definitions of width and depth that are adopted.

8.6.1 Sand-bed Channels

Several equations have been developed for fixed sand-bed channels, and the most comprehensive of these were formulated by Blench (1966) and Simons and Albertson (1963). Relatively little information is available regarding the effect of sediment transport on channel morphology and reliance has to be placed on equations derived by Blench who, on the basis of flume data, modified his original equations to take account of bedload transport. The effect of differences in bed and bank material is accommodated by means of a bed and a side factor.

Blench's Equations

Database and range of application
- Discharge (Q): 0.03–2800 $m^3\ s^{-1}$
- Sediment discharge (Q_s): 30–100 ppm
- Bed material size (D): 0.1–0.6 mm
- Bank material type: cohesive
- Bedforms: ripples–dunes
- Bank vegetation: not specified
- Valley slope: not specified
- Planform: straight
- Profile: uniform

Equations:

Mean width:

$$W^* = \frac{F_{bc}^{0.5}}{F_s^{0.5}} Q^{0.5} \text{ (m)} \tag{8.1}$$

Bed depth:

$$d^* = \frac{F_s^{0.33}}{F_{bc}^{0.66}} Q^{0.33} \text{ (m)} \tag{8.2}$$

Slope:

$$S = \frac{F_{bc}^{0.833} F_s^{0.083} \nu^{0.25}}{3.63\, g\, Q^{0.166}\left(1 + \dfrac{Q_s}{2330}\right)} \tag{8.3}$$

where g is the acceleration due to gravity (9.81 m s^{-2}), ν the kinematic viscosity (m^2 s^{-1}), and Q_s the bedload charge in parts per 100 000 by weight.

The bed factor F_{bc} and side factor F_s are defined by:

$$F_{bc} = F_b(1 + 0.012Q_s) \tag{8.4}$$

where

$$F_b = \frac{V^2}{d^*} \text{ (m s}^{-2}\text{)} \tag{8.5}$$

and

$$F_s = \frac{V^3}{W^*} \text{ (m}^2\text{ s}^{-3}\text{)} \tag{8.6}$$

where V is the mean velocity.

Values of F_b can be approximated by:

$$F_b = 0.58\, D_{50}^{0.5} \text{ (metric)} \tag{8.7}$$

where D_{50} is the median particle size (mm) from sieve analysis of the bed material. The side factor (F_s) is dependent on the bank material and has the following values:

$$\begin{aligned} F_s &= 0.009 \text{ (metric) for loam of very slight cohesiveness} \\ &= 0.018 \text{ (metric) for loam of medium cohesiveness} \\ &= 0.027 \text{ (metric) for loam of very high cohesiveness} \end{aligned}$$

This is the only method which makes explicit allowance for the effect of sediment transport on channel dimensions. Care should be exercised in the use of this method for transport rates in excess of 100 ppm. The conversion factors used to modify the original equations are based on flume studies; they were not derived from the original field data.

For meandering channels Blench indicated that the right-hand side of the slope equation (Eqn 8.3) should be multiplied by a factor k. He indicated that the value of k varied from 2 for well developed meanders to 1.25 for straight channels with alternate bars. This was considered to account for the increased energy loss in meandering channels which required an additional gradient. However, as meandering channels have a lower gradient than straight channels at a given site, it questions this assumption.

8.6.2 Gravel-bed Channels

Considerable research has been carried out on the morphology of gravel-bed rivers by Nixon (1959), Kellerhals (1967), Charlton et al. (1978), Bray (1982), Hey (1982b), Andrews (1984), and Hey and Thorne (1986). Most ignore the effect of bedload transport on channel morphology, or assume that its value is zero or trivial. Only Kellerhals' equations actually apply for low or zero load, while Hey and Thornes' were the first to explicitly include sediment load, albeit predicted values.

The latter showed that bedload transport affected both channel width and depth but, as the exponents were so small, its effect on cross-sectional shape could safely be ignored. Equations for width and depth, excluding bedload transport, were derived and these produce practical point estimates for design purposes. These equations, which are given below, define the mean bankfull width, depth and maximum depth based on data from four cross-sections (pool–riffle–pool–riffle) taken over a full meander wavelength.

Hey and Thorne's Equations

Database and range of application
- Bankfull discharge (Q): 3.9–424 $m^3\,s^{-1}$
- Bankfull sediment discharge (as defined by Parker et al.'s (1982) bedload transport equation) (Q_s): 0.001–14.14 $kg\,s^{-1}$
- Median bed material size (D_{50}): 0.014–0.176 m
- Bank material: composite, with cohesive fine sand, silt and clay overlying gravel
- Bedforms: plane
- Bank vegetation: Type I, 0% trees and shrubs; Type II, 1–5%; Type III, 5–50%; Type IV, >50%
- Valley slope (S_v): 0.00166–0.0219
- Planform: straight and meandering
- Profile: pools and riffles

Equations:

Bankfull width (reach average):

$$W = 4.33Q^{0.5}\ \text{(m)} \qquad \text{Vegetation Type I} \qquad (8.8a)$$

$$W = 3.33Q^{0.5}\ \text{(m)} \qquad \text{Vegetation Type II} \qquad (8.8b)$$

$$W = 2.73Q^{0.5}\ \text{(m)} \qquad \text{Vegetation Type III} \qquad (8.8c)$$

$$W = 2.34Q^{0.5}\ \text{(m)} \qquad \text{Vegetation Type IV} \qquad (8.8d)$$

Bankfull mean depth (reach average):

$$d = 0.22Q^{0.37}D_{50}^{-0.11}\ \text{(m)} \qquad \text{Vegetation Types I–IV} \qquad (8.9)$$

Bankfull slope:

$$S = 0.087Q^{-0.43}D_{50}^{-0.09}D_{84}^{0.84}Q_s^{0.10} \qquad \text{Vegetation Types I–IV} \qquad (8.10)$$

Bankfull maximum depth (reach average):

$$d_{\mathrm{m}} = 0.20Q^{0.36}D_{50}^{-0.56}D_{84}^{0.35}\ (\mathrm{m}) \qquad \text{Vegetation Types I–IV} \tag{8.11}$$

Meander arc length:

$$z = 6.31W\ (\mathrm{m}) \qquad \text{Vegetation Types I–IV} \tag{8.12}$$

Sinuosity:

$$p = S_{\mathrm{v}}/S \qquad \text{Vegetation Types I–IV} \tag{8.13}$$

Bankfull riffle width:

$$RW = 1.034W\ (\mathrm{m}) \qquad \text{Vegetation Types I–IV} \tag{8.14}$$

Bankfull riffle depth:

$$Rd = 0.951d\ (\mathrm{m}) \qquad \text{VegetationTypes I–IV} \tag{8.15}$$

Bankfull riffle maximum depth:

$$Rd_{\mathrm{m}} = 0.912d_{\mathrm{m}}\ (\mathrm{m}) \qquad \text{Vegetation Types I–IV} \tag{8.16}$$

Bankfull pool width:

$$PW = 0.966W \qquad \text{Vegetation Types I–IV} \tag{8.17}$$

Bankfull pool depth:

$$Pd = 1.049d \qquad \text{Vegetation Types I–IV} \tag{8.18}$$

Bankfull pool maximum depth:

$$Pd_{\mathrm{m}} = 1.088d_{\mathrm{m}} \qquad \text{Vegetation Types I–IV} \tag{8.19}$$

where D_n is the grain diameter of surface bed material (m) for which n per cent of the sample is finer (grid by number sample (Wolman, 1954)). As no direct observations were available regarding bedload transport rates, independent estimates had to be obtained to enable mobile-bed regime equations to be derived. Before applying these equations it is important to ensure that bedload transport rates can be defined by Parker et al.'s (1982) equation.

These equations can be applied to predict the planform as well as the cross-sectional shape of mobile gravel-bed rivers, and pools and riffles can also be incorporated in the design using Eqns 8.14–8.19.

Significantly these equations compare favourably with those developed by Kellerhals (1967) for fixed-bed channels. The regime width and depth equations are very similar, as they are relatively unaffected by bedload transport, and the slope equations only differ due to the presence/absence of bedload (Hey and Thorne, 1986). Equally, Richards (1987) has demonstrated that dimensionless forms of the Hey and Thorne width and depth equations are comparable to Andrews' (1984) dimensionless regime equations for gravel-bed rivers in Colorado. Greater diversity is shown between the slope equations, probably reflecting the fact that Andrews did not explicitly include bedload transport in his equations.

8.7 APPLICATION OF REGIME EQUATIONS

For rivers, channel morphology can be determined from discharge, bedload transport rate, calibre of bed and bank material, and valley slope. For newly formed reaches of channel, where initially there will be no trees or shrubs on the banks, the effect of vegetation can be ignored. The design discharge and transport rate are the bankfull discharge and its associated bedload transport. For river diversions or for the restoration of canalised rivers, they can best be obtained from bankfull flow conditions at neighbouring natural sections of stable channel. If these data are unavailable or inappropriate due to the river being either heavily engineered or unstable, then for sand- and gravel-bed rivers the flow and sediment transport associated with the 1.5 year-flood (annual maximum series) can be used for this purpose as this approximates the return period of the flow transporting most sediment (Section 8.2).

Only Hey and Thorne included meandering channels within their database and their equations enable the planform geometry of the channel to be determined. This requires the prediction of sinuosity (Eqn 8.13), given the valley slope and predicted channel slope, and meander arc length (Eqn 8.12), given the predicted channel width. This approach could be used with Blench's equations to determine the required planform geometry for sand-bed channels. Any attempt to impose a straight channel, effectively fixing the channel slope, can result in the redesigned section becoming unstable if the imposed slope does not correspond to that predicted by the regime slope equation.

Within any river reach there are local variations in width and depth due to the presence of pools and riffles. All available equations enable the average reach width and depth to be determined, but only the Hey and Thorne equations (Eqns 8.14–8.19) enable the local variability about the reach average to be established.

Regime equations can also be used to predict the morphology of mobile-bed canals. The best practical way is to predict channel dimensions from the discharge, sediment load, bed and bank material and valley slope. In many canal systems discharge and sediment load do not vary significantly over time. These constant values can be used as the design values in the regime equations, but for canals experiencing a range of flows and transport rates, it is necessary to define dominant values for design purposes. This can be achieved by determining the discharge, and associated sediment load, which will transport most sediment over a period which is representative of the imposed flow regime (Smith, 1977).

8.8 REFERENCES

Ackers, P. 1964. Experiments on small streams in alluvium. *Journal of the Hydraulics Division, American Society of Civil Engineers*, **90**(HY4), 1–37.

Ackers, P. and Charlton, F.G. 1970. Meander geometry with varying flows. *Journal of Hydrology*, **11**, 230–252.

Ackers, P. and White, W.R. 1973. Sediment transport: new approach and analysis. *Journal of the Hydraulics Division, American Society of Civil Engineers*, **HY11**, 2041–2060.

Andrews, E.D. 1980. Effective and bankfull discharges of streams in Yampa basin, Colorado and Wyoming. *Journal of Hydrology*, **46**, 311–330.

Andrews, E.D. 1984. Bed material entrainment and hydraulic geometry of gravel-bed rivers in Colorado. *Geological Society of America Bulletin*, **95**, 371–378.

Bettess, R. and White, W.R. 1987. Extremal hypothesis applied to river regime. In: Thorne, C.R., Bathurst, J.C. and Hey, R.D. (Eds), *Sediment Transport in Gravel-Bed Rivers*. Wiley, Chichester, 767–789.

Blench, T. 1966. *Mobile-Bed Fluviology*. University of Alberta Press, Edmonton, Canada.

Bray, D.I. 1982. Regime equations for gravel-bed rivers. In: Hey, R.D., Bathurst, J.C. and Thorne, C.R. (Eds), *Gravel-Bed Rivers*. Wiley, Chichester, 517–542.

Brebner, A. and Wilson, K.C. 1967. Determination of the regime equation from relationships for pressurized flow by use of the principle of minimum energy degradation. *Proceedings of the Institution of Civil Engineers*, **36**, 47–62.

Brookes, A. 1988. *River Channelization*. Wiley, Chichester, 326pp.

Carling, P. 1988. The concept of dominant discharge applied to two gravel-bed streams in relation to channel stability thresholds. *Earth Surface Processes and Landforms*, **13**, 355–367.

Chang, H. 1979. Geometry of rivers in regime. *Journal of the Hydraulics Division, American Society of Civil Engineers*, **106**, 691–706.

Chang, H. 1980. Stable alluvial canal design. *Journal of the Hydraulics Division, American Society of Civil Engineers*, **106**(HY5), 873–891.

Charlton, F.G., Brown, P.M. and Benson, R.W. 1978. *The Hydraulic Geometry of Some Gravel-bed Rivers in Britain*. Report IT 180, Hydraulics Research Station, Wallingford.

Davies, T.R.H. and Sutherland, A.J. 1983. Extremal hypothesis for river behaviour. *Water Resources Research*, **19**(1), 141–148.

Ferguson, R.I. 1986. Hydraulics and hydraulic geometry. *Progress in Physical Geography*, **10**(1), 1–31.

Griffiths, G.A. 1984. Extremal hypotheses for river regime: an illusion of progress. *Water Resources Research*, **20**(1), 113–118.

Hey, R.D. 1975. Design discharge for natural channels. In: Hey, R.D. and Davies, T.D. (Eds), *Science, Technology and Environmental Management*. Saxon House, UK, 73–88.

Hey, R.D. 1982a. Gravel-bed rivers: process and form. In: Hey, R.D., Bathurst, J.C. and Thorne, C.R. (Eds). *Gravel-Bed Rivers*. Wiley, Chichester, 5–13.

Hey, R.D. 1982b. Design equations for mobile gravel-bed rivers. In: Hey, R.D., Bathurst, J.C. and Thorne, C.R. (Eds), *Gravel-Bed Rivers*. Wiley, Chichester, 553–574.

Hey, R.D. 1987. Regime stability. In: *River Engineering – Part I*. Water Practice Manual No. 7, Institution of Water Engineers and Scientists, London, 139–147.

Hey, R.D. 1988. Mathematical models of channel morphology. In: Anderson, M.G. (Ed.), *Modelling Geomorphological Systems*. Wiley, Chichester, 99–125.

Hey, R.D. and Heritage, G.L. 1988. Dominant discharge in alluvial channels. *Proceedings of an International Conference on Fluvial Hydraulics*, Budapest, 143–148.

Hey, R.D. and Thorne, C.R. 1986. Stable channels with mobile gravel beds. *Journal of Hydraulic Engineering, American Society of Civil Engineers*, **112**(8), 671–689.

Inglis, C.C. 1946. *Meanders and Their Bearing on River Training*. Maritime Paper No. 7, Institution of Civil Engineers, London.

Inglis, C.C. 1949. *The Behaviour and Control of Rivers and Canals*. Research Publication No. 13, Central Water Power, Irrigation and Navigation Research Station, Poona, India.

Kellerhals, R. 1967. Stable channels with gravel-paved beds. *Journal of the Waterways and Harbors Division, American Society of Civil Engineers*, **93**(WW1), 63–83.

Kirkby, M.J. 1977. Maximum sediment efficiency as a criterion for alluvial channels. In: Gregory, K.J. (Ed.), *River Channel Changes*. Wiley, Chichester, 429–442.

Lane, E.W. 1937. Stable channels in erodible material. *Transactions of the American Society of Civil Engineers*, **102**, 123–194.

Nixon, M. 1959. A study of the bankfull discharges of rivers in England and Wales. *Proceedings of the Institution of Civil Engineers*, **12**, 157–174.

Parker, G., Klingeman, P.C. and McClean, D.G. 1982. Bedload and size distribution in paved gravel-bed streams. *Journal of the Hydraulics Division, American Society of Civil Engineers*, **108**(HY4), 544–571.

Purseglove, G. 1988. *Taming the Flood*. Oxford University Press, Oxford, 307pp.

Richards, K. 1987. Fluvial geomorphology. *Progress in Physical Geography*, **11**(3), 432–457.

Simons, D.B. and Albertson, M.L. 1963. Uniform water conveyance channels in alluvial material. *Transactions of the American Society of Civil Engineers*, **128**(1), 3339, 65–167.

Smith, K.V.H. 1977. Sediment balance method for irrigation canal design. *Proceedings of the Institution of Civil Engineers*, Part 2, **63**, 411.

White, W.R., Paris, E. and Bettess, R. 1980. The frictional characteristics of alluvial streams. *Proceedings of the Institution of Civil Engineers*, Part 2, **69**, 737–751.

White, W.R., Bettess, R. and Paris, E. 1982. Analytical approach to river regime. *Journal of the Hydraulics Division, American Society of Civil Engineers*, **108**(HY10), 1179–1193.

Williams, G.P. 1978. Bankfull discharge of rivers. *Water Resources Research*, **14**, 1141–1154.

Wolman, M.G. 1954. A method of sampling coarse river bed material. *Transactions of the American Geophysical Union*, **35**(6), 951–956.

Wolman, M.G. and Leopold, L.B. 1957. *River flood plains: some observations on their formation*. US Geological Survey, Professional Paper 282-C, US Government Printing Office, Washington DC.

Wolman, M.G. and Miller, J.P. 1960. Magnitude and frequency of forces in geomorphic processes. *Journal of Geology*, **68**, 54–74.

Yang, C.T. 1976. Minimum unit stream power and fluvial hydraulics. *Journal of the Hydraulic Engineering, American Society of Civil Engineers*, **102**(HY7), 919–934.

Yang, C.T., Song, C.C.S. and Woldenberg, M.J. 1981. Hydraulic geometry and minimum rate of energy dissipation. *Water Resources Research*, **17**(4), 1014–1018.

9 Styles of Channel Change

J. M. HOOKE
Department of Geography, University of Portsmouth, UK

9.1 INTRODUCTION

Analysis of the channel morphological system indicates that the channel can adjust in various ways but, because of the number of variables and possibilities involved, channel changes are usually indeterminate and cannot be predicted exactly (e.g. Hey, 1978). Theory predicts the direction of change in variables whilst case studies show how channels respond to different conditions, but few quantitative models yet predict the exact style or rate of channel change to be expected.

The morphological characteristics of channels are commonly analysed in two-dimensional slices, i.e. pattern (planform), cross-sectional form, and longitudinal form. The close associations between each have been demonstrated in Chapter 7. Most work has been on lateral instability and alterations of planform, partly because these are obvious changes and documentation on such changes is most readily available, but arguably also because it is in planform that the greatest adjustments take place.

Various causes of change have been identified with a basic division being made into autogenic and allogenic causes. Although shift in river channel position is inherent, it was long inferred that changes in characteristics of channel morphology were due to external causes. These may be either natural, such as climate change, or human-induced, such as land-use change and urbanisation (Table 9.1). There is a basic division between indirect channel changes, e.g. resulting from catchment alteration, and direct, deliberate modification of the channel, e.g. channelisation. It is essential to realise that both can result in further changes to the channel. Case studies of responses to channelisation have helped to identify the nature of adjustment (Brookes, 1988) and, very importantly, to predict the effects of works elsewhere.

It is now widely accepted that changes may be complex in spatial and temporal pattern and that sudden changes can occur without change in external conditions. Assumption of attainment and stability of equilibrium forms must also be questioned as the evidence grows of continued evolution of patterns over centuries rather than adjustments in decades. However, morphological changes do occur in response to changes in input of discharge and/or sediment and the expected directions of change were originally set out by Schumm (1969) (Table 9.2). The nature of the channel response depends on the inherent instability, the freedom to adjust and thus the sensitivity of different environments and channel reaches. In any engineering scheme

Applied Fluvial Geomorphology for River Engineering and Management, Edited by C. R. Thorne, R. D. Hey and M. D. Newson.

Table 9.1 Examples of anthropogenic drainage basin changes. Many of these can instigate changes of stream flow and of sediment and solute production and subsequently result in modifications of channel geometry (g), channel pattern (p) or drainage network (n) (Gregory and Welling, 1973)

		Form affected		
Direct changes				
Drainage network changes:	irrigation networks		n	
	drainage schemes		n	
	agricultural drains		n	
	ditches		n	
	road drains		n	
	storm water sewers		n	
Channel changes:	river regulation	g	p	
	bank stabilisation, protection	g	p	
Water and sediment balance:	abstraction of water	g		
	return of water	g		
	waste disposal	g		
Indirect causes				
Land use:	cropland	n	p	g
	building construction		p	g
	urbanisation	n	p	g
	afforestation	n	p	g
	reservoir construction		p	g
Soil character:	drainage	n		
	ploughing	n	p	
	fertilisers			

it is essential that, prior to work on design, any natural or pre-existing instability is identified. There is also a need to monitor responses to alterations.

9.2 EVIDENCE AND MEASUREMENT OF CHANGE

Various types of evidence and measurements can be used to analyse channel changes. The choice depends on the purpose of study and the rates of change. Where changes are rapid or the focus is on the processes of change and controls of the mechanisms, then instrumentation of erosion and deposition and measurement of flow properties and sediment transport are appropriate (see Chapters 4, 5 and 6). If the focus is on rates of change or on overall changes in form, then widespread monitoring of erosion or deposition would be required. Frequently, the rate of change is too slow or the period of assessment of instability is too short to allow direct measurement so the use of field or documentary evidence is needed. Historical sources provide readily accessible sources at an appropriate timescale to detect changes, fluctuations and trends. The most widely used sources are historical maps and aerial photographs, which enable analysis of planform changes. There are several limitations over the use of such data, the main one being that they only provide

Table 9.2 Examples of river metamorphosis (after Schumm, 1969)

Equations			Examples of change
Qs^{+}, Qw^{++}	$\approx$	$S^{-}, d50^{+}, D^{+}, W^{+}$	Long-term effect of urbanisation. Increased frequency and magnitude of discharge. Channel erosion (increasing width and depth)
Qs^{-}, Qw^{+} Qs^{--}, Qw	$\approx$ $\approx$	$S^{-}, d50^{+}, D^{+}, W^{-}$ $S^{-}, d50^{+}, D^{+}, W^{*}$	Intensification of vegetation cover through afforestation and improved land management reduces sediment loads
Qs^{-}, Qw^{+} Qs^{o}, Qw^{+}	$\approx$ $\approx$	$S^{-}, d50^{+}, D^{+}, W^{*}$ $S^{-}, d50^{+}, D^{+}, W^{+}$	Diversion of water into a river
Qs^{+}, Qw^{+} Qs^{-}, Qw^{-}	$\approx$ $\approx$	$S^{*}, d50^{*}, D^{*}, W^{+}$ $S, d50^{*}, D^{*}, W^{-}$	Parallel changes of water and sediment discharge with unpredictable changes of slope, depth and bed material
Qs^{++}, Qw^{+}	$\approx$	$S^{+}, d50^{-}, D^{-}, W^{+}$	Land-use change from forest to crop production. Sediment discharge increasing more rapidly than water discharge. Bed changes from gravel to sand, wider shallower channels
Qs^{o}, Qw^{-} Qs^{-}, Qw^{--}	$\approx$ $\approx$	$S^{+}, d50^{-}, D^{-}, W^{-}$ $S^{+}, d50^{-}, D^{-}, W^{-}$	Abstraction of water from river resulting in narrower stream
Qs^{+}, Qw^{-} Qs^{+}, Qw^{o}	$\approx$ $\approx$	$S^{+}, d50^{-}, D^{-}, W^{*}$ $S^{+}, d50^{-}, D^{-}, W^{-}$	Increased sediment supply and constant or reduced water discharge (e.g. hydraulic mining activity)

Qs = sediment discharge, Qw = water discharge, D = flow depth, W = flow width, S = slope, $d50$ = median diameter of bed material, o = no change, $\pm$ = increase or decrease respectively, $++/--$ = indicates a change of considerable magnitude, * = unpredictable

snapshots; care must be taken in inferring changes between those dates and in generalising rates. Secondly, the accuracy and scale of the sources may limit the quantitative detail which can be derived. Guidance on sources available, methods of checking accuracy, techniques of data compilation and types of data and analysis which may be applied are explained in Tables 9.3, 9.4, 9.5 and 9.6 and in Hooke and Kain (1982) and Hooke and Redmond (1989b). Figure 9.1 is an example of a compilation from such historical sources. Historical sources are generally of much greater value and reliability in analysing styles of change in meandering rivers than on braided rivers, because change in the former is much more progressive whereas changes are much more rapid and irregular in the latter and appearance is very much affected by stage of flow of mapping. Historical information on cross-sections or slope and longitudinal form are less readily available, though some valuable sources such as early profile surveys exist, particularly for navigable rivers e.g. Hawkesbury River, New South Wales, and where practical problems have been severe. Some details of slope profiles can be derived from aerial photography of a suitable scale and standard by photogrammetric methods.

Field evidence can complement documentary evidence and is needed to extend the timescale and provide corroboration of changes. The state of the river banks can indicate whether erosion is still active (Chapter 4) and the degree of degradation (lack of freshness) can indicate the relative time since erosion was active. A summary guide to help in rapid

9.3 Details of historical maps available for Britain

Map source	Date	Scale	Form	Location	Coverage	Accuracy	References
Estate	Some late C16th Majority 1700→	Usually 1:2376	MS	PRO Scottish Record Office British Library Bodleian, Oxford National Library of Wales LROs	Good for England and Scotland Over 20 000 available for consultation	Variable	Harley (1972)
Enclosure	C18th and C19th		MS	PRO Libraries CROs	England and Wales	Variable	Tate (1978)
Tithe	1840s	Usually 1:2376 or 1:4752	MS	CROs PRO	3/4 of England and Wales	Variable	Kain and Prince (1985)
Deposited plans	Post-1794	Min. scale 1:15 840 after 1807	MS	CROs House of Lords Records Office	Britain	Variable	Harley (1972)
County	C18th and C19th	1:63 360 1:31 680	MS and published	CROs Museums	England and Wales	Variable	Skelton (1970) Rodger (1972)
Military Survey of Scotland (Roy)	1747–1755	1:36 000	MS		Scotland	Reasonably accurate	Moir (1973)
Ordnance Survey Old Series 1st ed.	1795–1873	1″ = 1 mile 1:63 360	Printed	British Library-Museums	England and Wales	Poor	Harley (1965)

New Series 2nd ed.	1840–1893	1:63 360 1″ = 1 mile	Printed	British Library Museums	England and Wales	Moderate	Harley (1965)
3rd ed.	1901–1913	1″ = 1 mile	Printed	British Library Museums	England and Wales	Moderate	
4th ed.	1913–1926	1″ = 1 mile	Printed	British Library Museums	England and Wales	Moderate	
County Series	1840–1893	1:10 560	Printed	Museums Libraries	England and Wales	Good	Harley (1965)
1st ed.	1853–1893	1:2500		British Library CROs			
1st revision	1891–1914						
2nd revision	1904–1923						
National Grid Series	1948–1973	1:10 560 6″ = 1 mile (1:10 000 since 1969) 1:25 000	Printed	Libraries British Library	Great Britain	Good bank lines ± 1 m	Harley (1975)

Table 9.4 Errors and limitations of historical maps

Limitations	Checks
1. Nature of historical sources	
Only particular dates documented so no sampling control	Field checking
Difficult to infer nature and timing of changes between map dates	Consistency, documentation of floods
Use only in repository	
2. Nature of maps of rivers	
Methods of survey Cartographic representation Perception of change	Surveyor, instruments, cartographer, conventions, originality, purpose
3. Individual maps	
Shrinkage and stretching	Measurements
Planimetric accuracy	Measurements
Content accuracy	Documentary and field corroboration

field appraisal and inferences to be drawn on the nature of channel changes taking place is presented in Table 9.7. Presence of bars will indicate sites of deposition and height and amount of vegetation will indicate relative timescales of stability. The presence of palaeoforms indicates where the channel once flowed. Its relative freshness and amount of infilling will often indicate relative age of the channel but dating is required to build up a chronology of movement (e.g. from documentary evidence, lichenometry in rocky channels, vegetation and tree-ring dating, radiometric methods and archaeology and artifacts). Comparisons of the morphology of old channels will demonstrate whether changes in dimensions have taken place and the position will indicate the amount of lateral and vertical movement relative to the present day. Vegetation assemblages can indicate relative ages where there is a definite colonisation sequence, and age of trees can indicate minimum age of the surface. In some cases, particular morphological features called meander scrolls, deposited as ridges as a meander migrates, are of such regularity and clarity as to be used for plotting horizontal movement (Hickin, 1974; Figure 9.4).

Alternative approaches used where historical sources are not available and field evidence is unclear, e.g. for cross-sections, on 20–30 year timescale and in relation to specific human impacts, include the substitution of space for time. Comparisons are made between similar catchments or trends and distributions of characteristics are analysed

Table 9.5 Method of compilation of historical map data

1. Identify sources and dates available
2. Check background and general reliability of sources
3. Locate maps needed, e.g. in archives or libraries
4. Measure accuracy of individual maps
5. Abstract course by tracing, digitising
6. Superimpose maps by careful matching of common detail
7. Field check changes indicated
8. Make measurements and analysis

Table 9.6 Information that can be derived from historical maps

Data	Map compilation	Analysis technique
Location and distribution of instability and old channels	Superimposed or sequence	Visual inspection, map abstract
Length of change Sinuosity, degree of braiding	Separate or superimposed	Direct measurement Comparison for amount of change
Distances and rates of movement	Superimposed	Direct measurement
Areas of movement	Superimposed, shaded	Frame-grabber and image analysis
Direction of movement	Superimposed	Direct measurement, vector plots
Type of change	Sequence or superimposed	Classification
Change in characteristics	Digitised courses	Bend or reach definition Calculation
Floodplain ages	Superimposed	Mapping and zoning

within catchments. This approach assumes equilibrium forms and statistical analysis of characteristics and does have severe limitations (Gregory et al., 1992). Spatial and temporal analysis of changes resulting from alteration of inputs or presence of a particular input such as sediment, and from adjustments to channelisation both within and downstream of channelised reaches helps considerably in understanding the impact of such phenomena (Figure 9.2).

9.3 MEANDERING/SINGLE CHANNELS

9.3.1 Methods of Analysis

Once evidence is assembled on position and pattern of the river course at various dates, several methods of analysis can be used to elucidate and quantify the characteristics of change. Hooke (1984) has outlined various approaches to analysis (Figure 9.3). Figure 9.3A,B,C involve measurement of characteristics on the course of each date then comparison of the results. Figure 9.3D involves direct comparison of the courses and measurement of the changes and Figure 9.3E involves modelling the changes.

Meander patterns have conventionally been described by morphological variables, mainly meander wavelength (*ML*), meander breadth (*MB*), radius of curvature (*R*) and *ML*/*MB*,*R*/*W*. Comparison of bend statistics can indicate how the morphological characteristics have changed, but inconsistencies and problems arise in the use of statistical parameters and several variables are needed to characterise the patterns adequately.

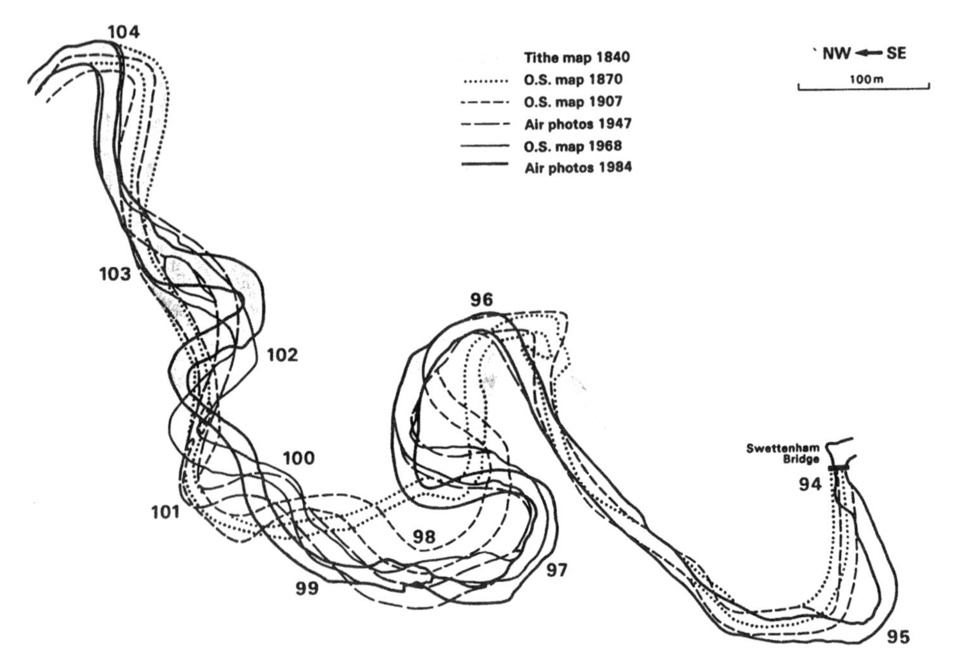

Figure 9.1 Historical compilation of river courses: River Dane, Cheshire, UK (from Hooke, 1987)

Even then, such statistics do not necessarily convey the nature and distance of movement. Measurement of most meander variables is subjective and assumes smooth, symmetrical forms. An alternative is digitising the course and objectively defining loops by points of inflection. Criteria still have to be established for minimum length and curvature of a bend but statistics can then be derived. Mathematical curve-fitting can also be applied to digitised data. Experiments have been made with spectral analysis to define the meandering characteristics but, although the general form of the power spectrum indicates the nature of the channel pattern, spectral analysis is insufficiently sensitive to measure characteristic wavelengths and detect changes.

9.3.2 Types of Meander Change

Classifications and general types of meander behaviour have been identified which characterise the nature of movement in a bend or in a reach and these are summarised in Table 9.8. In the same way as for morphology (Chapter 7), various classifications have been proposed with varying degrees of rigour, but no single standardised one has been adopted. Nevertheless, certain general terms and types of behaviour in individual meanders are widely recognised. The characteristics of each of these will now be discussed.

Table 9.7 Field indicators of channel instability (after Lewin et al., 1988)

Flood evidence	Instability tendency and probable sequence*
Floodplain morphology	
1. Inactive multichannel system on floodplain	(a) (i), (iii)→(ii), (iv) adjustment timescale measured in decades or (b) Possible channelisation
2. Terraces on valley-floor alluvium	(a) (i)→(ii) adjustment timescale measured in decades or (b) (ii)→tendency of British rivers to incise during the Holocene
Channel morphology	
1. Undermined bank protection structures, bridge footings, exposed tree roots on bankside	(a) (ii) or (b) Faulty design
2. Buried structures, contracting bridge openings	(a) (i) or (b) Over-designed channel
3. Erosion of both channel banks, increase in river width	(a) (iii) or (b) Under-designed channel
4. Channel, infilling and contraction	(a) (iv) or (b) Over-designed channel
5. All flood flows conveyed within channel	(a) (ii), (iii) or (b) Designed channel enlargement works
Floodplain sedimentary sequence	
1. Buried soils within alluvium	(a) (i) or (b) Made ground
2. Textural changes in vertical sequences of alluvium e.g. thick coarse within-channel sediments, overlying out-of-channel fines; e.g. fine sediment overlying terraced coarse within-channel deposits	(a) (i) or (b) Made ground (a) (ii) and/or (b) (iv)

* Instability tendencies
Vertical: (i) aggradation
(ii) incision
Lateral: (iii) enlarging
(iv) contracting

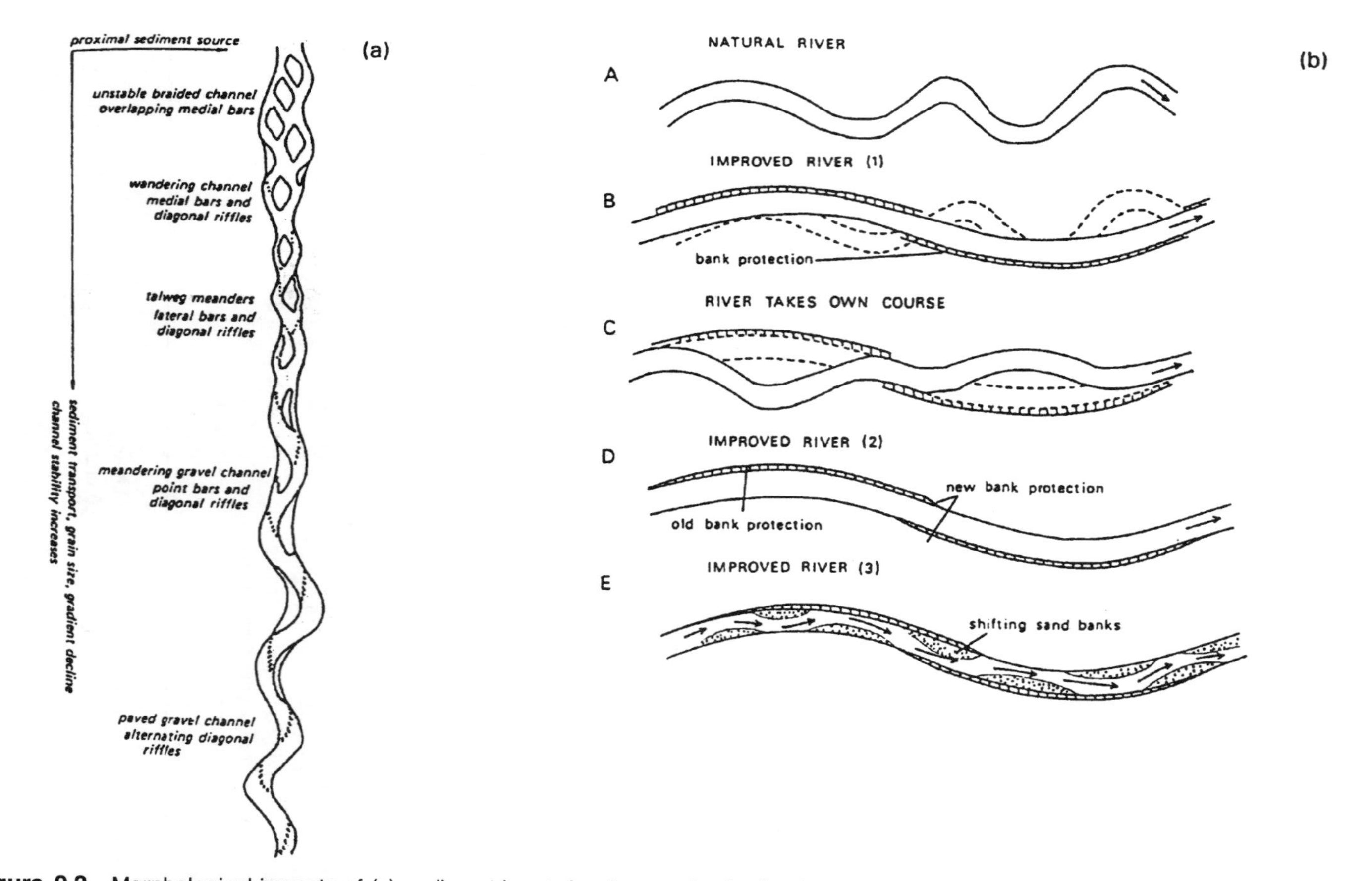

Figure 9.2 Morphological impacts of (a) sediment input showing proximal–distal occurrence of channel bars downstream from a sediment source (from Church and Jones, 1982) and (b) channelisation involving meander cutoffs (from Ryckborst, 1980)

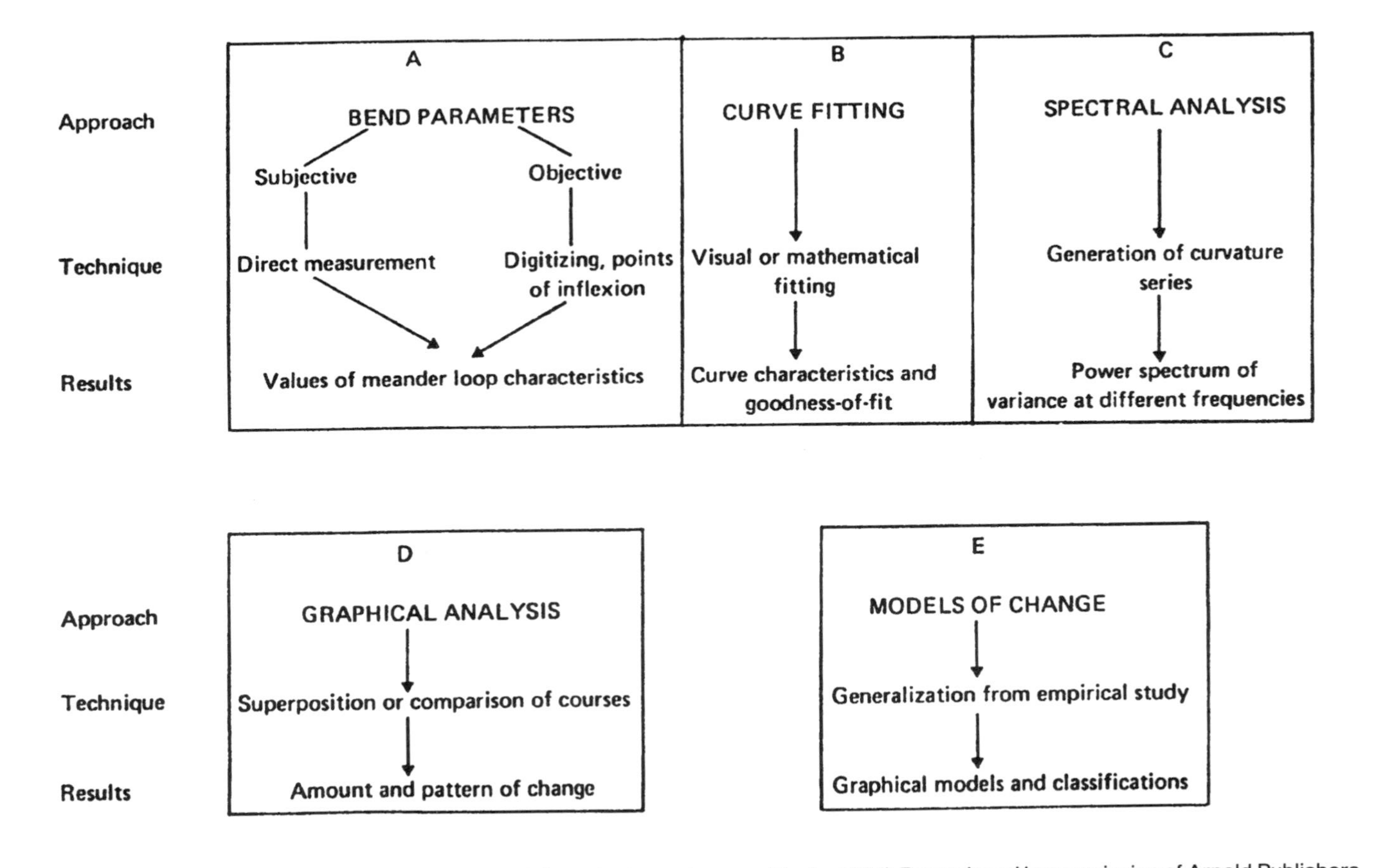

Figure 9.3 Approaches and techniques of analysis of river planform changes (Hooke, 1984). Reproduced by permission of Arnold Publishers

Table 9.8 Classifications or typologies of change

Author	Model/classification
Kondrat'yev (1968)	Graphical model of change from simple bend through to cutoff.
Daniel (1971)	Fitted a sine-generated curve to meanders. Identified five types of movement involving expansion, rotation and translation.
Keller (1972)	Processes of change and relationship to bedforms in a five-stage model.
Brice (1974)	Classification of morphology defined by circle combinations and analysis of sequences of change.
Hickin (1974)	Pattern of meander change developed from analysis of erosion pathlines.
Hickin and Nanson (1975)	Predictive model for determining rates of channel bend migration.
Kellerhals et al. (1976)	Identified six types of lateral activity: –downstream progression –progression and cutoffs –mainly cutoffs –entrenched loop development –irregular lateral activity –avulsion
Hooke (1977)	Used Daniel's (1971) primary elements of movement in double and triple combinations to compose a suite of 70 models of movement. Primary elements: –extension –translation –rotation –enlargement –lateral movement –complex change
Hooke and Harvey (1983)	Seven categories of change defined: –simple migration –confined migration –growth (extension) –lobing and compound growth –retraction and cutoff –complex changes (islands, abandonment etc.) –stable bends, no change
Brice (1984)	Classified changes in meander form according to mode of bank erosion

Migration

It has been shown by many flume and field studies that, in an unconstrained situation, meanders in alluvium gradually migrate in the downstream direction (e.g. Tiffany and Nelson, 1939; Friedkin, 1945; Ackers and Charlton, 1970). Doubt is now being cast on this assumption; for example, Neill (1970) in commenting on Carey's (1969) paper, points out that 'free meanders enlarge laterally then cutoff'. However, Carey's example of the Red River does show that migration takes place, and Brice (1977) found on the Sacramento River that the most common type of movement is migration. It may be that most migration is, in fact, confined but in most meanders with moderate curvature the maximum erosion is downstream of the apex, producing a major component of migration.

Even in wide alluvial floodplains, composition of sediments is not consistent, which means that migration is usually not symmetrical, even if it otherwise would be, and so there is unequal limb movement and this results eventually in cutoffs. Indeed, Hickin (1977) suggests that free meanders migrating through easily erodible material are more likely to develop severe distortions in response to small heterogeneities than are meanders developed in more resistant material.

Confined Migration

Very commonly, river channels are not free to migrate over a broad alluvial floodplain, but are restricted by features or deposits in the valley. This may impede both the migration and growth of meanders so bends become modified. Matthes (1941), Tanner (1955) and Lewin and Brindle (1977) suggested several factors which may be responsible for the anomaly of 'stacked' or deformed meanders including locally soft banks, resistant lenses, various non-alluvial formations, geological structure, engineering structures and stream confluences. The factors vary in the number of bends they are likely to affect. Schattner (1962) found that in the River Jordan flat loops tend to occur where the valley narrows, and on the River Dane in Cheshire, UK, a flattening or squaring of bends, often leading to double peaked curvature and double pools, is common where the channel impinges on older terraces or the valley wall (Hooke and Harvey, 1983). Yarnykh (1978) suggested that elongate, acute apex bends are due to confinements and found that one-fifth of meanders in a study reach of the Irtysch River and one-third of the bends on a study reach of the River Don are of this type. Much of the literature, particularly that emanating from flume studies and mathematical modelling, recognises a distinction between mobile channels in which the bends move and confined channels in which the bars move (Rhoads and Welford, 1991; Hasegawa, 1989).

Growth and Compound Development

The terms growth and extension are used for increase in bend amplitude and development perpendicular to the valley axis. Obviously bends must grow in their early stages from a straight long channel, but it was long assumed that, under stable conditions, the channel would reach an equilibrium form, thought to approximate a sine-generated curve. In 1947, Matthes could suggest that undue lengthening of a bend appears to be a deformity or abnormality rather than a normal characteristic of stream meanders.

Evidence has now emerged that continued growth and considerable elongation is a common phenomenon on rivers and that this often involves development of asymmetry and compound forms (e.g. Brice, 1974; Hooke and Harvey, 1983). Most of the case studies and derived models substantiate this. Brice (1974) traced the evolution of bends frommaps and aerial photographs and presented 16 forms (Figure 9.4a); the common sequence is increase in height of a simple symmetrical loop, then development of asymmetry by growth of a second arc, then evolution into a compound loop. Not all loops become compound and growth does not continue indefinitely. Eventually cutoffs take place because of the diverse orientations and different directions of growth of the loops. The early stages of Keller's (1972) model (Figure 9.4b) concur with many of the flume studies, showing development of shoals then erosion of the banks. Growth continues

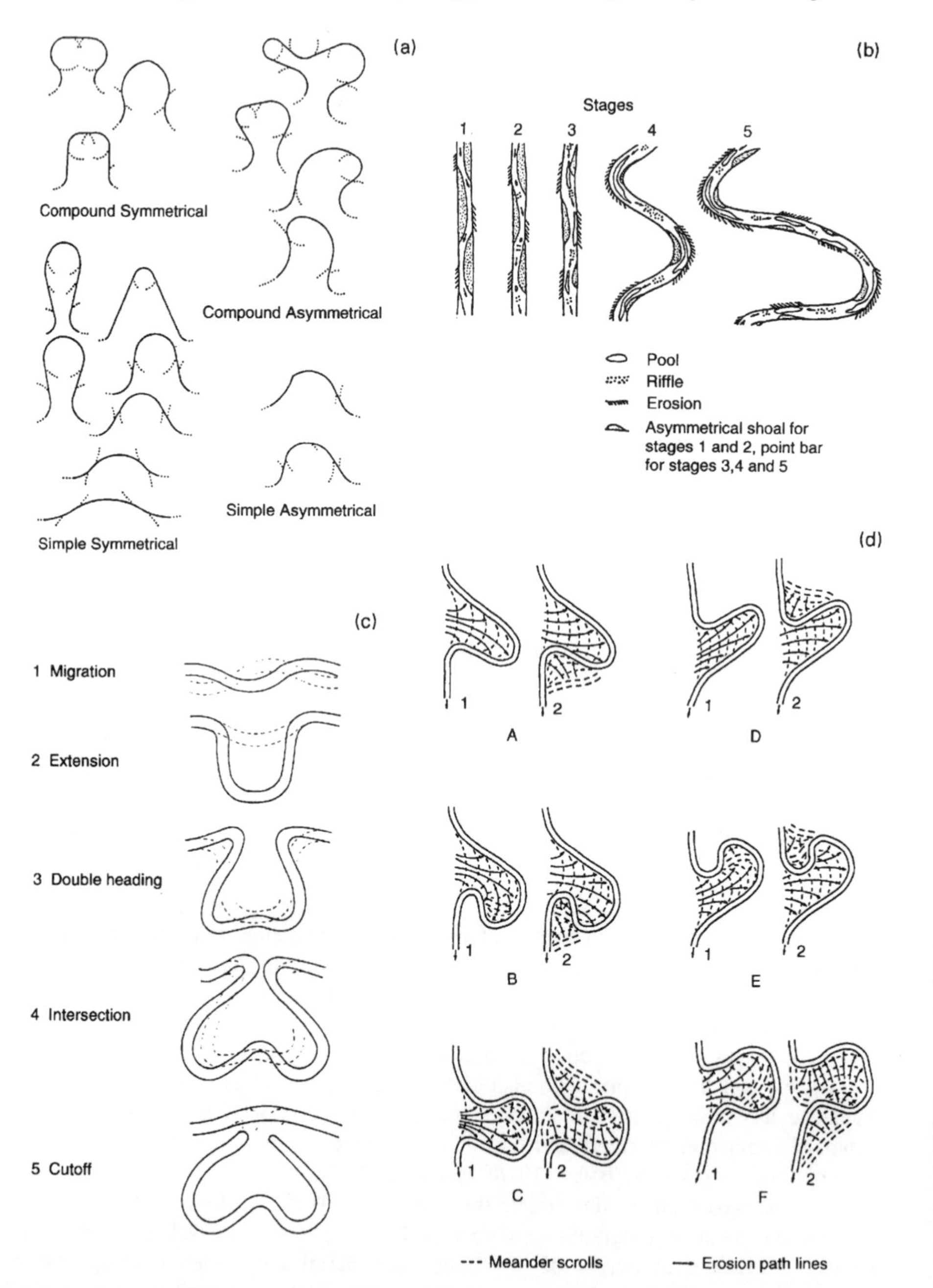

Figure 9.4 Case studies and models of sequences of meander change, after: (a) Brice (1974); (b) Keller (1972); (c) Hooke (1991); (d) Hickin (1974) (a, b, d, reproduced from *Progress in Physical Geography*. Reproduced by permission of Arnold Publishers)

and in the fifth stage Keller suggests that path length increases sufficiently that an extra pool and riffle are formed. This has similarities to the type of development which Hooke and Harvey (1983) found on bends on the River Dane in Cheshire, in which path length was found to increase and eventually a secondary arc or lobe was formed on the bend (Figure 9.4c); this continued to grow until a second inflection and true compound bend was formed with which was associated an extra pool and riffle. There are some differences in the details and modes of development; for example, in Keller's (1972) model, increase in the path length takes place in the limbs whereas Hooke and Harvey (1983) find the increase takes place in the apex region. In many models the growth is terminated by a cutoff (e.g. Kondrat'yev, 1968; Hickin and Nanson, 1975; Lewin, 1976; Thorne and Lewin, 1979; Hooke, 1991), but on some rivers, such as the River Dane, the secondary lobes tend themselves to develop and eventually form independent loops which continue to grow. Consequently the River Dane is becoming more sinuous and the number of bends is increasing (although occasionally a bend is overtaken or squeezed out). On this river there is no sign of a stable form and some of the most rapidly growing bends are those which approximate to a sine-generated model. It may be that the equilibrium model is theoretically correct, but in reality many rivers are not in equilibrium at the scale of meander development. It is possible that many are adjusting to changes on a larger timescale than decades, e.g. major climatic fluctuations, or it may be that such behaviour is inherent (Hooke and Redmond, 1992).

Cutoffs

There are divergent views on cutoffs in the literature. Even in 1983 Hickin, in reviewing meandering, says that: 'Until the last few decades our preoccupation with "equilibrium" forms has led to the dismissal of cutoffs as transient disturbances of the fluvial system', and remarks that they are frequent enough that 'many river reaches may be in a transient state permanently!'. Yet the occurrence and importance of cutoffs has long been recognised, especially with the early seminal work on the Mississippi and the spectacular nature of some of the cutoffs. There have been some case studies of cutoffs occurring (e.g. Johnson and Paynter, 1967; Mosley, 1975b) with some remarks on changes before and after. Other workers have remarked on the incidence of cutoffs (e.g. Lathrap, 1968), or the lack of them, as Sundborg (1956) found on the River Klaralven from the evidence of the past 300 years.

Various types of cutoff have been distinguished, the most common ones being chute and neck cutoffs. Laczay (1977) gives examples of neck cutoffs on the River Hernad in Hungary, and Brice (1977) found that both straight and diagonal cutoffs occur on the Sacramento River. Thorne and Lewin (1979) highlight the importance of chute cutoffs on the steep gravel-bed rivers in Wales. Lewis and Lewin (1983) identified five types of cutoff: chute, neck, mobile bar, multiloop chute and multiloop neck. They found that 55% of the 92 cutoffs studied were chute cutoffs. The distribution is variable but the overall rate for the period 1880–1900 was one in every five years and nearly one in every two years in the period 1950–1970 on 1000 km of river in Wales and the borderland. Long bypasses rather than individual cutoffs are cited by Kulemina (1973) on incomplete meandering reaches and some of these bypasses are several kilometres in length. A formula for calculating the frequency of cutoff that must occur to maintain equilibrium, given certain rates and pattern of development, was developed by Kondrat'yev (1968). Weilhaupt (1977) looked briefly at the morphology of ox-bow lakes and produced 16 basic models of the form, and the relationship between rate of infilling and type of cutoff has also been examined (e.g. Lewis and Lewin, 1983).

Conclusions

Two types of meander model have been generated over the past 20 years or so. The first generation of models were inductive and were primarily generalisations of the changes elucidated by the evidence and analysis discussed above. Most of these models were based on channels in one basin or region.

The main conclusion to emerge from this research on a variety of types of river all over the world is that, although migration is an important component of movement on many rivers, active rivers also exhibit a trend of increasing asymmetry and complexity over time. Empirical evidence, now supported by theoretical models and experiments, suggests that occurrence of cutoffs and decreasing sinuosity may be late stages of a sequence of meander evolution. The extent to which instability of meander characteristics is inherent is a matter of current research.

The second generation of meander evolution models are those developed largely since the early 1980s, which are computer simulations of meander behaviour. Most of these are based on theory of flow and sediment transport and are discussed in Chapter 10. These can now produce complex changes and appear to mimic some changes found in nature. However, they still need far more testing against historical data of real courses before they become applicable predictive models. Their development is providing some insight into controls and processes of meander change.

The style of channel change depends on the location and amount of erosion and deposition in each section and the effects of propagation of that change up- and downstream. The distribution of processes obviously depends ultimately on the shear stresses on bed and banks relative to the resistance. Much of the discussion in terms of meander changes has focused on whether the major control is near-bank velocity, difference between average and near-bank velocity or the pattern of secondary flows (e.g. Thompson, 1986; Figure 9.5). Various patterns of flow and of sediment movement in meander bends have also emerged from field studies (e.g. Dietrich, 1987).

9.4 BRAIDED/DIVIDED CHANNELS

Braided channels are generally thought of as being sections in which there is an excess of sediment load and deposition is dominant. Braided sections have been identified as occurring under conditions of high variability of discharge, high sediment load, wide valley floor, erodible bank materials, and steep slope. Although braided reaches tend to be primarily aggrading, processes of both erosion and deposition take place and both influence the type and form of bars in such sections and thus the styles of change. Braided rivers are regarded as inherently unstable and are dynamic, but the braided state may be an equilibrium condition. No matter what the style of braiding and the role of erosion, sediment must be available and deposited within such a reach. The conditions and

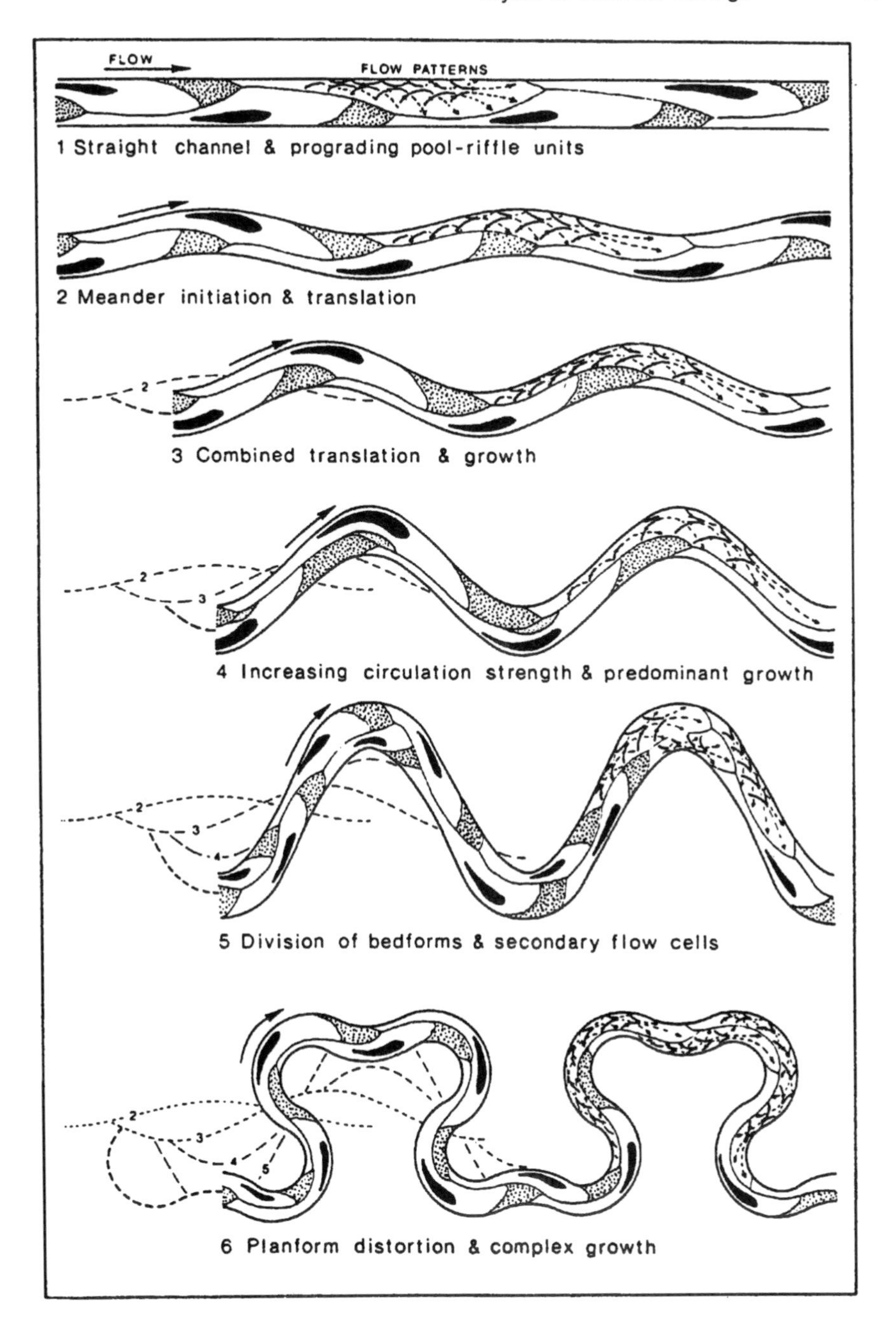

Figure 9.5 Model of secondary flow patterns in meander evolution sequence (from Thompson, 1986)

mechanics of deposition have already been discussed (Chapter 5). Many different types of bars have been identified in the literature with an accompanying proliferation and confusion of terminology. The best reviews are provided by Miall (1977), Lewin (1978), Smith (1978) and Church and Jones (1982) and illustrations of styles of bars and braiding are given in Figure 9.6. Basic criteria of classification include degree of elongation, symmetry and presence or absence of distal avalanche faces. Church and Jones (1982) show the evolutionary transformation between different types of bars (Figure 9.7).

Greater problems relate to the use of evidence of channel change in braided reaches, than for single channel reaches, especially at the historical timescale. This is because of the more dynamic nature of the reaches, with major morphological change taking place even within single events and with the nature and direction of change much more variable, and because the actual morphology and parameters of change are much more difficult to define. Various braiding indices have been suggested but the degree of braiding perceived depends very much on the stage of flow. Ferguson (1992) has warned of the dangers of making assumptions about processes from the morphology, and stresses the need to get in the channels and measure what is going on. With braided channels there has therefore been much greater emphasis on short-term changes and field measurement. Werritty and Ferguson (1980) have demonstrated, however, how this timescale can be meshed with the historical and longer timescales and the need for analysis at each timescale to understand the changes (Figure 9.8). The interrelated nature of the channels and bars and the need to examine subsurface changes has prompted the concept of pool–bar units, analogous to the pool–riffle unit of Thompson (1986) in meandering rivers.

The distinctive features of braided channels are the channel confluences and bifurcations. The channels tend to be curving and the bars to have a range of forms with various patterns of sediment distribution, often complex, indicating compound origin of bars. Coherent models of change in braided reaches are much less developed than for meandering channels and no overall review of styles of change exists. Ferguson (1993), however, concentrating on how channel bifurcations develop around mid-channel bars, has identified five modes of braiding development in gravel-bed rivers (partly based on Ashmore (1991)).

1. The traditional explanation of formation of *mid-channel* bars is that of *stalling* of material in the centre of a wide, shallow channel. Ashmore found that it was uncommon in his laboratory experiments, occurring at the threshold for bedload movement, as did Thompson (1987) on a stream in northern England.
2. Some bars seem to have a clearer link with pools and chutes a little way upstream. The term *chute and lobe* has been applied to such features but they mainly occur in very steep channels with coarse material and are ephemeral features. Longer-lasting depositional lobes occur in braided rivers which are less steep and shallow and their formation is termed *transverse bar conversion* by Ashmore. This was found to be common in experimental runs above the threshold for bed movement and where depth was two to three times D_{90}. In these conditions pools at sites of flow convergence can scour appreciably, generating enough sediment for substantial deposition where flow diverges, sometimes in sufficient depth for an avalanche face to develop. Thin bedload sheets then stall on top of the front of the lobe, which eventually emerges. Flow is then deflected off the edges where a pair of diagonal slip faces can develop.

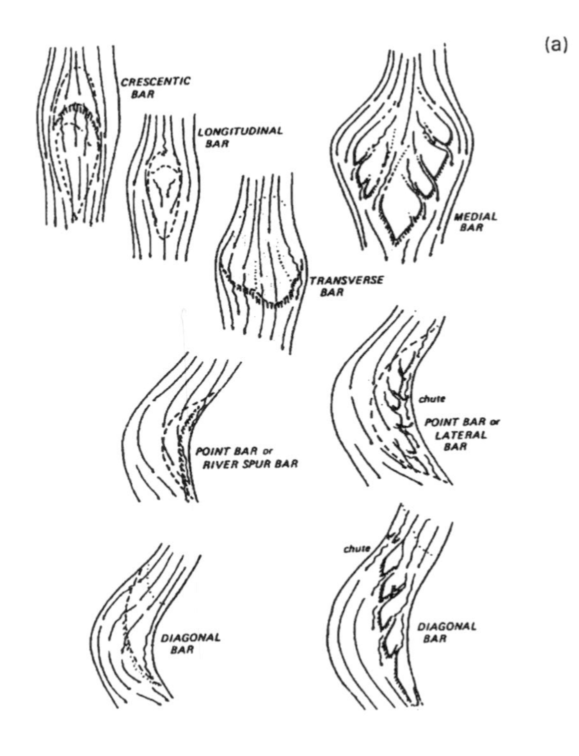

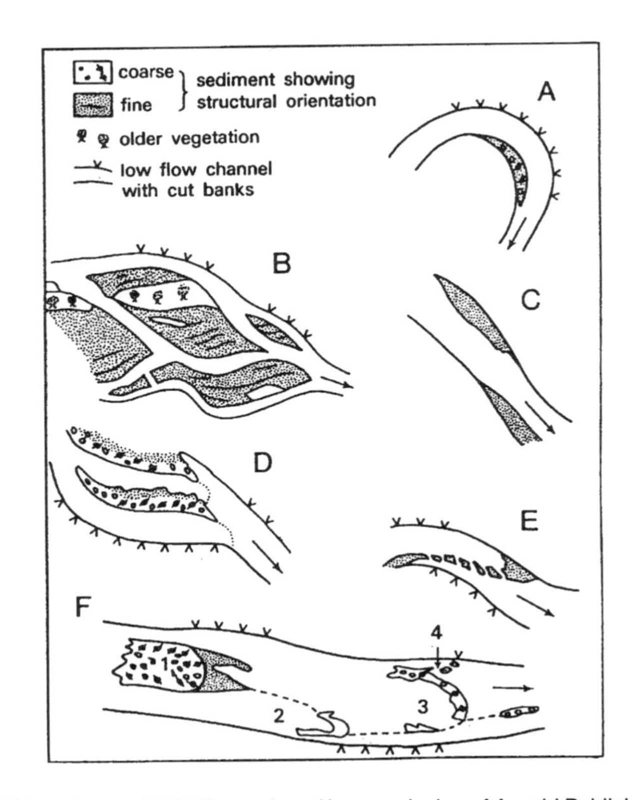

Figure 9.6 Types of channel bars: (a) from Church and Jones (1982); (b) from Lewin (1978). Reproduced by permission of Arnold Publishers

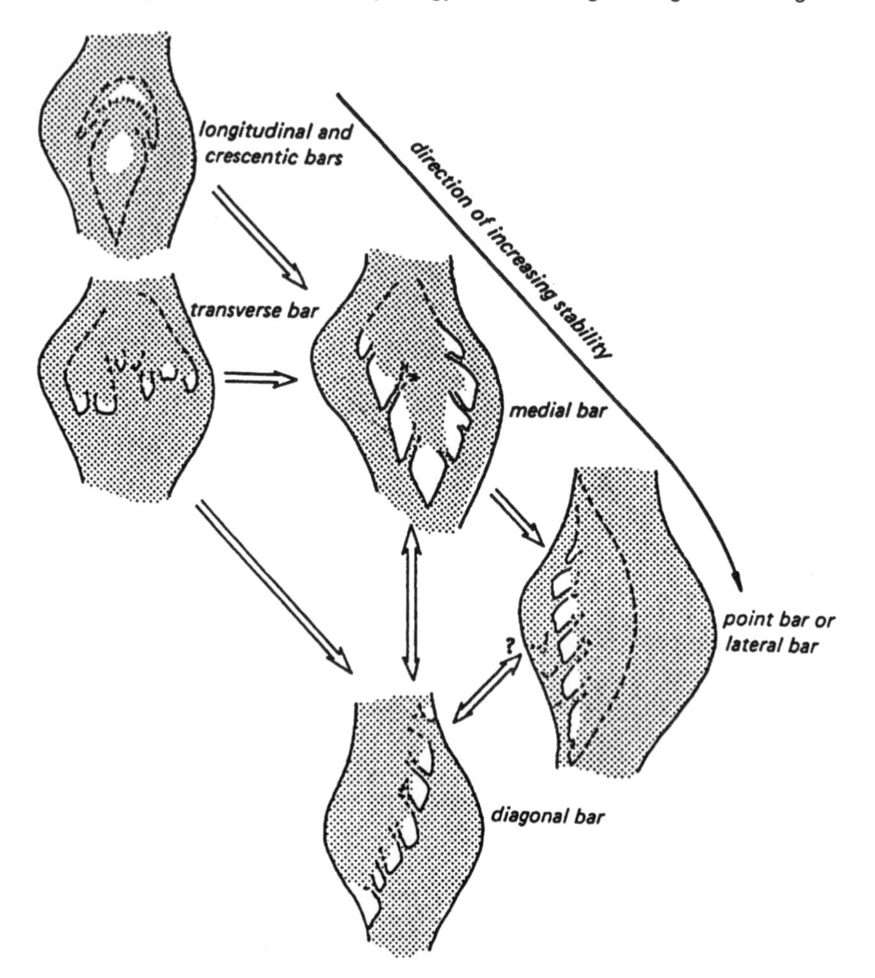

Figure 9.7 The main evolutionary transformations of gravel bars (from Church and Jones, 1982)

3. Once diagonal bar forms exist, with flow obliquely across a side shallow channel, conditions are set for another main mode of braid development, *chute cutoff*. This is an erosional process, also observed on gravel-bed meander point bars, with headward incision by flow taking a short cut across the bar. Although erosion is involved, the increase in bartop flow to initiate incision is often due to aggradation of the outer channel or bar rim.
4. *Multiple dissection of lobes* is also an erosional process. These have been described in the field, e.g. from New Zealand (Rundle, 1985), but were uncommon in experiments.
5. *Avulsion* is large-scale switching of the main flow. Often ponding occurs behind an aggrading bar leading to overflow, often on the outside of a curve. Overflow into an adjacent channel can cause headward erosion and stream capture. Bank erosion can also intersect an alternative alignment (Carson, 1986).

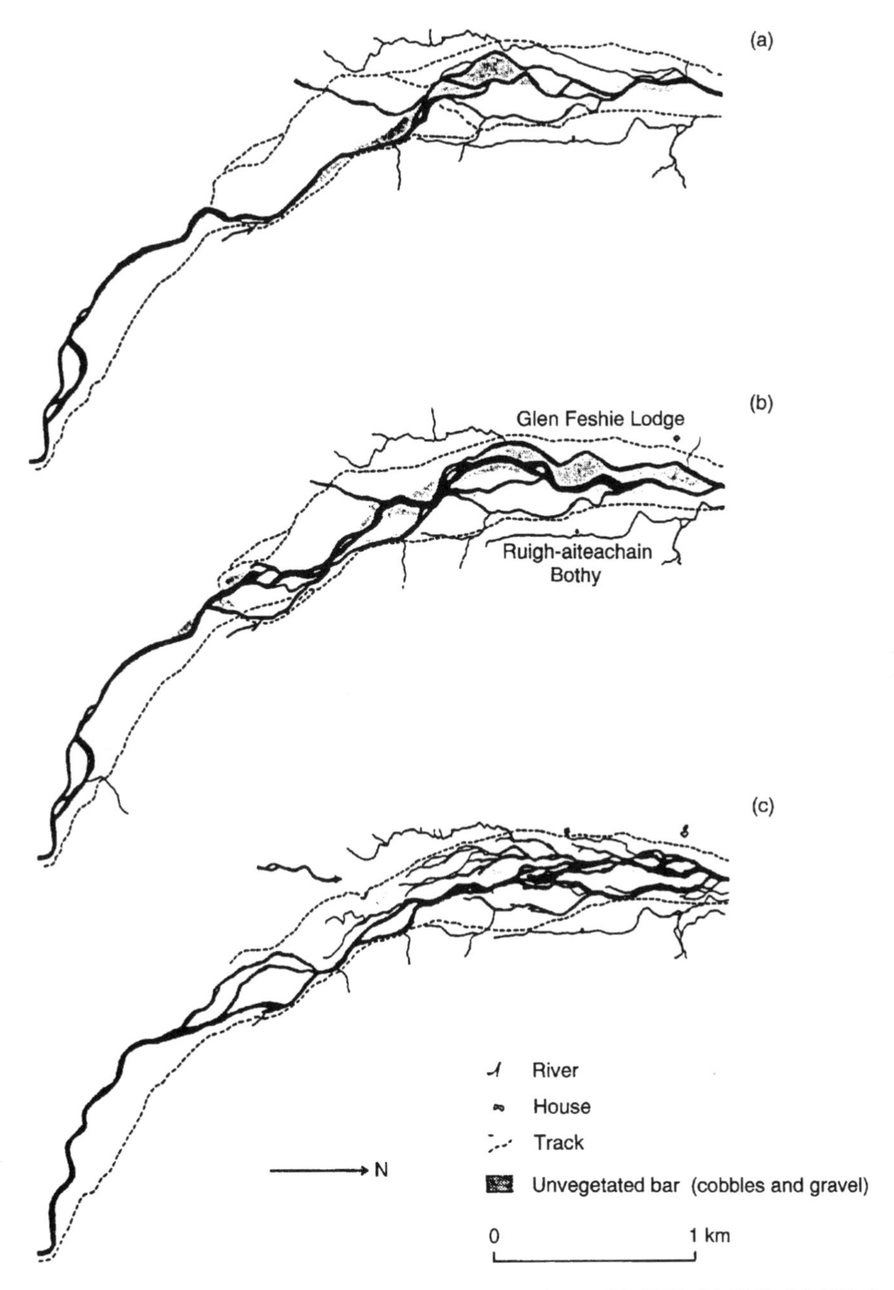

Figure 9.8 Pattern of braiding in Glen Feshie at three dates: (a) 1869; (b) 1899; (c) 1971 (from Werritty and Ferguson, 1980)

Ferguson ascribes the instability in braided channels and rapid variations to:

- high stream power and therefore high transport rates, with the possibility of rapid erosion or deposition where transport capacity increases or decreases downstream because of flow convergence or divergence;
- low relative depth which leads to rapid and sensitive feedback from local aggradation or degradation to flow and transport capacity;
- the changing share of discharge between distributaries as they erode or aggrade with knock-on effects downstream.

In braided rivers, changes are closely related to occurrence of flows competent to move the available material. Morphology of the bars and degree of braiding in a reach is closely related to the preceding sequence of flows. Similarly, climatic fluctuations at a longer timescale can produce differences in degree and style of braiding. Degree of braiding may also be influenced by conditions for stabilisation of bars by vegetation as shown by occurrence of drought in marginal conditions increasing braiding, increased moisture or invasion by new species stabilising bars, and by presence of toxic materials from mining sediments inhibiting vegetation growth and stabilisation.

9.5 RATES AND FREQUENCIES OF PLANFORM CHANGE

On most meandering rivers migration and growth are relatively continuous processes, movement taking place in relatively frequent events, i.e. below bankfull discharges. Erosion on active rivers may take place several times a year and channel shift is not solely related to large floods. Cutoffs and avulsions may be associated with overbank flows which cut across the floodplain, particularly chute cutoffs, but neck cutoffs are usually the culmination of continued erosion at apices of bends. Rivers on which meander scrolls are evident usually have a marked seasonal flow regime. The time taken for completion of the sequence of bend development from initial bend to cutoff varies from a few years to several centuries.

A non-linear relationship between form and rate of change has been found in meanders. Originally this was suggested by Hickin (1978) and is now substantiated by more evidence (Hickin and Nanson, 1984; Biedenharn et al., 1989; Hooke, 1987) (Figure 9.9). This relationship may help to assess the rate and type of change which is likely in a meander and indicates how rates may rapidly increase in the growth phase, though it has been criticised as a statistical artifact (Frobish, 1988).

Many figures for rates of channel shift are quoted as rates of erosion (see Chapter 4). Overall, rates and frequencies of channel shift vary very much with discharge regime (climate and land use), bank resistance and stream power (Lewin, 1983; Figure 9.10). Many of the most laterally active streams in humid temperate areas experience effective events one to several times a year. In semi-arid areas change is much more episodic and may lead to periods of distinct behaviour. Catastrophic change on a timescale of decades or centuries may have a dominating influence in certain environments, (e.g. Nanson and Erskine, 1988; Nanson, 1986; Wolman and Gerson, 1978).

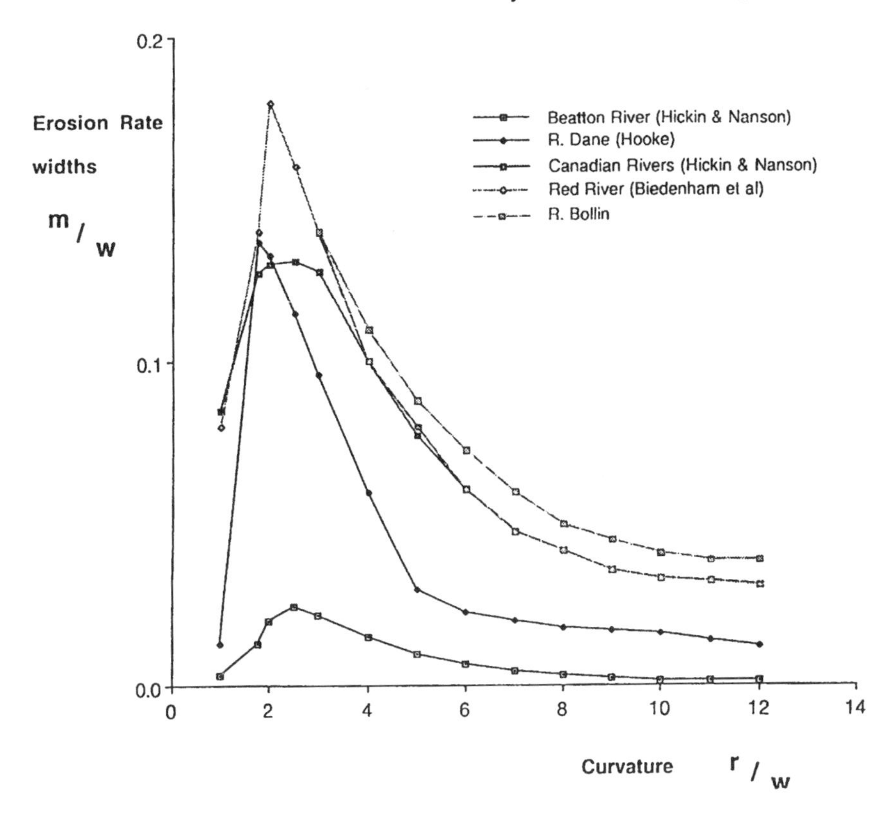

Figure 9.9 Relationship between erosion rate (migration rate) and bend curvature (from Hooke, 1991)

9.6 CROSS-SECTIONAL CHANGES

Analysis of these changes is much less common than for changes of pattern, mainly because of the need for repeated surveys of profiles. Most of the work has been associated with analysis of the response of channels to human activities. The close association between planform and cross-sectional form has already been indicated (Chapter 7) and some changes can, of course, be inferred from the changes in width indicated by the planform evidence. Changes in width, depth and channel capacity (area) can be investigated, and increase or decrease in each of these can occur.

Channel enlargement takes place where the dominant or effective discharges have increased. Most of the increase appears to take place by bank erosion and increase in width. Such cases have particularly been identified downstream of urban areas. Responses have been documented over the timescale of decades. Evidence of enlargement includes erosion on both sides, structures and trees undermined or trees in the channel (Gregory et al., 1992), and such changes obviously cause practical problems (e.g. Douglas, 1985).

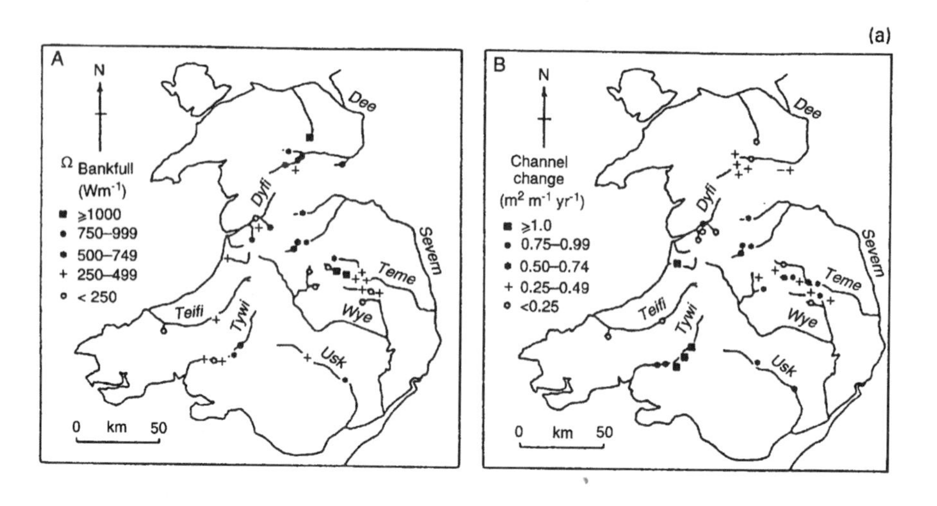

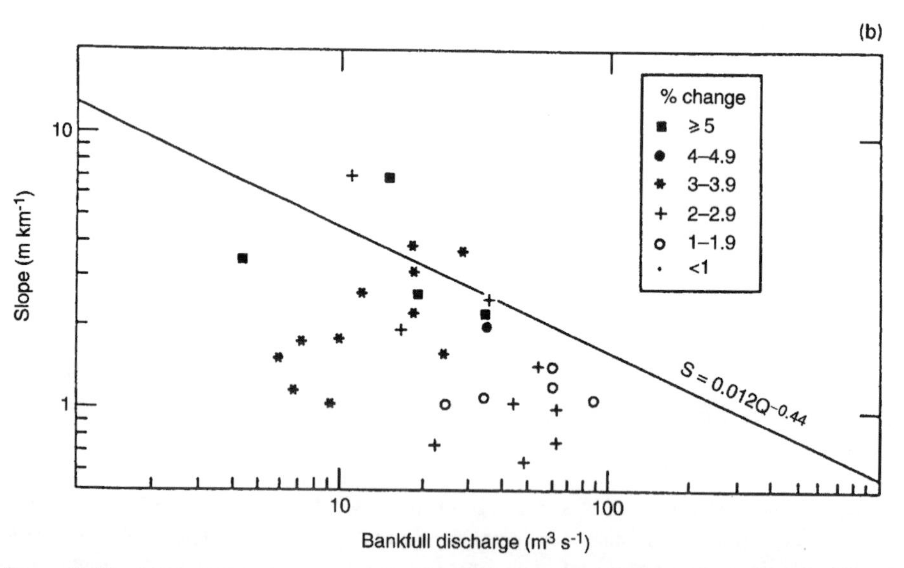

Figure 9.10 (a) Bankfull stream power and averaged rates of lateral channel shift for selected reaches of Welsh rivers (from Lewin, 1983). (b) Channel change rates in relation to slope and discharge (from Lewin, 1983). Reproduced by permission of Arnold Publishers

Decreases in channel capacity are rather easier to examine since the deposition is evident and can often be dated by vegetation, lichens, artifacts or structures. Deposition in the form of berms or benches along channels has been described, e.g. downstream of reservoirs (Petts, 1979) and in engineered reaches (Brookes, 1988). Other channel narrowing, for example in the transformation of braided channels in the SW USA in the last 50 years or so, has been by mid-channel and point bar deposition. The degree to which vegetation is a key factor in stabilisation, even encouraging sedimentation, and to what extent it is merely a post-depositional coloniser is debated.

Changes in cross-sectional form and size on the timescale of decades have taken place on streams in New South Wales, Australia, in response to fluctuations in climate. The restricted nature of the channels means that less adjustment is possible through planform but the scale of changes is large, particularly in sandy reaches (Erskine and Warner, 1988).

9.7 LONGITUDINAL CHANGES

Changes in the longitudinal direction can take place by incision or aggradation. Again, these normally have concomitant changes in pattern and cross-sectional form. Longitudinal changes have particularly been studied in the context of gullies and arroyos but more rarely in alluvial streams. Some work has focused on the issue of 'grade' and the nature of adjustment, though much of this work is at the timescale of thousands of years. Once incision occurs then it tends to work back upstream. A key question is how far the effect is transmitted. Some workers have shown that adjustment tends to be local and that much of the channel adjustment is taken up by variables other than slope (e.g. Leopold and Bull, 1979). Others demonstrate that adjustment is transmitted right the way up through the system (e.g. Begin et al., 1981). Pronounced and rapid incision upstream of channel cutoffs or straightening is common (e.g. Mosley, 1975b; Brookes, 1988). The actual nature of adjustment depends much on sediment supply, as Thoms and Walker (1992) show in analysing the response of the Murray River, Australia, to the construction of weirs. Some sections are aggrading and others degrading and adjustment is still working through the system 54 years after construction of the weirs (Figure 9.11). Complex response of channels, with aggradation and incision varying both spatially and temporally, has been demonstrated through flume studies (Schumm, 1977) and field studies (e.g. Haible, 1980).

9.8 FLOODPLAIN CHARACTERISTICS

The style and rate of channel changes has a great effect upon the morphology and sedimentology of floodplains. The relative contribution of lateral and vertical accretion and the continuity of deposition influence the composition of the floodplain. Basic differences in stratigraphy have long been identified. Facies models have been proposed representing streams in different types of environment with different sediment loads and with different degrees of mobility. Such two-dimensional models have severe limitations for interpreting three-dimensional assemblages. In addition, it has been found that each new situation demands a new facies model (Miall, 1985) negating the synthesising role of the facies model concept (Brierley, 1989). Brierley derived new river planform facies models for three different types of pattern on the Squamish River, British Columbia, but found that even this more quantitative approach could not be used to differentiate depositional suites and concluded that use of such data at the scale of channel planform is inapplicable. Similarly, Lewin (1983) has pointed out the complexity of floodplain sedimentation patterns and noted that these are not uniquely related to the type of

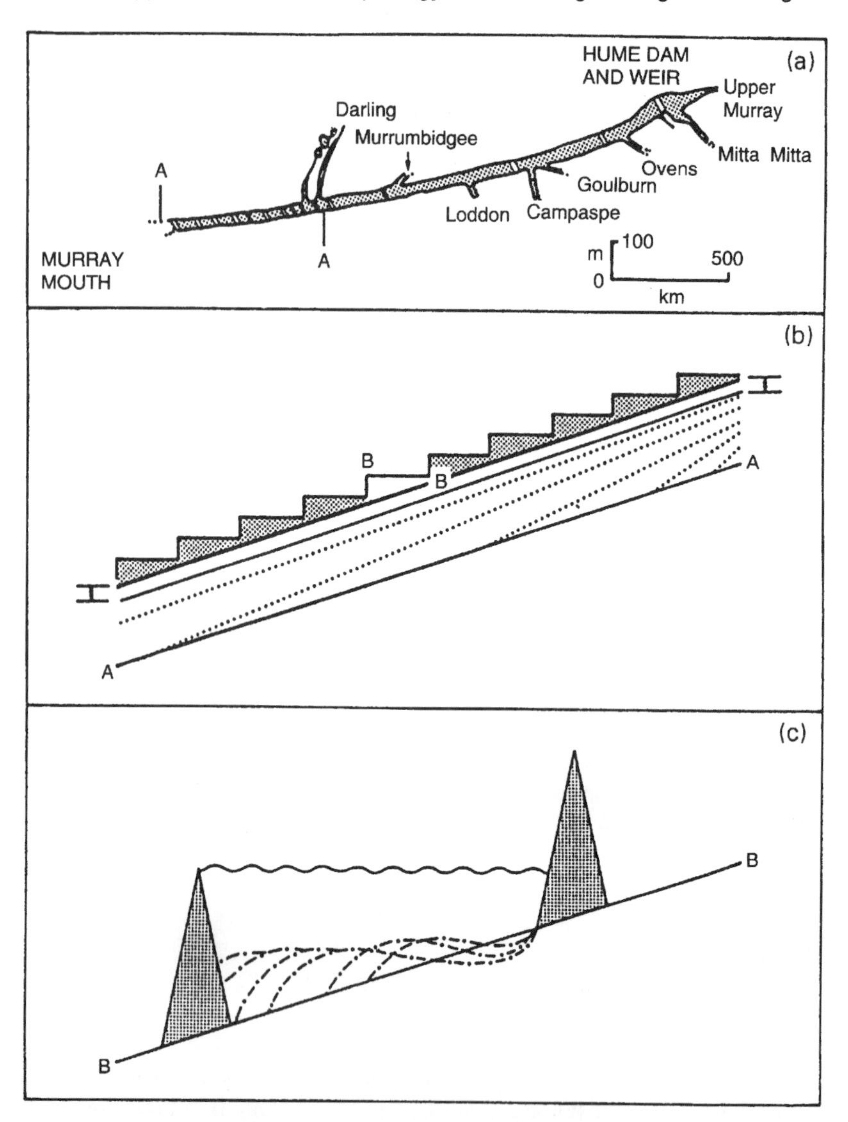

Figure 9.11 Schematic diagrams of the response of the river bed over time to flow regulation, River Murray. (a) Long profile; (b) channel profile of lowland region; (c) Adjustment of bed slope between weirs (from Thoms and Walker, 1992)

pattern. Processes such as counterpoint bar sedimentation and catastrophic stripping of floodplains have been recognised in certain locations (Lewin, 1983; Nanson, 1986). One of the best reviews of floodplain morphology exemplifying the range of channel morphologies is still that of Lewin (1978), in which he shows the association with style of channel change (Figure 9.12).

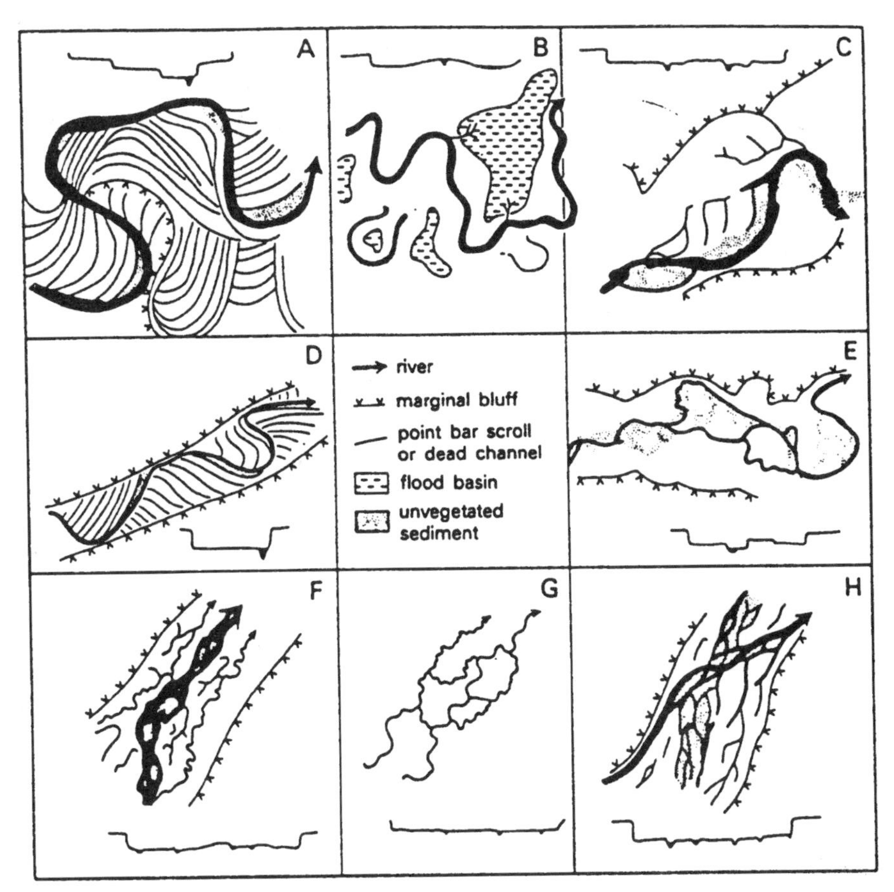

Figure 9.12 Selected floodplain types: (A) meandering, incising river; (B) meandering river with levees; (C) migrating meandering/braided river; (D) confined meandering stream; (E) braided/ meandering river with islands; (F) braided main channel with migrating bars and meandering accessory channels; (G) stable anastomosing channel; (H) braid-bar plain with channel switching (from Lewin, 1978)

9.9 CHANNEL SENSITIVITY

9.9.1 Natural Streams

The pattern of distribution of instability within drainage basins also varies. On some channels small isolated sections of active erosion and deposition occur with minor movement, usually in the apices of bends. On other rivers, mobile sections alternate with stable sections, frequently in relation to gross controls of valley slope, valley width or resistance of bank materials. Within drainage basins as a whole there is a tendency for lateral mobility to be greatest in the middle sections (Bull, 1979), particularly in piedmont sections where valleys open out from upland areas but slope and stream power are still high (Hooke and Redmond, 1989a).

The sediment supply can be seen to be a major control on channel activity both at the scale of a few metres within a meandering or braided section and at the larger scale of occurrence of unstable sections and their characteristics. It is a major cause of change within a river channel from meandering to braided pattern and vice versa, both temporally and spatially, and likewise affects vertical behaviour of erosion or aggradation. Harvey (1986) has demonstrated how the influx of large amounts of sediment resulting from a catastrophic rainstorm in the Howgill Fells transformed a meandering stream to a braided channel. Gradually over the decade since that event, the channel has adjusted back as the supply has diminished. Similarly, the braided reach discussed by Thompson (1987) is downstream of a major valley scar supplying sediment.

Large floods will often be the trigger to push a system over the pattern threshold, though it must be near it for that to occur (Schumm, 1979). The dominance of braiding or meandering in a reach will vary with the sequence of discharge events. Thompson (1987) showed that long periods of gradual change favour development of an increasingly complex braided pattern, whilst periods of rapid change in response to more frequent flooding encourage the persistence of simpler meandering patterns.

The distinction between the meandering and braided patterns is not a clear one, meandering channels often having mid-channel bars (e.g. Hooke, 1986), but the transition between the two types is relatively narrow. Cases of change between meandering and braided channels over time are particularly common in semi-arid areas where full-scale metamorphosis of channels has taken place.

9.9.2 Channelised Rivers

The nature of response in channelised rivers depends very much on the pre-existing characteristics and the degree and type of control incorporated in the channelisation. Basically, rivers tend to meander and need sites for deposition along the channel. Thus bars frequently form in channelised sections and banks may erode. Adjustment may also be

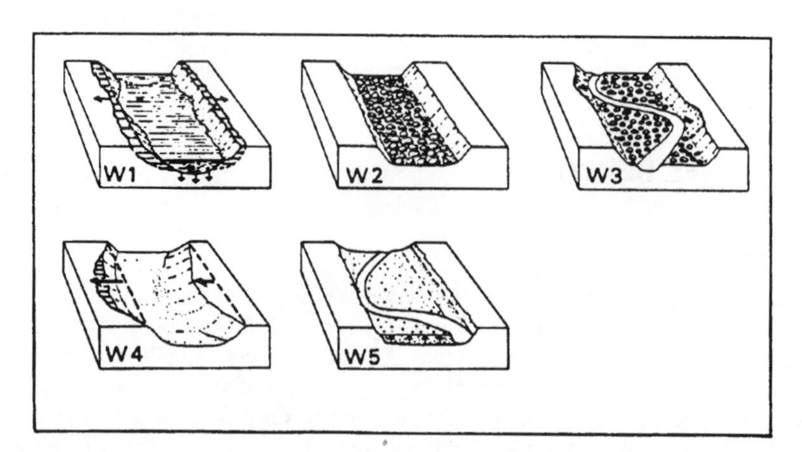

Figure 9.13 Principal types of adjustment in straightened river channels in Denmark (from Brookes, 1987b)

made vertically, particularly due to the steeper slope resulting from straightening, and the calibre and roughness of the channel bed may be altered (Brookes, 1987; Figure 9.13). Understanding of pre-existing instability is an essential prerequisite for effective design of schemes and will influence the degree of control necessary to maintain stability. However, a better alternative is to recognise the natural characteristics of the stream and to design and manage the channel for instability. For this and other ecological and aesthetic reasons, meanders are now being put back into straightened streams (Brookes, 1988). In examining responses of channels downstream of engineering works and channelisation schemes, Brookes (1987a,b) has demonstrated a general relationship between adjustment and stream power, with lower power channels generally being stable.

9.10 REFERENCES

Ackers, P. and Charlton, F.G. 1970. Meander geometry arising from varying flows. *Journal of Hydrology,* **11,** 230–252.

Ashmore, P. 1991. How do gravel-bed rivers braid?. *Canadian Journal of Earth Sciences,* **28**(3), 326–341.

Begin, Z.R., Meyer, D F. and Schumm, S.A. 1981. Development of longitudinal profiles of alluvial channels in response to base-level lowering. *Earth Surface Processes and Landforms,* **6**, 49–68.

Biedenharn, D.S., Combs, P.G., Hill, G.J., Pinkard, C.F. and Pinkston, C.B. 1989. Relationship between channel migration and radius of curvature on the Red River. In: Wong, S.S. (Ed.), *Sediment Transport Modelling*. American Society of Civil Engineers, 536–541.

Brice, J.C. 1974. Evolution of meander loops. *Geological Society of America Bulletin*, **85**, 581–586.

Brice, J.C. 1977. *Lateral Migration of the Middle Sacramento River, California*. US Geological Survey Report, WRD/WR1-77/052.

Brierley, G.J. 1989. River planform facies models: the sedimentology of braided wandering and meandering reaches of the Squamish River, British Columbia. *Sedimentary Geology*, **61**, 17–35.

Brookes, A. 1987a. River channel adjustments downstream from channelization works in England and Wales. *Earth Surface Processes and Landforms*, **12**, 337–351.

Brookes, A. 1987b. The distribution and management of channelized streams in Denmark. *Regulated Rivers*, **1**, 3–16.

Brookes, A. 1988. *Channelized Rivers: Perspective for Environmental Management.* Wiley, Chichester, 326pp.

Bull, W.B. 1979. Threshold of critical power in streams. *Geological Society of America Bulletin*, **90**, 453–464.

Burkham, D.E. 1972. *Channel Changes of the Gila River in Stafford Valley, Arizona, 1946–1970.* US Geological Survey Professional Paper 665-G.

Carey, W.C. 1969. Formation of flood plain lands. *Proceedings of American Society of Civil Engineers, Journal of the Hydraulics Division*, **95**, 981–994.

Carson, M.A. 1986. Characteristics of high-energy 'meandering' rivers: The Canterbury Plains, New Zealand. *Geological Society of America Bulletin*, **97**, 886–895.

Carson, M.A. and Lapointe, M.F. 1983. The inherent asymmetry of river meander planform. *Journal of Geology*, **91**, 41–45.

Church, M. and Jones, D. 1982. Channel bars in gravel-bed rivers. In Hey, R.D., Bathurst, J.C. and Thorne, C.R. (Eds), *Gravel-bed Rivers*. Wiley, Chichester, 291–338.

Dietrich, W.E. 1987. Mechanics of flow and sediment transport in river bends. In: Richards, K.S. (Ed.), *River Channels: Environment and Processes*. Blackwell, Oxford, 179–227.

Douglas, I. 1985. Hydrogeomorphology downstream of bridges: one mechanism of channel widening. *Applied Geography*, **5**, 167–170.

Erskine, W.D. and Warner, R.F. 1988. Geomorphic effects of alternating flood- and drought-dominated regimes on NSW coastal rivers. In: Warner, R.F. (Ed.), *Fluvial Geomorphology of Australia*. Academic Press, Sydney, 223–244.

Ferguson, R.I. 1993. Understanding braiding processes in gravel-bed rivers: progress and unsolved problems. In: Best J.L. and Bristow C.S. (Eds), *Braided Rivers*, Geological Society Special Publication No. 75, 73–87.

Ferguson, R.I. and Werritty, A. 1983. Bar development and channel changes in the gravelly River Feshie, Scotland. In: Collinson, J. and Lewin, J. (Eds), *Modern and Ancient Fluvial Systems*. International Association of Sedimentologists, Special Publication no. 6, 181–193.

Friedkin, J.F. 1945. *A Laboratory Study of the Meandering of Alluvial Rivers*. US Waterways Experiment Station, Vicksburg, Mississippi.

Frobish, D.J. 1988. River-bend curvature and migration: How are they related?. *Geology*, **16**, 752–755.

Graf, W.L. 1981. Channel instability in a braided, sand bed river. *Water Resources Research*, **17**, 1087–1094.

Gregory and Walling. 1973. Drainage Basin Form and Process. Arnold, London.

Gregory, K.J., Davis, R.J. and Downs, P.W. 1992. Identification of river channel change due to urbanization. *Applied Geography*, **12**, 299–318.

Haible, W.W. 1980. Holocene profile changes along a California coastal stream. *Earth Surface Processes*, **5**, 249–264.

Harvey, A.M. 1986. Geomorphic effects of a 100 year storm in the Howgill Fells, Northwest England. *Zeitschrift für Geomorphologie*, **30**(1), 71–91.

Hasegawa, K. 1989. Studies on qualitative and quantitative prediction of meander channel shift. In: Ikeda, S. and Parker, G. (Eds), *River Meandering*. American Geophysical Union, Washington, 485pp.

Hey, R.D. 1978. Determinate hydraulic geometry of river channels. *Proceedings of the American Society of Civil Engineers, Journal of the Hydraulics Division*, **104**, 869–885.

Hickin, E.J. 1974. The development of meanders in natural river channels. *American Journal of Science*, **274**, 414–442.

Hickin, E.J. 1977. The analysis of river planform responses to changes in discharge. In: Gregory, K.J. (Ed.), *River Channel Changes*. Wiley, Chichester, 249–264.

Hickin, E.J. 1978. Hydraulic factors controlling channel migration. In: Davidson-Arnott, R. and Nickling, W. (Eds), *Research in Fluvial Systems*. Geo Abstracts, Norwich, 59–66.

Hickin, E.J. 1983. River channel changes: retrospect and prospect. In: Collinson, J.D. and Lewin, J. (Eds), *Modern and Ancient Fluvial Systems*. Blackwell, Oxford, 61–83.

Hickin, E.J. and Nanson, G. 1975. The character of channel migration on the Beatton River, northeast British Columbia, Canada. *Geological Society of America Bulletin*, **86**, 487–494.

Hickin, E.J. and Nanson, G. 1984. Lateral migration rates of river bends. *ASCE, Journal of Hydraulic Engineering*, **110**, 1557–1567.

Hooke, J.M. 1977. The distribution and nature of changes in river channel pattern. In: Gregory, K.J. (Ed.), *River Channel Changes*. Wiley, Chichester, 265–280.

Hooke, J.M. 1984. Changes in river meanders: a review of techniques and results of analysis. *Progress in Physical Geography*, **8**, 473–508.

Hooke, J.M. 1986. The significance of mid-channel bars in an active meandering river. *Sedimentology*, **33**, 839–850.

Hooke, J.M. 1987. Changes in meander morphology. In: Gardiner, V. (Ed.), *International Geomorphology*. Wiley, Chichester, 591–609.

Hooke, J.M. 1991. *Non-linearity in River Meander Development: Chaos Theory and its Implications*. Department of Geography, University of Portsmouth, Working Paper No. 19, 22pp.

Hooke, J.M. and Harvey, A.M. 1983. Meander changes in relation to bend morphology and secondary flows. In: Collinson, J.D. and Lewin, J. (Eds), *Modern and Ancient Fluvial Systems*. Blackwell, Oxford, 121–132.

Hooke, J.M. and Kain, R.J.P. 1982. *Historical Change in the Physical Environment: A Guide to Sources and Techniques*. Butterworths, Sevenoaks, 236pp.
Hooke, J.M. and Redmond, C.E. 1989a. River channel changes in England and Wales. *Journal of the Institution of Water and Environmental Management*, **3**, 328–335.
Hooke, J.M. and Redmond, C.E. 1989b. Use of cartographic sources for analysing river channel change with examples from Britain. In: Petts, G.E. (Ed.), *Historical Change of Large Alluvial Rivers: Western Europe*. Wiley, Chichester, 79–93.
Hooke, J.M. and Redmond, C.E. 1992. Causes and nature of river planform change. In: Billi, P., Hey, R.D., Thorne, C.R. and Tacconi, P. (Eds), *Dynamics of Gravel-bed Rivers*. Wiley, Chichester, 549–563.
Johnson, R.H. and Paynter, J. 1967. The development of a cut-off in the River Irk at Chadderton, Lancashire. *Geography*, **52**, 41–49.
Keller, E.A. 1972. Development of alluvial stream channels: a five-stage model. *Geological Society of America Bulletin*, **83**, 1531–1540.
Kellerhals, R.M., Church, M. and Bray, D.I. 1976. Classification and analysis of river processes. *Proceedings of the American Society of Civil Engineers, Journal of the Hydraulics Division*, **102**, 813–829.
Kondrat'yev, N.Ye 1968. Hydromorphic principles of computations of free meandering and signs and indexes of free meandering. *Soviet Hydrology*, **4**, 309–335.
Kulemina, N.M. 1973. Some characteristics of the process of incomplete meandering of the channel of the upper Ob River. *Soviet Hydrology*, **6**, 562–565.
Laczay, I.A. 1977. Channel pattern changes of Hungarian river: the example of the Hernad River. In: Gregory, K.J. (Ed.), *River Channel Changes*. Wiley, Chichester, 184–192.
Lathrap, D.W. 1968. Aboriginal occupation and changes in river channel on the central Ucayali, Peru. *American Antiquity*, **33**, 62–79.
Leopold, L.B. and Bull, W.B. 1979. Base level, aggradation and grade. *Proceedings of the American Philosophical Society*, **123**, 168–202.
Lewin, J. 1972. Late-stage meander growth. *Nature – Physical Science*, **240**, 116.
Lewin, J. 1977. Channel pattern changes. In: Gregory, K.J. (Ed.), *River Channel Changes*. Wiley, Chichester, 167–184.
Lewin, J. 1978. *Progress in Physical Geography*, **2**, 416.
Lewin, J. 1983. Changes of channel patterns and floodplains. In: Gregory, K.J. (Ed.), *Background to Palaeohydrology*. Wiley, Chichester, 303–319.
Lewin, J. and Brindle, B.J. 1977. Confined meanders. In: Gregory, K.J. (Ed.), *River Channel Changes*. Wiley, Chichester, 221–234.
Lewin, J., Macklin, M.G. and Newson, M.D. 1988. Regime theory and environmental change – irreconcilable concepts?. In: White, W.R. (Ed.), *International Conference on River Regime*. Wiley, Chichester, 431–445.
Lewis, G.W. and Lewin, J. 1983. Alluvial cutoffs in Wales and the Borderlands. In: Collinson, J.D. and Lewin, J. (Eds), *Modern and Ancient Fluvial Systems*. Blackwell, Oxford, 145–154.
Matthes, G.H. 1941. Basic aspects of stream meanders. *Transactions of the American Geophysical Union*, **32**, 632–636.
Miall, A.D. 1977. A review of the braided-river depositional environment. *Earth Science Reviews*, **13**, 1–62.
Miall, A.D. 1985. Architectural-element analysis: a new method of facies analysis applied to fluvial deposits. *Earth Science Reviews*, **22**, 261–308.
Mosley, M.P. 1975a. Channel changes on the River Bollin, Cheshire, 1872–1973. *East Midland Geographer*, **6**, 185–199.
Mosley, M.P. 1975b. Meander cutoffs on the River Bollin, Cheshire in July 1973. *Revue de Geomorphologie Dynamique*, **24**, 21–32.
Nanson, G.C. 1986. Episodes of vertical accretion and catastrophic stripping: a model of disequilibrium floodplain development. *Bulletin of the Geological Society of America*, **97**, 1467–1475.

Nanson, G.C. and Erskine, W.D. 1988. Episodic changes of channels and floodplains on coastal rivers in New South Wales. In: Warner, R.F. (Ed.), *Fluvial Geomorphology of Australia*. Academic Press, Sydney, 201–222.

Nanson, G.C. and Hickin, E.J. 1983. Channel migration and incision of the Beatton River. *Proceedings of the American Society of Civil Engineers, Journal of the Hydraulics Division*, **109**, 327–337.

Neill, C.R. 1970. Formation of flood plain lands (discussion). *Proceedings of the American Society of Civil Engineers, Journal of the Hydraulics Division*, **96**, 297–298.

Petts, G.E. 1979. Complex response of river channel morphology subsequent to reservoir construction. *Progress in Physical Geography*, **3**, 329–362.

Rhoads, R.L. and Welford, M.R. 1991. Initiation of river meandering. *Progress in Physical Geography*, **15**, 127–156.

Ryckborst, H. 1980. Geomorphological changes after river-meander surgery. *Geologie en Mijnbouw*, **59**, 121–128.

Schattner, I. 1962. *The Lower Jordan Valley*. Publications of Hebrew University, Jerusalem, Magness Papers.

Schumm, S.A. 1969. River metamorphosis. *Proceedings of the American Society of Civil Engineers, Journal of the Hydraulics Division*, **95**, 255–273.

Schumm, S.A. 1977. *The Fluvial System*. Wiley, Chichester.

Schumm, S.A. 1979. Geomorphic thresholds. *Transactions of the Institute of British Geographers*, **4**, 485–515.

Smith, N.D. 1978. Some comments on terminology for bars in shallow rivers. In: Miall, A.D. (Ed.), *Fluvial Sedimentology*. Canadian Society of Petroleum Geologists, 85–88.

Sundborg, A. 1956. The River Klaralven, a study of fluvial processes. *Geografiska Annaler*, **38**, 127–316.

Tanner, W.F. 1955. Geological significance of stacked meanders. *Geological Society of America Bulletin*, **66**, 1698.

Thompson, A. 1986. Secondary flows and the pool–riffle unit: a case study of the processes of meander development. *Earth Surface Processes and Landforms*, **11**, 631–641.

Thompson, A. 1987. Channel response to flood events in a divided upland stream. In: Gardiner, V. (Ed.), *International Geomorphology*. Wiley, Chichester, 691–709.

Thoms, M.C. and Walker, K.F. 1992. Channel changes related to low-level weirs on the River Murray, South Australia. In: Carling, P.A. and Petts, G.E. (Eds), *Lowland Floodplain Rivers: Geomorphological Perspectives*. Wiley, Chichester, 235–250.

Thorne, C.R. and Lewin, J. 1979. Bank processes, bed material movement and planform development in a meandering river. In: Rhodes, D.D. and Williams, G.P. (Eds), *Adjustments of the Fluvial System*. George Allen and Unwin, London, 117–137.

Tiffany, J.B. and Nelson, G.A. 1939. Studies of meandering of model-streams. *Transactions of the American Geophysical Union*, **20**, 644–649.

Weihaupt, J.G. 1977. Morphometric definitions and classifications of oxbow lakes, Yukon River Basin, Alaska. *Water Resources Research*, **13**, 195–196.

Werritty, A. and Ferguson, R.I. 1980. Pattern changes in a Scottish braided river over 130, and 200 years. In: Cullingford, R.A., Davidson, D.A. and Lewin, J. (Eds), *Timescales in Geomorphology*. Wiley, Chichester, 53–68.

Wolman, M.G. and Gerson, R. 1978. Relative scales of time and effectiveness of climate in watershed geomorphology. *Earth Surface Processes and Landforms*, **3**, 189–208.

Yarnykh, N.A. 1978. Channel and bank deformations in an acute meander. *Soviet Hydrology, Selected Papers*, **17**, 278–281.

10 Prediction of Morphological Changes in Unstable Channels

KEITH S. RICHARDS AND STUART N. LANE
Department of Geography, University of Cambridge, UK

10.1 INTRODUCTION

There has been considerable emphasis in fluvial geomorphology on identifying and defining the geometric properties in planform, cross-section and longitudinal profile of equilibrium alluvial channels, and their mutual adjustments to prevailing discharge and sediment transport regimes. Much of this has developed from classic work on hydraulic geometry by Leopold and Maddock (1953), which explicitly acknowledged a debt to the river engineering tradition of regime theory (see Chapter 8). A regime channel was defined by Blench (1969, p.2) as one adjusted to:

> ...average breadths, depths, slopes and meander sizes that depend on (1) the sequence of water discharges imposed on them, (2) the sequence of sediment discharges acquired by them from catchment erosion, erosion of their own boundaries, or other sources, and (3) the liability of their cohesive banks to erosion or deposition.

One characteristic of the geomorphological approach to river management is that it adopts a holistic view of the catchment and river system, and interprets channel equilibrium and channel change in relation to basin processes (Gardiner, 1991). For example, the average dimensions and shape of the channel in a given reach are dependent on the prevailing discharge regime, the sediment yield imposed by the hydrological and denudational processes in the upstream catchment, and the sedimentology of the immediate valley fill environment. However, Chapter 9 shows that there can be continual dynamic fluctuating behaviour about the steady average state that reflects the 'regime' condition, and in the short term this may obscure any tendency towards adjustment in the medium term.

In general, the regime approaches outlined in Chapter 8 can be used to predict the new equilibrium channel form to be expected given environmental changes within the drainage basin. Such predictions only indicate the eventual equilibrium state after complete adjustment to a specified and finite set of environmental changes (altered discharge and sediment regimes, for example). Disturbance to a river reach as a result of changes in environmental controls usually involves continual dynamic adjustment processes that are reflected in a series of transient states. The nature of these transient channel patterns and geometries depends on the type of disturbance; Figure 10.1 suggests that, in addition to progressive changes, stepped and impulse influences and associated responses (often

Applied Fluvial Geomorphology for River Engineering and Management, Edited by C. R. Thorne, R. D. Hey and M. D. Newson.

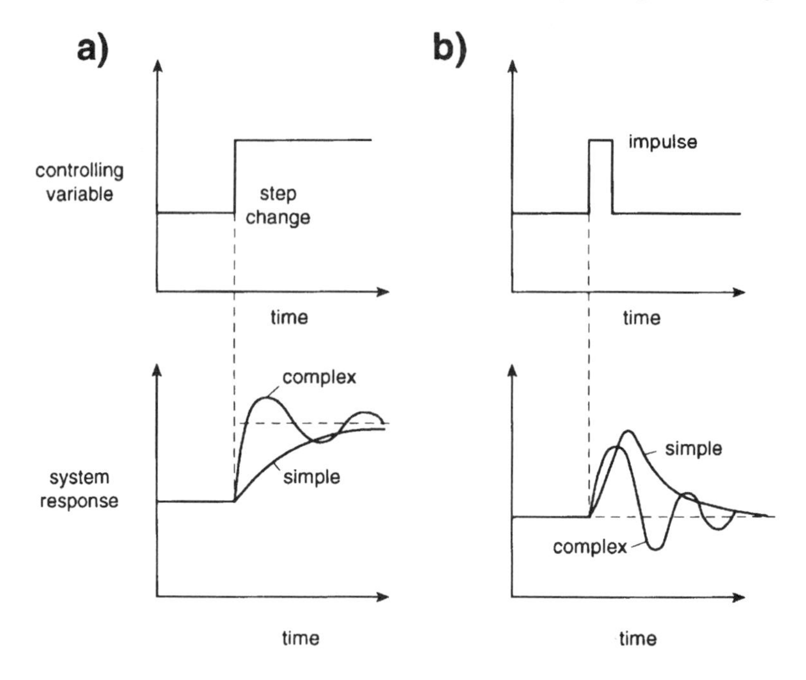

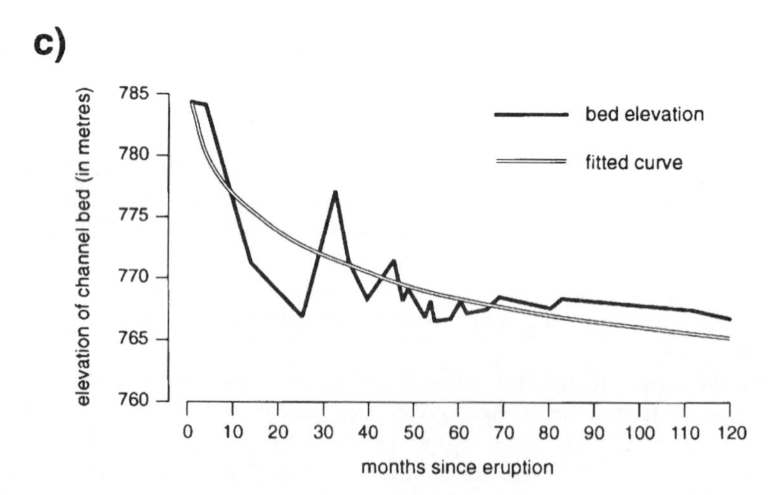

Figure 10.1 Types of externally imposed change, and responses: (a) shows possible responses to a step-function change, (b) to a single impulse, and (c) illustrates an example of a river channel response to the impulse of massive material deposition as a result of the eruption of Mount St Helens, November 1980 (after Simon, 1992. By permission of Elsevier Science – NL)

complex) may occur. The timescales of these adjustments reflect the 'relaxation time' of the river reach in question. Furthermore, these transient states involve process–form feedbacks and linkages through which the change in river channel morphology conditions the response to future environmental change.

The purpose of this chapter is to consider the nature and prediction of such dynamic behaviour, which becomes a significant issue in river management if the transient states persist for significant periods of time, or if the complex oscillatory responses to stepped or impulse changes result in potentially damaging overshoots beyond the likely final equilibrium state. In addition, the problem of indeterminacy in the prediction of equilibrium channel geometry is considered.

10.2 CHANNEL CHANGES: NATURAL AND IMPOSED ADJUSTMENT OF REGIME

River responses to environmental change have been summarised in qualitative terms by Schumm (1969). This is shown in Table 9.2, which lists the adjustments anticipated for specified directions of change in discharge and sediment load, and distinguishes between dependent channel properties whose response is relatively predictable and those whose response is indeterminate. This model was quantified empirically by Rango (1970), in the form of a series of multiple regression equations designed to predict stream response to increased discharges resulting from cloud seeding. However, no attempt was made to simulate the impact of simultaneous changes in discharge and sediment load. Consequently, such simple models suffer from a failure to incorporate mutual adjustment and interdependence amongst variables. More recently, Miller (1984) has suggested that a simultaneous equation model can be used to predict different adjustments under constrained and unconstrained conditions (Box 10.1). However, even this approach is limited by its reliance on empirical data from channels known to be in regime (to define the component relationships), and by the necessary restriction of its application to channels similar to those in the initial data set. A more general approach to the modelling of transient changes from a particular initial condition requires a methodology directly based on the governing process laws, as outlined in the following sections.

10.3 ONE-DIMENSIONAL FEEDBACK: THE DOWNSTREAM LONG PROFILE

The dynamics of channel change, both autogenic (within-equilibrium) and allogenic (disequilibrium), arise because of the feedback that occurs amongst the channel shape, the flow pattern, and the processes of sediment transport. Hey (1979) described this qualitatively for the one-dimensional (long-profile) case, arguing that an external stimulus such as a sudden lowering of the local base level (for example, caused by channelisation) would be transmitted upstream, and that each location along the affected reach would experience alternate phases of bed degradation and aggradation of decreasing amplitude. Degradation reflects the drawdown, flow acceleration and increase in bed material transport occurring at the local increase in bed slope at the knickpoint initiated by the bed lowering. Aggradation reflects a backwater influence caused when the products of this

Box 10.1 A simultaneous equation model of channel adjustment

The Problem

A river reach experiences flooding, and dredging is considered in order to increase the channel capacity. However, this change in river depth is not entirely independent of other variables that will control the channel capacity and stability (such as slope and width).

Conceptualisation

These interrelationships can be considered in terms of the following multivariate logarithmic relationships:

$$\log W = \log B_{10}\, B_{11} \log D\, B_{12} \log S\, B_{13} \log Q$$
$$\log D = \log B_{20}\, B_{21} \log W\, B_{22} \log S\, B_{23} \log Q$$

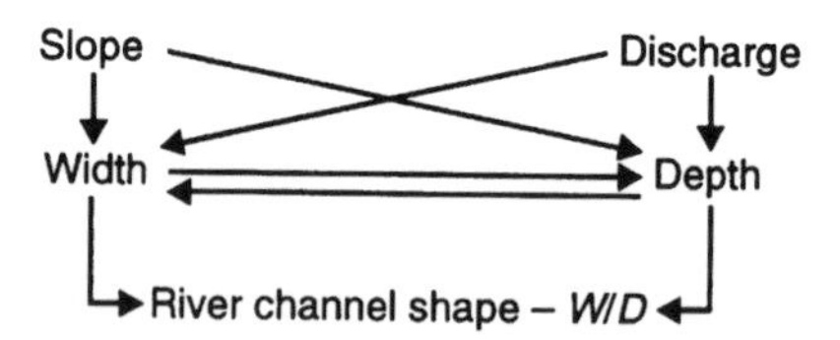

Figure 1 A simultaneous equation model of channel adjustment

The existence of mutual adjustment between variables precludes direct solution of these equations, but with simultaneous solution for $\log W$ and $\log D$ they reduce to:

$$\log W = n_{10} + n_{11} \log S + n_{12} \log Q$$
$$\log D = n_{20} + n_{21} \log S + n_{22} \log Q$$

where the n_{ab} are functions of the values B_{ij}.

Unfortunately there are too many unknowns in the full equations to obtain the values of B_{ij} through solution of the reduced equations for $\log W$ and $\log D$ but this can be overcome through the use of indirect least squares regression. With calculation of the values of B_{ij} and n_{ab}, two sets of equations result, the reduced equations measuring the direct effects of the external variables (in this case slope and discharge) on the internal dependent variables, width and depth.

An Example

Miller (1984) applied this model to an extensive data set and found that the discharge coefficient (n_{22}) in the reduced depth equation was 0.14, although when using the full depth equation, B_{23} was 0.66. The fact that these two coefficients were different suggests that there was a significant indirect effect of discharge on channel depth through the mutual adjustment mechanism of width and depth.

Application

The utility of the mutual adjustment analysis is that it can be used in two different scenarios in which channel dredging occurs. The reduced form should be used when the river is permitted to behave in an unconstrained manner (i.e. with no bank protection) and when, as a result, the river may also adjust its width in response to dredging. The full equation for depth may be used if the potential for width change can also be constrained (through existing or potential bank protection measures). This form of analysis therefore allows a choice of the most appropriate form of solution for a given type of channel management scenario.

erosion are deposited on the lower gradient reach below the retreating knickpoint. This model was quantified by Pickup (1975, 1977), adopting the structure illustrated in Box 10.2. Here, the feedback loops are explicitly defined. The relationship between the bed profile and the (one-dimensional) flow pattern is represented by the gradually varied flow equation:

$$\frac{\mathrm{d}y}{\mathrm{d}x} = \frac{s_o - s_f}{1 - (aQ^2 w/gA^3)} \tag{10.1}$$

where the terms are defined in the Appendix to this chapter. The relationship between the local flow and the bedload transport rate is represented by the Meyer-Peter and Muller (1948) bedload transport equation:

$$\gamma_w(Q_s/Q)(k_b/k_r)^{3/2} R s_f = 0.047(\gamma_s - \gamma_w)\mathrm{d}_m + 0.25(\gamma_w/g)^{1/3} G^{2/3} \tag{10.2}$$

Finally, the relationship between local bedload transport and scour and fill of the channel bed is represented by the sediment continuity equation:

$$\frac{\mathrm{d}z}{\mathrm{d}t} = \frac{1}{\gamma_s}\frac{\mathrm{d}G}{\mathrm{d}x} + i \tag{10.3}$$

Thus, each step in the feedback can be simulated numerically using finite-difference approximations of these physical relationships, and the channel dynamics can be predicted as a recursive process. Both the instantaneous flow and sediment transport variation along the reach, and the time-dependent upstream migration of the erosion wave can be simulated, as shown in Figure 10.2a and b. The complexities of such models can be increased in order to include the effects of gravel armour development on the bed following the selective entrainment and removal of the finer bed material; this is a common process downstream from dams which limits the tendency of a river to post-impoundment bed degradation.

One limitation of a model such as Pickup's is that it assumes that the morphological adjustments occur only in terms of long-profile change; for example, it implies that the channel banks remain stable during the process of bed degradation. However, it is likely that vertical instability caused by bed lowering will trigger bank erosion and destabilisation, and therefore generate channel widening (see Chapter 6). In this case, bank instability is commonly a result of the increase of bank height above the limiting height for stability against wedge (Culmann) failure. The result is that the sediment continuity equation (Eqn 10.3) must be evaluated in the context of a positive lateral sediment input term, *i*. This may be difficult to evaluate because a significant fraction of the volume of sediment generated

Box 10.2 A one-dimensional model for bed adjustment

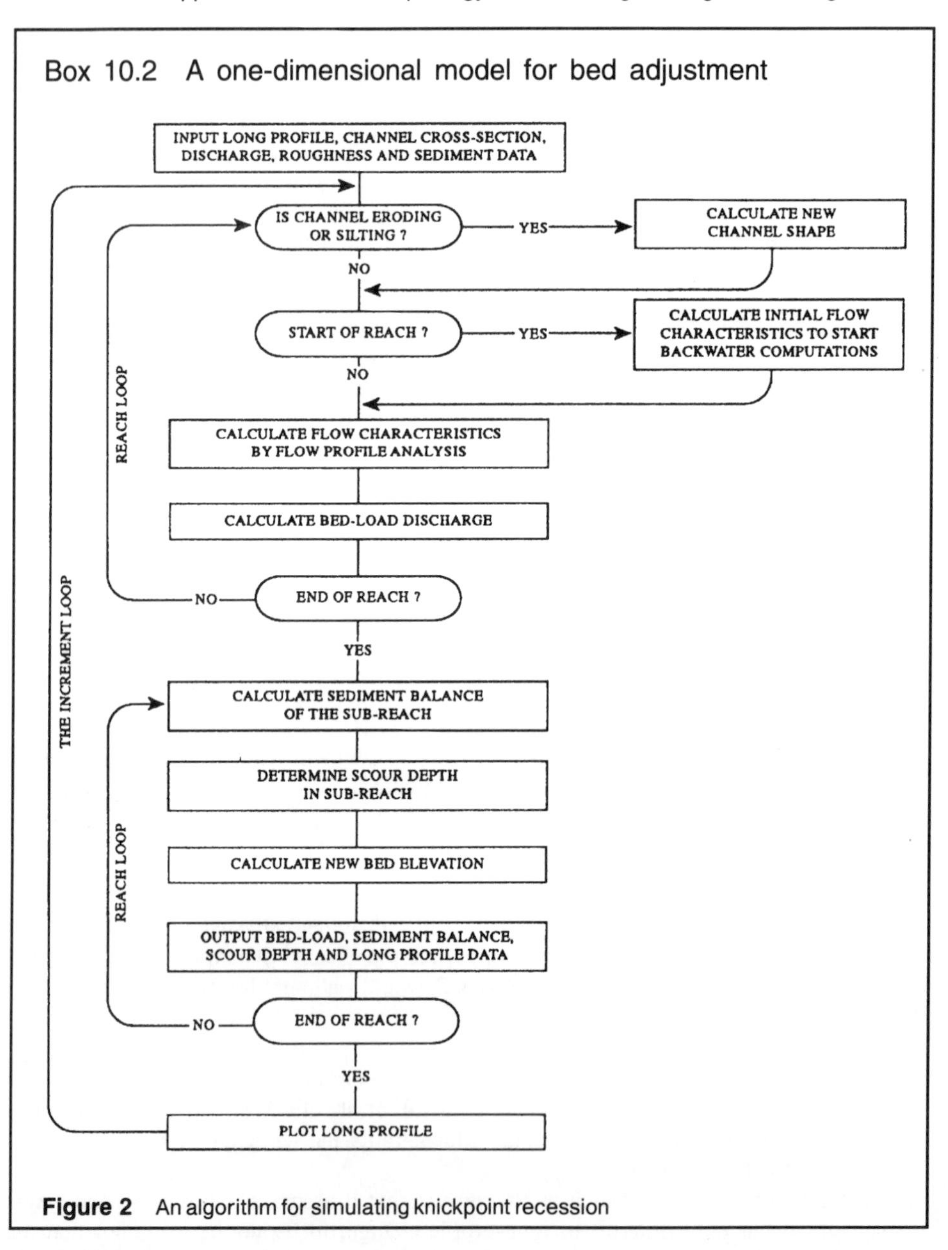

Figure 2 An algorithm for simulating knickpoint recession

by bank failure may be fine-grained (silt and clay) material that is transported as wash load, rather than forming an additional contribution to the bed material load. When an additional bed material load supply is produced by bank erosion during bed degradation, the problem becomes two-dimensional, and after knickpoint recession the equilibrium gradient may be steeper than the initial gradient because of the adjustment of channel width that accompanies the incision. Successful modelling of both the transient and equilibrium

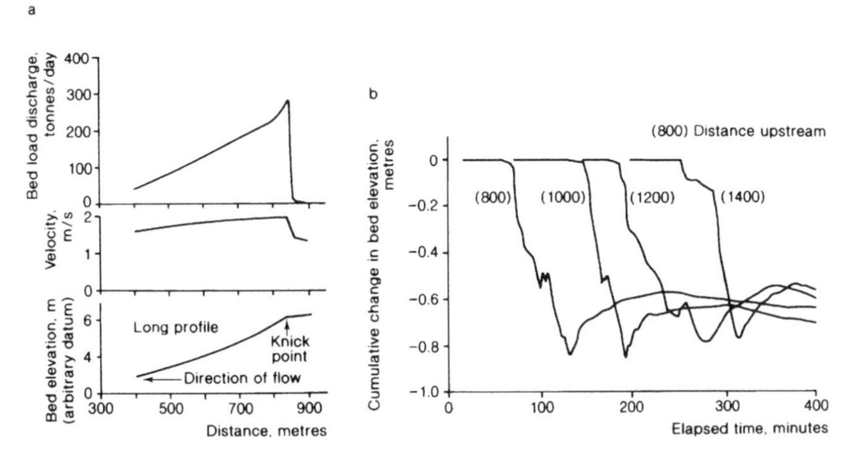

Figure 10.2 (a) Downstream variations in flow velocity and bedload transport rate over a knick-point, for inflow discharge of 2 $m^3 s^{-1}$ (after Pickup, 1975. By permission of Gebrüde Borntraeger). (b) Cumulative simulated change in bed elevation with time at four locations along a reach (distance upstream, in parentheses), during the upstream passage of a knickpoint (after Pickup, 1977. By permission of John Wiley & Sons Ltd)

channel responses to disturbance therefore await physically realistic models of the bank erosion and width adjustment processes, and, particularly in the transient case, these will have to simulate the porewater pressure dynamics within the bank in order that the timing of bank failure and sediment supply is predicted at the correct stage during a hydrograph. Darby and Thorne (1996a) illustrate one such modelling approach based on simplified bank and phreatic surface geometries. However, bank failures may occur at any point during a hydrograph depending on: (i) wetting front dynamics in the partially saturated bank during the rising stage; (ii) the extent of basal scour and undercutting during peak stage; and (iii) the rate of drawdown on the falling stage. This dynamic behaviour is difficult to capture in simple models.

Two-dimensional effects may be parameterised in one-dimensional models in order to resolve some of the limitations of one-dimensional analysis. For example, Begin et al. (1981) have employed a model of knickpoint recession which incorporates the effects of bank erosion. In this model, the sediment continuity equation is modified by an approximation of the river bank contribution to sediment supply, using:

$$i = B_0 e^{-bx} \tag{10.4}$$

These authors demonstrated that where a channel entrenches into homogeneous, isotropic material following an instantaneous lowering of base level, it is possible to assume that the sediment discharge is proportional to bed slope, and to show that the additional supply of sediment as a result of bank erosion reduces the degree of bed lowering, and results in a steeper final gradient, both of which are consistent with the results of flume experiments. Alternative numerical approaches to the modelling of bank instability following bed degradation are provided by Osman and Thorne (1988) and Darby and Thorne (1996b).

An understanding of the transient and equilibrium responses to bed degradation is clearly relevant to any river management which involves the range of channelisation options liable to increase stream power per unit bed area (or per unit depth), and thereby increase bed material transport capacity (e.g. deepening by dredging, or channel straightening). When the increase in power is sufficient to mobilise the bed material, the bed degradation will be transmitted upstream, resulting in channel maintenance problems beyond the reach originally targeted (Parker and Andres, 1976; see also Chapter 11). Brookes (1991) illustrates one case where straightening a small stream in Britain more than doubled the stream power from 35 $\mathrm{W\,m^{-2}}$ to 80 $\mathrm{W\,m^{-2}}$, necessitating further channel maintenance and construction of drop structures and flow deflectors to induce energy losses. The environmental context of channel management determines the potential for upstream and downstream effects, and rivers with sufficiently high stream powers exhibit an inevitable tendency to recover their natural condition after channelisation. Leeks et al. (1988) reported that straightening and deepening the Afon Trannon during the nineteenth century caused subsequent engineering problems which reflected the high rates of bed material movement in the steepened stream, availability of mobile sediments, and heterogeneous floodplain materials which prevented identification of a stable route for the relocated channel. Similarly, Brookes (1991) describes a continual increase of sinuosity in a reach of the Upper Severn over 125 years after straightening in the 1850s, as this dynamic river recovered its meandering state. These examples serve to underline the limitations of piecemeal, local channelisation schemes; rivers tend to restore natural characteristics, which an appraisal of their Holocene behaviour would reveal, and in doing so cause upstream and downstream channel stability and operational maintenance problems.

10.4 TWO-DIMENSIONAL FEEDBACK: PLAN TOPOGRAPHY

Quantitative modelling of the dynamic, transient response of channel planform to disturbance is difficult to achieve, as the bed topography also varies as the planform changes. However, the effect of the bed topography on the flow field is critical. Topographic skewing of the main flow in curved channels, both single-thread (Dietrich, 1987) and multithread (Whiting and Dietrich, 1991), is known to generate significant distortions to the velocity and shear stress fields. Point and medial bars act to force the flow laterally, resulting in strong topographically-induced lateral pressure gradients and convective accelerations, and a simple representation of the bed shear stress as the product of the unit weight of water, hydraulic radius and longitudinal energy slope, is therefore unable to control the spatial pattern of sediment transport correctly in a model designed to relate the bed topography to the flow and sediment transport fields. The terms in the downstream and cross-stream force balances that are due to the cross-stream bed slope and flow curvature effect are large and must be included for accurate estimation of the local bed shear stress field, or predictions of the pattern in the bedload transport rate dependent on the shear stress will be erroneous.

However, the basic conceptual model structure employed by Pickup (1975, 1977) and discussed in relation to the one-dimensional case can also be applied to the plan adjustment of a river reach. Box 10.3 suggests a qualitative feedback model applicable to coarse-grained river beds, in which the topographic changes of the bed are primarily controlled by bedload movement, and suspended bed material load is of relatively minor significance for the topography. In this conceptual model each ‘box’ represents not a single value, but a map of the attribute throughout the reach in question. This implies that as the

Box 10.3 Qualitative model of form–process interrelationships

The diagram below, based on Ashworth and Ferguson (1986), describes the feedback loops and process–response mechanisms associated with dynamic alluvial channels.

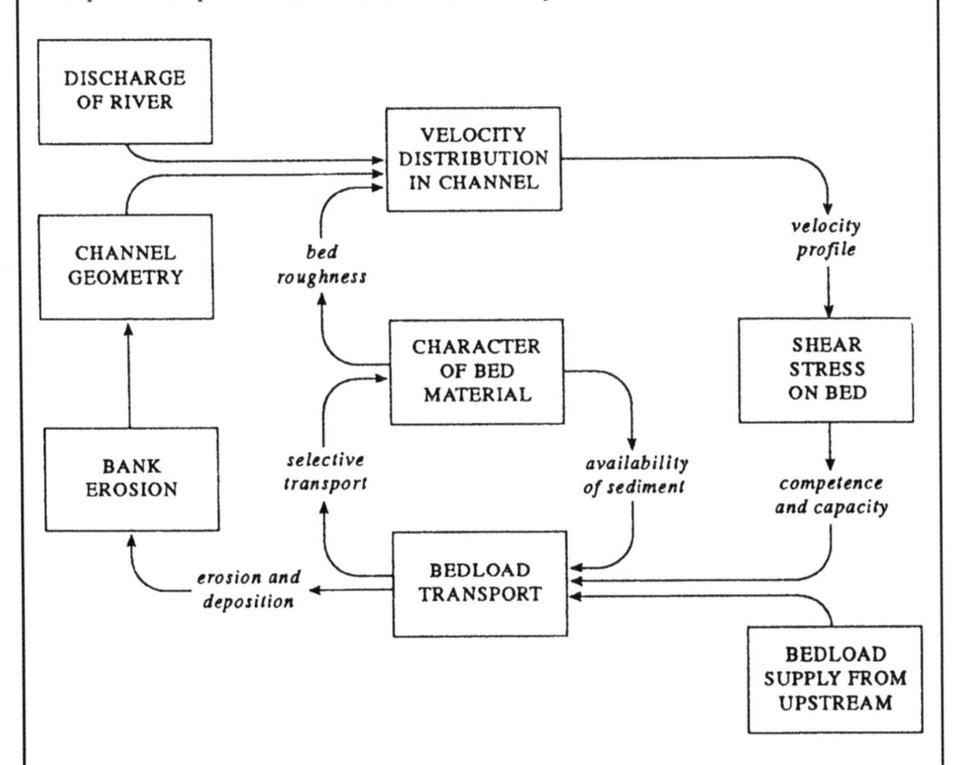

Figure 3 A feedback model of channel change: initial form, flow pattern, transport pattern and sediment mass conservation determine patterns of channel bed change

Entering the diagram at the top left, unsteady discharge through a river channel with a given topographic form and a given channel perimeter sedimentology, produces spatially variable velocity patterns and bed shear stresses both on the river bed and within each vertical profile. Interaction of these flow characteristics with sediment availability at the bed and the supply of sediment from upstream controls the sediment transport rate which, through erosion and deposition, controls bank erosion leading to a new channel geometry. Bedload transport itself feeds back to alter the character of the bed material, through the effects of selective transport. The changes in topography and bed roughness combine with the discharge (potentially different) to determine the flow and sediment transport conditions in the next time period.

Such a description provides a useful conceptual framework for viewing likely river channel response to various management activities. In a quantified form, it has the potential to allow the prediction of the dynamic adjustment of a river channel (across space and through time) to engineering solutions.

flow and bedload enter at the head of the reach, they are distributed across the existing bed topography and redistributed according to the pattern of flow vectors induced by that topography. The spatial pattern of bedload transport within the reach results in a pattern of local erosion and deposition in accord with the conservation of bedload mass represented by the continuity equation (Ashworth and Ferguson, 1986; Richards, 1988). As with Pickup's model, this has to be evaluated as a recursive process in which the bed topography and the distribution of sediment at the end of one time period define the initial conditions at the beginning of the next. The bed topography and the pattern of sediment sizes adjust to every new water and sediment input. When bed erosion occurs adjacent to a river bank, the critical height of the bank in stability terms may be exceeded, the bank may retreat and the channel either widens or migrates laterally (cf. Chapter 6), although this may be lagged in time depending on the rate of basal scour. This widening may be compensated by bank accretion elsewhere in a reach (e.g. Bridge, 1993) and this maintains a 'regime' width adjusted to the average throughput of sediment (among other things). However, this conceptual model also accommodates systematic adjustment of the reach when the flow and/or sediment supply to the reach alters. One implication of the model is that the reach character can no longer be simply represented by a set of nine variables (the degrees of freedom of Chapter 8), but has to be described in terms of a detailed topography and spatial distribution of bed material grain size. Although this is clearly very demanding of field data, Lane et al. (1995) have demonstrated that such a modelling approach is feasible for small gravel-bed rivers with complex geometries.

In principle, therefore, a numerical model of adjustment in a gravel-bed river may be formulated. As a first approximation, this involves a finite difference or finite element solution of the two-dimensional depth-averaged Navier–Stokes equations for the conservation of fluid momentum and mass given respectively as follows.

The x-component momentum equation:

$$\begin{aligned}\frac{\partial \overline{U}}{\partial t}+\overline{U}\frac{\partial \overline{U}}{\partial x}+\overline{V}\frac{\partial \overline{U}}{\partial y} &= -g\frac{\partial}{\partial x}(z_o+h)+\frac{1}{\rho h}\frac{\partial(h\overline{\tau_{xx}})}{\partial x}+\frac{1}{\rho h}\frac{\partial(h\overline{\tau_{xy}})}{\partial y}-\frac{\tau_{bx}}{\rho h}\\ &\quad+\frac{1}{\rho h}\frac{\partial}{\partial x}\int_{z_o}^{z_o+h}\rho(\overline{u}-\overline{U})^2\partial z\\ &\quad+\frac{1}{\rho h}\frac{\partial}{\partial y}\int_{z_o}^{z_o+h}\rho(\overline{u}-\overline{U})(\overline{v}-\overline{V})\partial z\end{aligned} \tag{10.5}$$

The y-component momentum equation:

$$\begin{aligned}\frac{\partial \overline{V}}{\partial t}+\overline{U}\frac{\partial \overline{v}}{\partial x}+\overline{V}\frac{\partial \overline{v}}{\partial y} &= -g\frac{\partial}{\partial y}(z_o+h)+\frac{1}{\rho h}\frac{\partial(h\overline{\tau_{xy}})}{\partial x}+\frac{1}{\rho h}\frac{\partial(h\overline{\tau_{yy}})}{\partial y}-\frac{\tau_{by}}{\rho h}\\ &\quad+\frac{1}{\rho h}\frac{\partial}{\partial x}\int_{z_o}^{z_o+h}\rho(\overline{u}-\overline{U})(\overline{v}-\overline{V})\partial z\\ &\quad+\frac{1}{\rho h}\frac{\partial}{\partial y}\int_{z_o}^{z_o+h}\rho(v-\overline{V})^2\partial z\end{aligned} \tag{10.6}$$

where

$$\overline{\tau_{ij}} = \text{turbulent shear stresses} = \upsilon_{ij}\left(\frac{\delta V_j}{\delta x_i} + \frac{\delta V_i}{\delta x_j}\right)$$

$$\overline{\tau_{bi}} = \text{bottom shear stresses} = C_f \rho \overline{V_i}\left(\overline{V_i^2} + \overline{V_j^2}\right)^{1/2}$$

The conservation of mass equation:

$$\frac{\partial h}{\partial t} + \frac{\partial}{\partial x}(h\overline{U}) + \frac{\partial}{\partial y}(h\overline{V}) = 0 \tag{10.7}$$

The eddy viscosity (υ_{ij}) may be specified or, preferably, modelled using transport and dispersion equations. A critical concern in the application of depth-averaged flow models is the degree to which the model adequately represents the process of turbulence (Clifford and French, 1993) and its effect on the turbulent shear stresses (τ_{ij}) and therefore on the mean flow properties. Box 10.4 indicates the increasing sophistication of alternative turbulence closure models used in flow simulations. These may result in significant differences in predicted mean velocities or bed shear stresses, although detailed assessment of the significance of the turbulence parameterisation in models of flow in natural channels is only currently in progress (the more complex two-equation models require parameters whose values have, to date, only been quantified in flume experiments). Furthermore, two-dimensional flow models involve a range of different assumptions. Some assume a rigid lid (that is, they specify the water surface), and only produce accurate estimates of flow vectors when the local Froude number is less than about 0.7; however, free surface models are now being successfully employed to study flood routing and floodplain inundation (Anderson and Bates, 1994). Equations 10.5 and 10.6 also contain correlations that represent the effects of secondary flow on the primary flow calculations (the $(\bar{V}_i - \bar{v}_i)(\bar{V}_j - \bar{v}_j)$ terms), and their effects on primary flow velocities must be modelled numerically as the spiralling effect associated with such channels can be a key element driving the flow. Such three-dimensional effects can be parameterised in two-dimensional models (e.g. Bernard et al., 1992), although in some cases it may be necessary to acquire detailed field data to confirm the existence of secondary flows in the channels to be simulated. Exploration of obviously three-dimensional flow structures such as those occurring at confluences requires the use of fully three-dimensional models, and the complex eddy patterns that occur in the shear zones between confluent flows are difficult to capture in time-averaged simulations; large eddy simulation methods of the kind employed in oceanographic and atmospheric modelling are required for this purpose.

However, the two-dimensional models have been shown to be capable of simulating the pattern of depth-averaged flow vectors even within a complex braided reach (Figure 10.3; Lane et al., 1995). Successful modelling of this kind demands rigorous definition of the boundary conditions, in terms of a detailed map of bed topography and spatially distributed roughness, and a suitable upstream or downstream boundary condition in the form of a depth–discharge relation which provides velocity information distributed across the section. The depth–discharge relation may be simplified in the case of a steady-state situation, although this will require specification of a water surface. In Figure 10.3, the water surface was fixed from information obtained from oblique (terrestrial) analytical photogrammetry. Model assessment through distributed sensitivity analysis (Lane et al., 1994) and distributed comparison of observed flow velocities with model predictions have

Box 10.4 Two-dimensional depth-averaged flow models: different approaches to the turbulence closure problem

Representation of Turbulence

In theory, to simulate a turbulent river flow it is only necessary to solve the three-dimensional Navier–Stokes equations, with appropriate boundary conditions, using any one of a number of numerical procedures for three-dimensional, non-linear, coupled differential equations (Younis, 1992). The outcome of such a simulation will be a complete description of the time–space variation in the flow. However, this is impractical as to solve the equations with sufficient accuracy a computational grid is required, with spacings smaller than the smallest eddy but large enough to cover the field of interest. The number of nodes required in such a mesh depends on the Reynolds number (Younis, 1992):

$$N = R_e^{2.75}$$

At the scale of turbulence production within a natural river channel, the relevant boundary conditions (the dimensions and distributions of all water particles) cannot be defined. The effects of turbulence on the mean properties of flow are therefore assessed using Reynolds averaging (Reynolds, 1895; Clifford and French, 1993). The instantaneous variables are decomposed into mean and fluctuating components, and these are introduced into the Navier–Stokes equations and then averaged. This introduces unknown turbulence correlations, the Reynolds stresses, which represent the rate at which momentum is transported by turbulent fluctuations. The problem with this method is that there are now more unknowns than equations to solve them – the problem of closure.

Modelling the Effects of Turbulence on Mean Flow Parameter

The solution to this problem is to model the consequences of Reynolds stresses using flow parameters that are either known or knowable, and this can be done with varying degrees of sophistication. The simplest method is based upon the Boussinesq (1877) eddy viscosity concept where, by analogy with laminar flows, the Reynolds stresses are modelled as proportional to the velocity gradient. The proportionality constant defines eddy (or turbulence) viscosity. Unlike molecular viscosity, the eddy viscosity is not a fixed function of the fluid but varies spatially within the flow. The eddy viscosity can be represented as follows.

(i) *Zero order model* – the eddy viscosity is either prescribed (as a constant value or as a function of flow parameters) or modelled as being proportional to a prescribed length scale (the Prandtl (1952) mixing length hypothesis).
(ii) *One equation model* – the eddy viscosity is thought to be proportional to both the length scale and a velocity scale (Prandtl, 1952), the latter being related to the kinetic energy of the turbulent motion. This is obtained from solution of a differential transport equation and extends the zero-order approach through the incorporation of some 'history' effect, but still prescribes the length scale rather than attempting to model it.
(iii) *Two equation model* – this turbulence model attempts to represent the length and velocity scales using both the kinetic energy of turbulence motion and the dissipation of turbulence energy due to viscous action. Although parameterising this model is different, it is widely used in commercial computational fluid dynamics applications (Knight, 1992; Younis, 1992).

Future Developments

Current research in modelling turbulence effects seeks to replace the Boussinesq eddy viscosity concept, and to calculate the Reynolds stresses directly from the solution of their own differential transport equations: five in the case of two-dimensional flows. As computational efficiency improves, such models are likely to become increasingly appropriate for modelling natural fluid flows, ultimately in three dimensions. However, the fundamental problem will remain of measuring the topographically complex geometry and heterogeneous nature of boundary roughness at the resolution required for modelling more sophisticated turbulence effects. Boussinesq approximations will thus be used for some time in studies of natural channels (Knight, 1992).

produced encouraging results, although further model improvement requires additional in-depth investigations of secondary flow processes and the controls of bed roughness (grain and bedform) in three dimensions.

A bedload transport equation may be coupled to this simulation of the flow and, given data on the grain size distribution of the bed material, and the pattern of depths and depth-averaged velocities predicted by the flow model, an estimate of the bed shear stress can be derived and used to predict the bedload transport rate as a function of the shear stress excess (above the threshold required to move the local bed material). Local transport

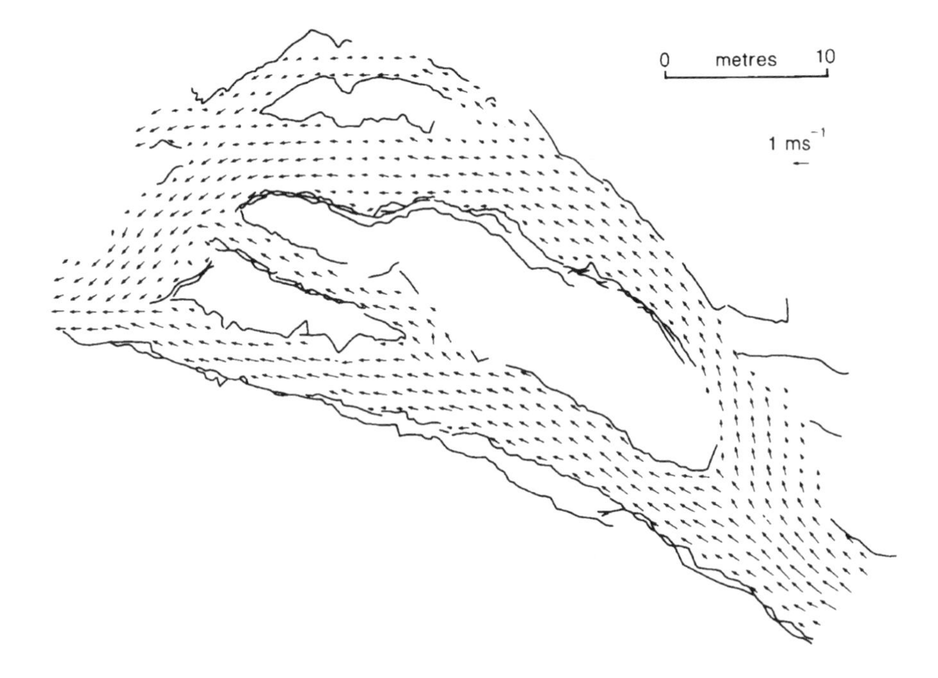

Figure 10.3 Depth-averaged flow vectors in a natural bar-head divergence in a braided reach of the Haut Glacier d'Arolla meltwater stream, Valais, Switzerland, derived using a two-dimensional, depth-averaged finite volume numerical solution of the shallow water equations, and a detailed map of the bed topography

continuity is then evaluated to assess the changes in bed elevation, and the revised bed topography becomes the initial condition for the next time step. At present it is not possible to progress beyond this first approximation in coarse-grained channels, because the threshold of motion for gravel is a complex function of the local bed structure (Richards and Clifford, 1991). However, there are several other applications for this depth-averaged flow model. For example, it provides the basis for predicting the local mean velocity in the vicinity of structures such as bridge piers, vanes and dykes, which can be compared with an estimate of the local threshold velocity for bed material motion in order to assess channel stability following construction of these features. In sand-bed streams where definition of the threshold of motion is less problematic, interaction between flow and the mobile bed has been modelled with some success. For example, van Rijn (1987) has combined a two-dimensional depth-averaged flow model, similar to that described above, with a logarithmic velocity profile to give a quasi-three-dimensional representation of the flow field (in a manner conceptually similar to the incorporation of bank erosion effects in the one-dimensional models discussed above), and has computed the associated sediment transport rates using a three-dimensional mass balance equation for the convection, diffusion and gravitational (settling) transport of suspended sediment. Figure 10.4 illustrates a prediction by this model of the pattern of bed scour around the nose of a spur projecting into the flow in a straight channel.

10.5 PREDICTING MORPHOLOGICAL CHANGE IN DYNAMIC SYSTEMS

The principles of non-linearity and indeterminacy underlie several approaches to the prediction of morphological change in geomophology, and indicate that eventual equilibrium states are likely to be unpredictable without the physical treatment outlined in the previous sections. Fluvial processes appear to be inherently non-linear, and to involve discontinuous responses to changes in the control variables. This view has been promulgated by Schumm (1973), who has argued that drainage basins demonstrate 'intrinsic' or 'geomorphic' thresholds (in addition to the familiar 'extrinsic' thresholds, such as the threshold of bed material motion exceeded when critical flow shear stresses are attained). Typically, an intrinsic threshold arises when sediment storage has continued for a sufficient period that the gradient of the aggrading surface becomes steep enough to trigger renewed erosion. The familiar threshold slope–discharge criterion discriminating meandering from braided rivers is an example of an intrinsic threshold (see Chapter 9). Given that this criterion is well known, it is perhaps surprising that little research has focused on the possibility that 'catastrophe theory' might describe fluvial system behaviour.

Graf (1988) outlines the basic theory of the 'cusp catastrophe'. If a simple physical system is defined by two control variables (a and b, which are likely to represent a 'force' and a 'resistance' variable) and one response variable (x), then these are related by an energy function $E(a, b, x)$, and for each (a, b) pair, a value of x minimises E. The system is at equilibrium when $\mathrm{d}E/\mathrm{d}x = 0$, and a map of the values of (a, b, x) which satisfy this criterion defines a three-dimensional, equilibrium surface. In some cases, the surface has a wrinkle, or fold, which means that there are certain (a, b) combinations where x may adopt more than one value which each locally satisfy the equilibrium criterion. Figure 10.5 shows the familiar channel pattern threshold interpreted in this way, with the control

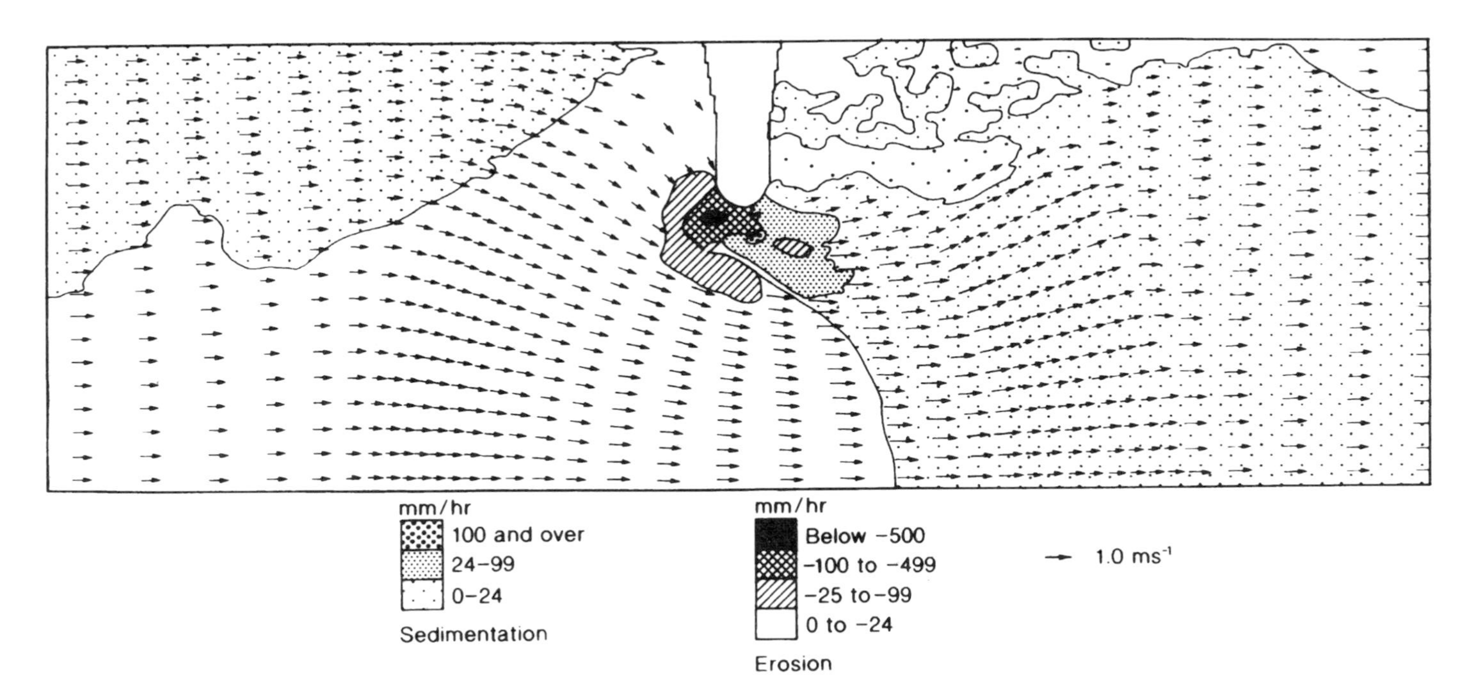

Figure 10.4 The pattern of sediment transport vectors and bed erosion and deposition simulated around a spur in a straight reach (after van Rijn, 1987. By permission of the author)

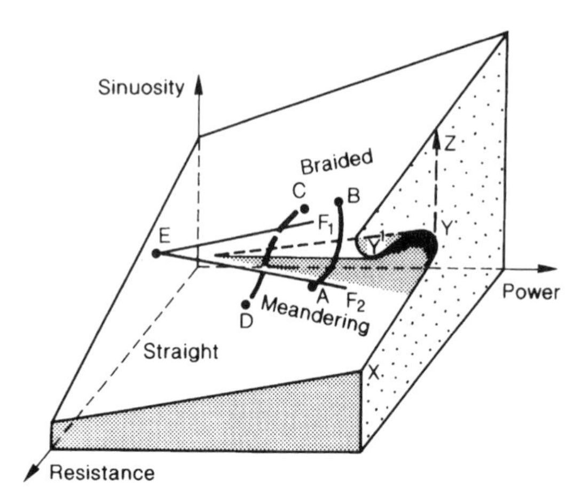

Figure 10.5 A qualitative model of non-linear change in channel pattern equilibrium, based on the cusp catastrophe

variables being stream power and bank erodibility, and the response variable being a measure of channel sinuosity defined as the total channel length per unit reach length. This provides a qualitative model which accounts for hysteretic behaviour of channel pattern change, in which the response of channel pattern to changes in the control variables differs according to whether the control variables are increasing or decreasing over time. For example, extreme flood events may rapidly convert a meandering stream to a braided form (the route A to B in the diagram), but the recovery is delayed by the gradual restoration of bank stability as fine sediments accumulate (and bank vegetation re-establishes). This is route C to D on the diagram. In addition, the cusp catastrophe indicates that quite divergent responses may be made by two similar systems undergoing change (routes E to F_1 and F_2). Graf (1988) suggests ways of fitting the cusp model to empirical data to quantify the equilibrium surface for a particular fluvial system, and applies these principles to studies of differential sedimentation at river junctions, and erosion of valley floor gullies.

The suggestion that multiple modes of channel adjustment may occur in response to disturbance has also been made by Phillips (1991), as an alternative to the deterministic 'degrees of freedom' model (Chapter 8), and in recognition that few of the physical laws required for a unique solution of the channel geometry are known. Predictable mutual adjustment of the nine degrees of freedom is also precluded by the different relative rates of change of the hydraulic variables. This implies that the equilibrium surface is multivalued, with several local energy minima (defined as $E(w, d, v, s, \ldots, Q, d_m, \ldots)$), any one of which could represent an attractor for the changing channel morphology. A similar conclusion is reached formally and quantitatively by Chang (1988), employing the hypothesis of minimum stream power per unit length in order to compensate for the lack of physical laws (for example, resistance and sediment transport laws) needed to account for all of the adjacent dependent variables. Beginning with the independent variables Q, G and d_m and a straight channel, Chang assumes a series of trapezoidal channel widths, and for each calculates the depth (from a friction equation) and the bed slope (from a bedload transport equation). The slopes vary non-linearly with width, and the stable channel geometry is that at which the slope (and therefore the power per unit length) is a minimum. This reflects the operation of counteracting factors: for larger widths a steeper slope is required to counter the reducing depth and maintain sediment transport, while for small widths the sloping banks become a larger proportion of the total width and a steeper bed slope is again required to counter the reduced downstream transport in the shallow flow over the banks. However, it is clear from this example that different stable geometries (minimum bed slopes) will occur as bank slopes vary in response to changes of bank material and, in addition, the freedom to develop meander bends of different curvature results in variable rates of bedload transport for a given bed slope. Thus, in the case of mutual multivariate adjustment, there are still several channel configurations which satisfy local minima in the energy criterion.

10.6 MORPHOLOGICAL CHANGE IN THE DRAINAGE BASIN

Sections 10.3 and 10.4 have described one- and two-dimensional feedback models of the dynamics of channel response to disturbance. In certain circumstances this feedback can extend throughout a channel network to create transient adjustments in the entire network – in effect, a large-scale, three-dimensional feedback, which is likely to be quantitatively indeterminate as mutual and interactive adjustment occurs of several channel properties. The now classic (or infamous) example of the Willow Drainage Ditch (Daniels, 1960) demonstrates that channelisation without drop structures (Brookes, 1991) can cause such a drainage network response when dredging and channel deepening initiate knickpoint recession which is transmitted headwards, and ultimately causes gully erosion in what were stable headwater hollows (Figure 10.6). During this erosional adjustment to the step-change lowering of base level, a 'complex response' ensues (Schumm, 1973). The bed degradation which begins in the main stream results in increased sediment yield, until the knickpoint recedes into the first tributary valleys, whose erosional products then fill the main channel (in extreme cases, resulting in overlapping sedimentary deposits). When a similar phase of aggradation occurs in these tributaries, the main valley experiences bed degradation again. The result is a damped oscillatory response of sediment yield to the single initial stimulus provided by the channelisation, a process similar to that described in Section 10.3, but not so readily modelled quantitatively at this scale. The example illustrates, however, first that river management must recognise the possibility of delayed reaction to an intervention or disturbance, and second that a river system may 'overshoot' in adjusting towards an anticipated new equilibrium channel form. When an initial period of adjustment appears complete, this may only be the first phase in an oscillatory adjustment; later periods of destabilisation may also occur, and the sediment production associated with the immediate post-disturbance period may not represent the total sediment production. Thus, lowering the level of a reservoir, for example, can result in a damped oscillatory response of sediment yield from the inflowing streams over several years (Born and Ritter, 1970), and the loss of storage capacity in the initial years will not then be the total loss attributable to the destabilisation.

Quantitative modelling of this scale of dynamic, transient response to disturbance is difficult to achieve. The two-dimensional, reach-scale models discussed in Section 10.4 impose extreme demands on the accurate description of boundary conditions, and to achieve this at the drainage basin scale is clearly even more difficult. As a result,

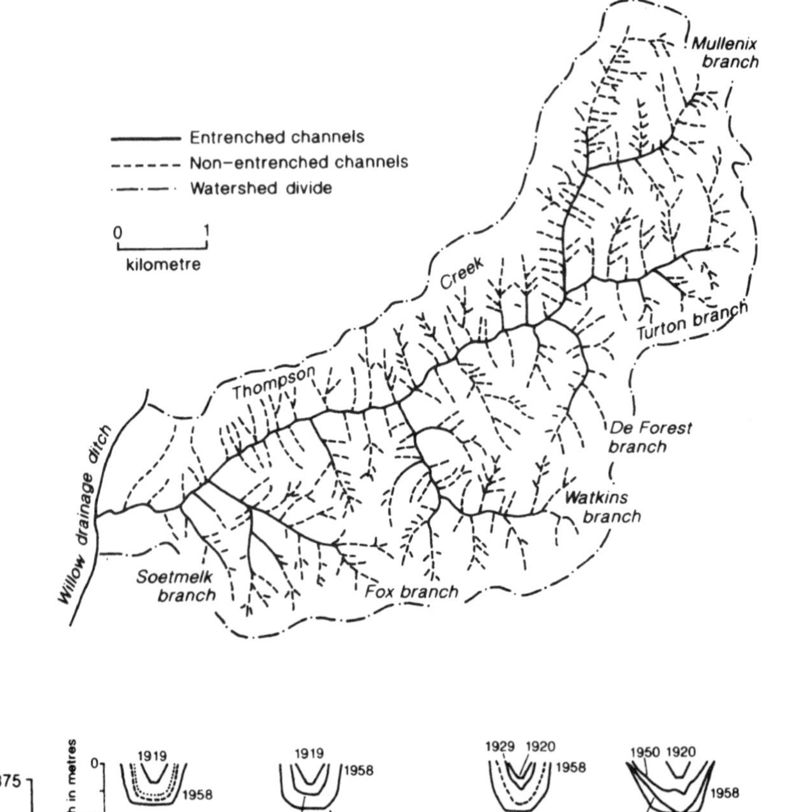

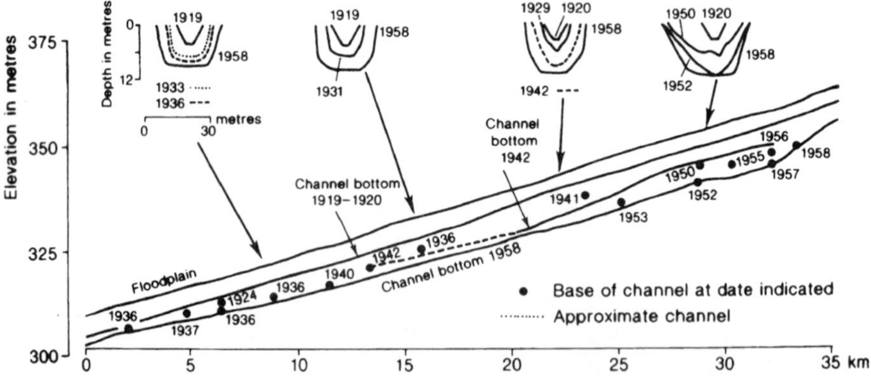

Figure 10.6 Gully formation in the Willow Drainage Ditch in response to channelisation and bed lowering (after Daniels, 1960. By permission of the *American Journal of Science*)

geomorphological approaches lack a secure physical basis. For example, Patton and Schumm (1973) have suggested that an erosional phase is triggered when the aggradational phase in a particular valley reach attains a critical (threshold) gradient (see Section 10.5), and the valley-fill sediments become unstable in relation to the shear stresses imposed by effective flows. This is an example of a so-called 'intrinsic' or 'geomorphic' threshold. In smaller valleys, the valley gradient effect appears to be obscured by the

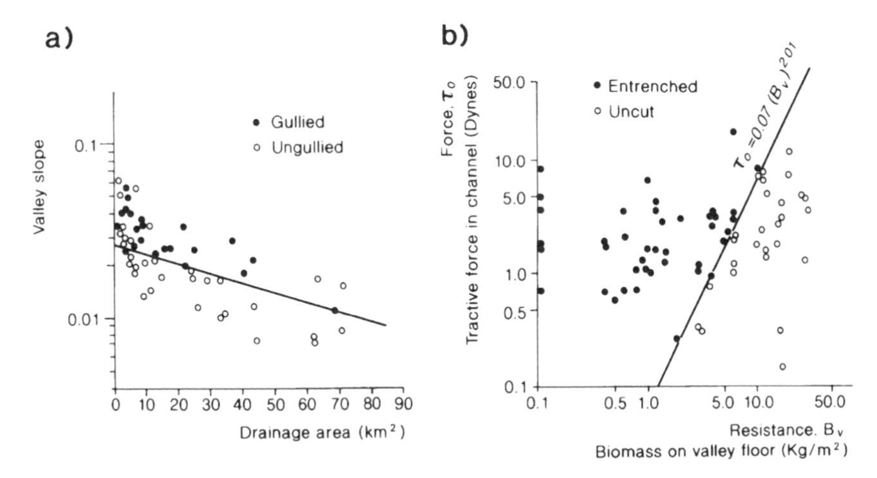

Figure 10.7 (a) Gully formation as a function of slope and basin area (after Patton and Schumm, 1975. By permission of the Geological Society of America). (b) Conditions for gully initiation as a function of shear stress in the 10-year flood and valley floor biomass (after Graf, 1979. By permission of John Wiley & Sons Ltd)

influence of the biomass (Graf, 1979), which determines the erodibility of the valley-floor sediments. These empirical threshold relationships are shown in Figure 10.7, but neither would be a reliable basis for the prediction of the initiation of damped oscillatory sediment yield responses. Furthermore, it is by no means clear that tributary sediment yield results in steepening of the main valley gradient, since much of the 'excess' sediment produced from a tributary is stored in alluvial bars and fans at the tributary junctions (Richards, 1993). From these temporary storages, intermittent releases of sediment occur when fantoe deposits are eroded by extreme events, or when the local downvalley gradient across them has become sufficiently accentuated (Small, 1973; Church, 1983). This conceptual model suggests that a more physically-based method, such as that developed by Pickup et al. (1983), may be used to model the discontinuous temporal behaviour of sediment yield from tributaries. Originally used to simulate the downstream translation of a wave of bedload derived from eroding mine tailings dumps adjacent to rivers, this is a finite-difference approximation of a kinematic-wave sediment routing, and for a single input of sediment results in predictions of wave translation such as those in Figure 10.2.

10.7 RIVER SYSTEM DEVELOPMENT

The example of complex response following the crossing of intrinsic thresholds discussed in the previous section raises an additional consideration for river managers: the need to understand the depositional context of river system development independently of the effects of environmental change. In Britain, there are no large rivers responsible for deposition in extensive sedimentary basins, and most British rivers are relatively stable. The British experience of river management cannot therefore be translated to other regions without careful assessment of the fluvial geomorphology (Richards et al., 1987). In large sedimentary basins, rivers often act as depositional systems, not as single independent channels. They convey sediment from eroding source areas, such as adjacent mountains, to the sedimentary basin, where deposition occurs preferentially along the alignment of the rivers. The basin, however, fills relatively uniformly over a longer timescale; the Gangetic Plain, for example, displays no more than 20 m of relief over 100 km of cross-sections, and this is comparable to the seasonal stage range in the trunk rivers. The implication is that the rivers in such a system constitute 'avulsive river systems' (Richards et al., 1993), in which some channels are dominant for a period, but gradually aggrade until a sufficient elevation difference develops for flood overspill to cause self-diversion into one or more subparallel secondary channels, some of which may be former courses that have atrophied and become (temporarily) 'underfit'. A similar process does occur in Britain, albeit on a much smaller scale: the local channel avulsions observed in a wandering gravel-bed river such as the River Feshie are examples (Ferguson and Werritty, 1983). Rivers such as the Rapti on the northern margin of the Gangetic Plain, the rivers of the Okavango Delta, the Hadejia-Nguru rivers of northern Nigeria, and the Columbia River in Canada are typical of the larger-scale avulsion-controlled systems elsewhere in the world. It is clearly important to understand the interactions between rivers in such systems, to assess the potential for avulsion during the planned life of a river engineering project, and to estimate the possible consequences of an alteration in the pattern of sediment supply as a result of impoundment (Richards et al., 1987). This requires application of a range of geomorphological techniques, such as: morphological mapping to identify the system of river channels and palaeochannels; stratigraphic investigation to assess the relationships between depositional units; both relative and absolute dating of channel changes from sedimentological, pedological and historical evidence; and attempts to estimate present rates of sedimentation. Numerical modelling frameworks for prediction of avulsive channel behaviour are lacking, but in essence would require a structure similar to that developed by Hooke and

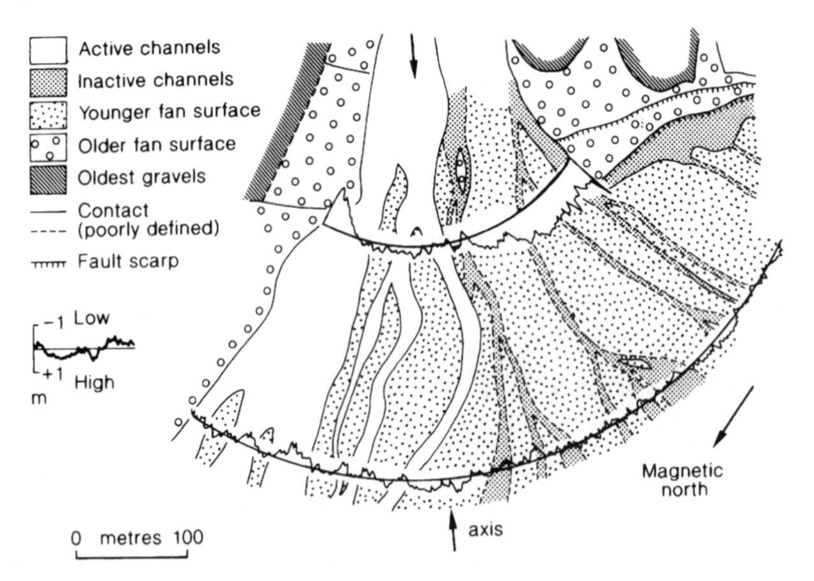

Figure 10.8 Shifting patterns of flow on the Mormon alluvial fan on the east side of Death Valley, California, indicating topographic deviations from an ideal conical surface (after Hooke and Rohrer, 1979. By permission of John Wiley & Sons Ltd)

Rohrer (1979) to simulate the shifts of channel alignment that occur on alluvial fan surfaces as lateral differences in elevation arise because of the temporary linear location of deposition on the fan radius occupied by the dominant channel (Figure 10.8). However, calibration of such a model requires the geomorphological data noted above (for example, on spatially differentiated rates of sedimentation), and therefore necessitates extensive field investigation of the system to be modelled. In this respect, therefore, evaluation for river management purposes of a large-scale fluvial adjustment process such as this is no different from evaluation of single reaches in one or two dimensions using more sophisticated and physically based numerical models.

10.8 CONCLUSIONS

Scale, environment and uncertainty may be the three themes that are central to understanding the problems of prediction of morphological change in unstable channels. Schumm and Lichty (1965) show how, as time or space scales change, the formative variables associated with river channel dynamics shift between being dependent, independent and irrelevant. For instance, at the longest timescales, river channel morphology may be dependent upon catchment hydrological conditions, vegetation, relief and sedimentology. Over shorter timescales, channel morphology switches from being dependent to being the independent control of flow characteristics. However, in reality, a feedback occurs across the timescales. This is well illustrated by the fact that, while valley gradient controls stream power and, therefore, determines the channel planform, given sufficient time the dynamics of that planform determine the construction of the valley fill and, therefore, the valley gradient.

The method required to predict morphological change is both determined and constrained by the temporal and spatial scale of enquiry. However, a consideration of time and space scale is insufficient without consideration of the sensitivity of river channel response to the local environment and the local alluvial history (cf. the cusp catastrophe example in Section 10.5). Different environments result in a variation in the relative time and space scales of channel adjustment. River channels in unvegetated semi-arid catchments are generally more unstable, less likely to display equilibrium forms, and more likely to experience substantial modification during extreme events. This example illustrates that there are significant variations in the uncertainty associated with the prediction of morphological change in different environments. Nevertheless, it should be noted that while absolute levels of uncertainty may increase as time and space scales expand, the *relative* magnitude of uncertainty may be unchanged and has to be measured against the scale of the system for which predictions are being made. Prediction of morphological change over longer timescales inevitably adopts a more qualitative approach which therefore appears to embody a considerable degree of uncertainty. However, quantitative predictions derived from physically based distributed models are also subject to considerable levels of uncertainty, for example in the specification of boundary conditions and parameter values. Often attached to the results of such simulation models is a spurious impression of accuracy which should remind us that these models, whilst they may be used for predictive purposes, may be little more than tools for probing the depths of our uncertainty.

10.9 REFERENCES

Anderson, M.G. and Bates, P.D. 1994. Evaluating data constraints on two-dimensional finite element models of floodplain flow. *Catena*, **22**, 1–15.

Ashworth, P.J. and Ferguson, R.I. 1987. Interrelationships of channel processes, changes and sediments in a proglacial river. *Geografiska Annaler*, **68A**, 361–371.

Begin, Z.B., Meyer, D.F. and Schumm, S.A. 1981. Development of longitudinal profiles of alluvial channels in response to base-level lowering. *Earth Surface Processes*, **6**, 49–68.

Bernard, R.S., Grimes, B., Howington, S., Schneider, M. and Turner, H. 1992. *STREMR Numerical Model for Depth-averaged Flow.* United States Army Corps of Engineers, Waterways Experiment Station, Vicksburg, 140pp.

Blench, T. 1969. *Mobile Bed Fluviology.* University of Alberta Press, Edmonton.

Born, S.M. and Ritter, D.F. 1970. Modern terrace development near Pyramid Lake, Nevada, and its geologic implications. *Geological Society of America, Bulletin*, **81**, 1233–1242.

Boussinesq, J. 1877. *Essai sur la theorie des eaux courantes.* Mem. Presentes Acad. Sci. Paris, 23, 46pp.

Bridge, J.S. 1993. The interaction between channel geometry, water flow, sediment transport and deposition in braided rivers. In: Best, J.L. and Bristow, C.S. (Eds), *Braided Rivers.* Geological Society, London, Special Publication 75, 13–71.

Brookes, A. 1991. Recovery and restoration of some engineered British river channels. In: Boon, P.J., Calow, P. and Petts, G.E. (Eds), *River Conservation and Management.* John Wiley & Sons, Chichester, 337–352.

Chang, H.H. 1988. *Fluvial Processes in River Engineering.* Wiley Interscience, New York, 432pp.

Church, M. 1983. Patterns of instability in a wandering gravel bed channel. In: Collinson, J.D. and Lewin, J. (Eds), *Modern and Ancient Fluvial Systems.* International Association of Sedimentologists, Special Publication No. 6, 169–180.

Clifford, N.J. and French, J.R. 1993. Monitoring and modelling turbulent flow: historical and contemporary perspectives. In: Clifford, N.J., French, J.R. and Hardisty, J. (Eds), *Turbulence: Perspectives on Flow and Sediment Transport.* John Wiley & Sons, Chichester, 1–34.

Daniels, R.B. 1960. Entrenchment of the Willow Drainage Ditch. *American Journal of Science*, **258**, 161–176.

Darby, S.E. and Thorne, C.R. 1996a. Development and testing of riverbank-stability analysis. *Journal of Hydraulic Engineering*, **122**, 443–454.

Darby, S.E. and Thorne, C.R. 1996b. Numerical simulation of widening and bed deformation of straight sand-bed rivers. I: Model development. *Journal of Hydraulic Engineering*, **122**, 184–193.

Dietrich, W.E. 1987. Mechanics of flow and sediment transport in river bends. In: Richards, K.S. (Ed.), *River Channels: Environment and Process.* Basil Blackwell, Oxford, 179–227.

Ferguson, R.I. and Werritty, A. 1983. Bar development and channel changes in the gravelly River Feshie, Scotland. In: Collinson, J.D. and Lewin, J. (Eds), International Association of Sedimentologists, Special Publication No. 6, 181–193.

Gardiner, J.L. 1991. *River Projects and Conservation: a Manual for Holistic Appraisal.* John Wiley & Sons, Chichester, 236pp.

Graf, W.L. 1979. The development of montane arroyos and gullies. *Earth Surface Processes*, **4**, 1–14.

Graf, W.L. 1988. Applications of catastrophe theory in fluvial geomorphology. In: Anderson, M.G. (Ed.), *Modelling Geomorphological Systems.* John Wiley & Sons, Chichester, 33–47.

Hey, R.D. 1979. Dynamic process-response model of river channel behaviour. *Earth Surface Processes*, **4**, 59–72.

Hooke, R.LeB. and Rohrer, W.L. 1979. Geometry of alluvial fans: effect of discharge and sediment size. *Earth Surface Processes*, **4**, 147–166.

Knight, D. 1992. Turbulent flows in a natural river. *Abstracts, International Association of Hydrology Research Conference, 'Is Turbulence Modelling of Any Use?'*, Institution of Civil Engineers, 9 April 1992.

Lane, S.N., Richards, K.S. and Chandler, J.H. 1994. Distributed sensitivity analysis in modelling environmental systems. *Proceedings of the Royal Society, Series A*, **447**, 49–63.
Lane, S.N., Richards, K.S. and Chandler, J.H. 1995. Within-reach spatial patterns of process and channel adjustment. In: Hicken, E.J. (Ed.), *River Geomorphology.* John Wiley & Sons, Chichester, 105–130.
Leeks, G.J., Lewin, J. and Newson, M.D. 1988. Channel change, fluvial geomorphology and river engineering: the case of the Afon Trannon, mid-Wales. *Earth Surface Processes and Landforms*, **13**, 207–213.
Leopold, L.B. and Maddock, T. 1953. *The Hydraulic Geometry of Stream Channels and Some Physiographic Implications*. United States Geological Survey, Professional Paper 252, 1–57.
Meyer-Peter, E. and Muller, R. 1948. Formulas for bedload transport. *International Association for Hydraulic Structures Research, Second Meeting*, Stockholm, 39–64.
Miller, T.K. 1984. A system model of stream-channel shape and size. *Geological Society of America, Bulletin*, **95**, 237–241.
Osman, A.M. and Thorne, C.R. 1988. Riverbank stability analysis. I: Theory. *ASCE Journal of Hydraulic Engineering*, **114**, 134–150.
Parker, G. and Andres, D. 1976. Detrimental effects of river channelization. *Proceedings of the Conference, Rivers '76*, American Society of Civil Engineers, 1248–1266.
Patton, P.C. and Schumm, S.A. 1975. Gully erosion, northern Colorado: a threshold phenomenon. *Geology*, **3**, 88–90.
Pickup, G. 1975. Downstream variations in morphology, flow conditions and sediment transport in an eroding channel. *Zeitschrift fur Geomorphologie*, **19**, 443–459.
Pickup, G. 1977. Simulation modelling of river channel erosion. In: Gregory, K.J. (Ed.), *River Channel Changes*. John Wiley & Sons, Chichester, 47–60.
Pickup, G., Higgins, R.J. and Grant, I. 1983. Modelling sediment transport as a moving wave – the transfer and deposition of mining waste. *Journal of Hydrology*, **60**, 281–301.
Phillips, J.D. 1991. Multiple modes of adjustment in unstable river channel cross-sections. *Journal of Hydrology*, **123**, 39–49.
Prandtl, L. 1952. *Essentials of Fluid Dynamics*. Hafner Publications, New York, 453pp.
Rango, A. 1970. Possible effects of precipitation modification on stream channel geometry and sediment yield. *Water Resources Research*, **6**, 1765–1770.
Reynolds, O. 1895. On the dynamical theory of incompressible viscous fluids and the determination of the criterion. *Philosophical Transactions of the Royal Society*, **186A**, 123–164.
Richards, K.S. 1988. Fluvial geomorphology. *Progress in Physical Geography*, **12**, 435–456.
Richards, K.S. 1993. Sediment transport into and through the network. In: Beven, K. and Kirkby, M.J. (Eds), *Channel Network Function*. John Wiley & Sons, Chichester, 221–254.
Richards, K.S. and Clifford, N. 1991. Structured beds in gravelly rivers. *Progress in Physical Geography*, **15**, 407–422.
Richards, K.S., Brunsden, D., Jones, D.K.C. and McCraig, M. 1987. Applied fluvial geomorphology: river engineering project appraisal in its geomorphological context. In: Richards, K.S. (Ed.), *River Channels: Environment and Process*. Basil Blackwell, Oxford, 348–382.
Richards, K.S., Chandra, S. and Friend, P. 1993. Avulsive channel systems: characteristics and examples. In: Best, J.L. and Bristow, C.S. (Eds), Geological Society, London, Special Publication, 75, 195–203.
Schumm, S.A. 1969. River metamorphosis. *Proceedings of the American Society of Civil Engineers, Journal of the Hydraulic Division*, **95**, 255–273.
Schumm, S.A. 1973. Geomorphic thresholds and complex response of drainage systems. In: Morisawa, M. (Ed.), *Fluvial Geomorphology.* Publications in Geomorphology, SUNY Binghampton, 69–85.
Schumm, S.A. and Lichty, R.W. 1965. Time, space and causality in geomorphology. *American Journal of Science*, **263**, 110–119.

Simon, A. 1992. Energy, time and channel evolution in catastrophically disturbed fluvial systems. In: Phillips, J.D. and Renwick, W.H. (Eds), *Geomorphic Systems*. Proceedings of the 23rd Binghampton Symposium, September 1992, 345–373.

Small, R.J. 1973. Braiding terraces in the Val d'Herens, Switzerland. *Geography*, **58**, 129–135.

van Rijn, L.C. 1987. *Mathematical Modelling of Morphological Processes in the Case of Suspended Sediment Transport*. Delft Hydraulics Communication No. 382, 127pp.

Whiting, P.J. and Dietrich, W.E. 1991. Convective accelerations and boundary shear stress over a channel bar. *Water Resources Research*, **27**, 783–796.

Younis, B. 1992. Some turbulence models. *Abstracts, International Association of Hydrology Research Conference, 'Is Turbulence Modelling of Any Use?'*, Institution of Civil Engineers, 9 April 1992.

APPENDIX: NOTATION

a	velocity distribution coefficient
A	cross-sectional area
b	exponent relating bank erosion to distance downsteam
B_{ij}	coefficient in multivariate channel geometry equations
B_o	constant in expression for downstream bank erosion
C_f	friction coefficient
d_m	mean bed material size
D	water depth
g	acceleration due to gravity
G	submerged weight bedload transport per second per unit width
h	water depth
i	sediment supply from bank erosion
k_b	Strickler's bed roughness coefficient
k_r	Strickler's particle roughness coefficient
n_{ab}	coefficients in Miller's (1984) reduced equations
N	number of computational nodes
p	pressure
Q	water discharge
Q_s/Q	bank friction correction factor
R	hydraulic radius
R_e	Reynolds number
s_f	friction slope
s_o	bed slope
t	time
u,v	instantaneous downstream (u) or vertical (v) velocity
$\bar{u}, \bar{v}$	Reynolds averaged downstream ($\bar{u}$) or vertical ($\bar{v}$) velocity
$\bar{U}, \bar{V}$	depth-averaged, Reynolds averaged downstream ($\bar{U}$) or vertical ($\bar{V}$) velocity
V	mean velocity
w	channel width
W	channel width
x	downstream distance
y	water surface elevation
z	bed elevation
z_o	bed elevation
γ_s	specific weight of sediment
γ_w	specific weight of water
ρ	water density
∇	gradient operator

11 River Dynamics and Channel Maintenance

ANDREW BROOKES
Environment Agency, London, UK

11.1 INTRODUCTION

This chapter examines the reasons for river channel maintenance and lists the most widely used and accepted methods of countering sedimentation and erosion problems. Attention is then given to alternative strategies which work with nature and to the benefits of incorporating geomorphological principles in channel design and operational maintenance.

11.2 RIVER MANAGEMENT PROBLEMS AND TRADITIONAL RESPONSES

11.2.1 Reasons for Channel Maintenance

River improvement works can result in morphological problems, requiring heavy maintenance intervention (Brookes, 1988). Morphological effects are often not considered at the planning stage of flood control projects, and it is only when a project becomes operational that substantial maintenance commitments and high costs may become apparent. If the regime of the river is affected then alterations in the capacity of a channel to transport sediment can lead to erosion or deposition. For example, adjustments of channel morphology have been noted within reaches which have been widened to convey peak flood discharges within banks (Box 11.1).

> **Box 11.1 Adjustments within enlarged channels**
>
> The problems of over-wide channels have been documented for the River Tame near Birmingham in central England (Nixon, 1966) and for various other schemes in the UK (Hydraulics Research, 1987). Widening a channel reduces the stream power per unit bed area, thereby decreasing the sediment discharge capacity. Sediment is deposited because of the reduced velocities and unit stream power in the artificially wide section. Vegetation may colonise the sediment deposits, leading to a further reduction of flood capacity unless maintenance is used to remove the shoals and vegetation.

Applied Fluvial Geomorphology for River Engineering and Management, Edited by C. R. Thorne, R. D. Hey and M. D. Newson.

Channel realignment or straightening has been a very common channel improvement practice, affecting many thousands of kilometres of watercourse. The length of a meandering channel is artificially reduced by straightening. This has the effect of increasing the channel slope and reducing energy losses in bendways, which in turn increase mean velocity, thereby reducing flood levels. However, the increase of channel slope elevates the sediment transport capacity so that it is greater than the supply at the upstream end of the straightened reach. The difference in sediment load is balanced through erosion of the channel bed and banks leading to destabilisation of the straightened reach. Siltation in the unaltered reach downstream sometimes leads to further problems.

An active maintenance programme is, therefore, often required to retain the characteristics of improved channels for navigation, flood defence or land drainage. The frequency and extent of maintenance will depend on the local degree of instability.

Other types of river improvements which often have maintenance implications include: the construction of flood walls and embankments to contain flood flows; the building of flood relief or bypass channels; and flood control reservoirs designed to store flood water temporarily. Local improvements to structures such as weirs, bridges and bank protection can lead to morphological impacts also requiring maintenance. In this respect, many conventional engineering solutions to river management problems are not sustainable.

Maintenance problems also arise from land-use changes in the catchment upstream such as urban development and agricultural or forestry practices. Table 11.1 lists some typical problems experienced by river managers. Gravel extraction, mining and river regulation may all potentially result in local and downstream maintenance problems.

11.2.2 Types of Maintenance Practice

Maintenance can be divided into several categories including dredging, weed control, tree clearance, and removal of rubbish.

Table 11.1 Examples of land-use problems

Independent parameters	Dependent parameters	Examples of change
$Q_w^+ Q_s^+$	$\sim b^{\pm} d^{+} L^{\pm} S^{-} P^{+} F^{-}$	Discharge increases as bedload decreases following urban development when sediment source areas have been paved over. Erosion of bed and banks. Fine suspended sediments arising from urban runoff may create a low-cost, low-frequency maintenance problem.
Q_s^+	$\sim b^{+} d^{-} L^{+} S^{+} P^{-} F^{+}$	Increased bed material load due to deforestation or intense arable farming. Moderate cost, low-frequency maintenance, particularly in lowland catchments.

A plus or a minus exponent is used to indicate how various aspects of channel morphology change with an increase or decrease in water discharge or bedload.
Q_w = water discharge (e.g. mean annual flood); Q_s =bed material load (expressed as a percentage of total load); b = width; d = depth; L = meander wavelength; S =channel slope; P = sinuosity; F = width/depth ratio.

Dredging

Dredging is undertaken to remove deposited sediments and maintain the specified cross-sectional area. A large range of machinery is available for dredging and its selection depends on the size of the channel, access and ground conditions, type of material being removed and method and distance of disposal of spoil. Increasingly, environmental constraints dictate the use of machinery which will be least environmentally damaging.

The simplest form of dredging is that in which the material is broken up, loosened and left to be transported out of a channel by the current. In large rivers the material may be pumped ashore or discharged into barges before being dumped at selected locations. The type of machinery used for dredging has been well documented (Joglekar, 1971; Brandon, 1987) and includes both mechanical dredgers which remove the material by bucket, grab or ladder devices, and hydraulic suction dredgers which pump the material from the bed in pipes. In smaller watercourses dredging is carried out by hydraulic excavators or draglines attached to land-based vehicles. Mini-tracked hydraulic excavators are increasingly used in various countries, including England and Wales. These are very small machines with a relatively short reach and are restricted to the removal of small quantities of material. However, because of their size, weight and mobility, access can be gained to very restricted areas on banks and in channel beds, limiting environmental disturbance such as the loss of trees.

Weed Control

The control of aquatic macrophytes is sometimes included within the definition of 'clearing and snagging' and the annual growth of these plants is one of the greatest problems in the maintenance of channel capacity in highly productive streams in both Britain and North America (Haslam, 1978). Plants physically reduce the capacity of a channel to a limited extent, but particularly increase the hydraulic roughness as well as promote the accumulation of silt, and in certain areas they have to be managed several times a year. Submerged, emergent or marginal bank plants are controlled by a variety of methods and these include cutting, chemical control, grazing and shading.

Cutting has been the traditional method of removing plants that constitute flood hazards in Britain and this has been carried out either by a scythe operated by an individual or a number of scythe blades linked together and hauled along the channel bed by operatives on each bank. Mechanised methods include the use of paddle-driven, flat-bottomed boats which have reciprocating cutting knives, and newer designs of boat have been introduced which can be adapted for use in rivers of various sizes. The cut weeds may be allowed to float downstream to a suitable collection point and a wooden boom may be laid across the river and the debris later removed by means of a power-driven elevator or a dragline. This type of weed control, if not selective, may have adverse impacts on the ecology and morphology of a channel by disrupting the natural processes of erosion and deposition.

During the last 20 years in various European countries, trees have been deliberately planted as a means of reducing aquatic plant growth, by shading. In Britain, trees and hedges have been planted along several watercourses (Haslam, 1978) and experimental studies have been undertaken by Dawson and Kern-Hansen (1978) and by Dawson (1981).

Clearly, this is a long-term solution since it may take several years for the trees to attain maturity.

Tree Clearance

This involves the removal of fallen trees and debris jams from a channel (Herbkersman, 1981) as well as the harvesting of timber from the banks and floodplain (Patric, 1975). This has been referred to in North America as 'clearing and snagging'. In Britain much 'pioneer tree clearance' work was undertaken during the first 10 years of the existence of Catchment Boards because many rivers were in a derelict state before the passing of the 1930 Land Drainage Act. Trees grew inwards from both banks of the channel with the result that when a river was in flood the branches trailing in the water would reduce the mean velocity and increase flood levels. Furthermore, trees which had been undermined tended to collapse across the river and collect further material, thereby obstructing the flow and also causing localised scour of the channel. Tree clearance, however, may increase stream temperatures (Little, 1973) and encourage excessive growth of any aquatic macrophytes that may be present in the channel. Removal of trees, especially the stumps, can cause instability by weakening the structural strength of the bank.

Rubbish Removal

The term 'clearing and snagging' may also be applied to the removal of debris such as rubbish, sediment and plant growth from urban channels. This type of work is particularly important to clear debris which would otherwise block culverts and bridges and cause a potential flooding problem. Much of this material is not natural, but is thrown into the river by riparian owners or vandals. Traditional approaches to channel maintenance include (see Brandon, 1987):

(a) planned preventative maintenance which involves a schedule of pre-planned maintenance to keep an asset such as a flood control channel in prime condition (e.g. regular dredging of a silt-clay river system);
(b) breakdown maintenance which involves intervention only when an asset fails (e.g. in gravel-bed rivers dominated by flood-dependent sediment transport);
(c) planned inspection which involves routine inspection of watercourses; action is only taken where a need is identified.

11.2.3 Constraints of Management Standards and Environmental Law

In addition to the more traditional maintenance techniques, a number of alternative strategies and procedures have been developed. These include: a minimalist approach of allowing a previously managed channel to recover naturally; a catchment approach which treats the cause of a maintenance problem rather than the symptoms; and techniques which allow more natural characteristics to be formed in a channel. In many countries, environmental-legislation dictates that consideration should now be given to more environmentally sound approaches. For example, in England and Wales, Section 16 of the Water Resources Act 1991 relates to the general environmental duties of drainage bodies which include the need to 'further the conservation and enhancement of natural beauty and the conservation of flora, fauna and geological or physiographical features of special interest' (HMSO, 1991, p. 13). Danish legislation ensures that streams have a diverse flora and fauna, that they are a natural part of the landscape, and that they are aesthetically and recreationally pleasing. The Danish Watercourse Acts of 1982/1995 state that river works are to be planned and undertaken with regard to the stream quality. Stream quality refers not only to pollution but also to the physical form of a river channel, including pools and riffles (Brookes, 1987; Nielsen, 1996). Some organisations now have budgets expressly for the environmental enhancement of watercourses.

However, maintenance is undertaken for very valid reasons such as flood defence and land drainage, and therefore in dealing with any one problem various issues need to be considered, including the standard of service to be provided to a customer.

11.2.4 Standard of Service

The frequency with which maintenance is undertaken depends on the cost, the standard or level of service to be provided, and experience (Brandon, 1987; Hydraulics Research, 1992c; Brookes, 1995a). The standard of service for a particular river channel in an urban area may be the 1:100 year level of flood protection, while for arable areas the design flood might be the 20 or 25 year event.

By comparing a target standard of service with the actual standard of service it is possible to determine whether a reach of river is over-serviced, under-serviced, or adequately serviced. Methodologies have been developed in several countries to define the changes in maintenance regime required to achieve the target standard of service (e.g. Hydraulics Research, 1992a).

11.3 LOCATIONAL CONSTRAINTS AND GEOMORPHOLOGICAL GUIDING PRINCIPLES

If a fluvial system is stable, evidence suggests that the natural channel morphology and floodplain can accommodate large floods without a major maintenance commitment. However, it has often been necessary to create washlands, riparian corridors and flood basins to regulate overbank flood flows.

Effective assessment of the potential morphological effects prior to the design of a preferred scheme for flood defence should lead to improved designs in the future (e.g. Hydraulics Research, 1992a,b). One option may be to design a two-stage channel where the low-flow channel remains as existing and the berms are widened and lowered to accommodate the design flood discharge. However, during a flood sediment will be deposited on the berms, particularly if they are heavily vegetated. Accurate calculation of the capacity of two-stage channels is a subject under much discussion and continuing research at present. Another option may be to construct a flood-relief channel, leaving the original channel intact.

For channels which need to be managed for flood defence or agricultural drainage purposes it is suggested that geomorphological understanding can be used in a variety of ways, as described below.

11.3.1 'Do Minimum' Option

In engineering terms, minimum maintenance involves gradual reduction of the maintenance frequency to a level where the asset just attains its prime objective (Brandon, 1987). Although cheap to operate, it is difficult to judge the correct level and still provide an adequate standard of service. However, in environmental terms lack of channel maintenance permits gradual stream recovery. For example, in the case of the River Tame in central England, referred to in Box 11.1, the channel reverted to its original capacity in less than 30 years (from 1930 to 1959) in the absence of maintenance (Nixon, 1966). This was due to the enlarged channel being adjusted to a particular design flow event, which made it out of equilibrium with the normal range of flows.

The nature and rate of adjustment following channel works depend not only on the available energy or stream power but also on the sediment supply from the catchment upstream or from channel erosion (Brookes, 1992). For a number of channels in England and Wales it has been shown that whilst recovery of the low-flow width had occurred through deposition of sediment, the actual cross-sectional area was reduced from the artificially enlarged size by up to a maximum of 35%. It was therefore concluded that the modified channels studied were not adversely affected by morphological adjustment. However, it is recommended that further hydraulic modelling studies be undertaken for channels which are recovering, in order to determine the significance of reformed deposits on flood conveyance capacity. Particularly where channels have been over-serviced in the past, the minimal maintenance approach, allowing continued recovery, may be acceptable.

11.3.2 Catchment Approach

For channels where the standards of service for flood defence cannot be achieved due to excessive sedimentation, a solution may be to appraise more accurately the cause of the problem. A traditional approach to channel maintenance has been to treat the symptom and not the cause. This has often led to the need for frequent and costly maintenance. From an environmental point of view, regular channel maintenance is also disruptive and suppresses recovery of natural morphological diversity and habitat value. It is suggested that a geomorphological contribution to channel maintenance is to identify and to answer the 'where', 'what type' and 'over which timescale' questions (Sear and Newson, 1991, 1993; Sear, 1996). In particular, the sediment system in a river should be regarded as a continuum of supply, transfer and storage, operating at both the catchment and reach scales. Figure 11.1 illustrates the 'knock-on' effect of a large input of sediment in the uplands and the effect of increased sediment transmission downstream. Geomorphology can be used to:

- correctly identify the cause and scale of a problem (e.g. local or catchment);
- identify the impacts of a proposed solution (e.g. upstream and downstream impacts on sediment supply);
- correctly target the solution (e.g. siting of gravel traps or bank protection).

The impacts of conventional solutions can be determined by using stream power or other indices of channel stability; by drawing on experience from similar options carried out in comparable environments; and by determining the sediment balance of the reach in relation to catchment processes and history. In particular, a catchment-wide approach to

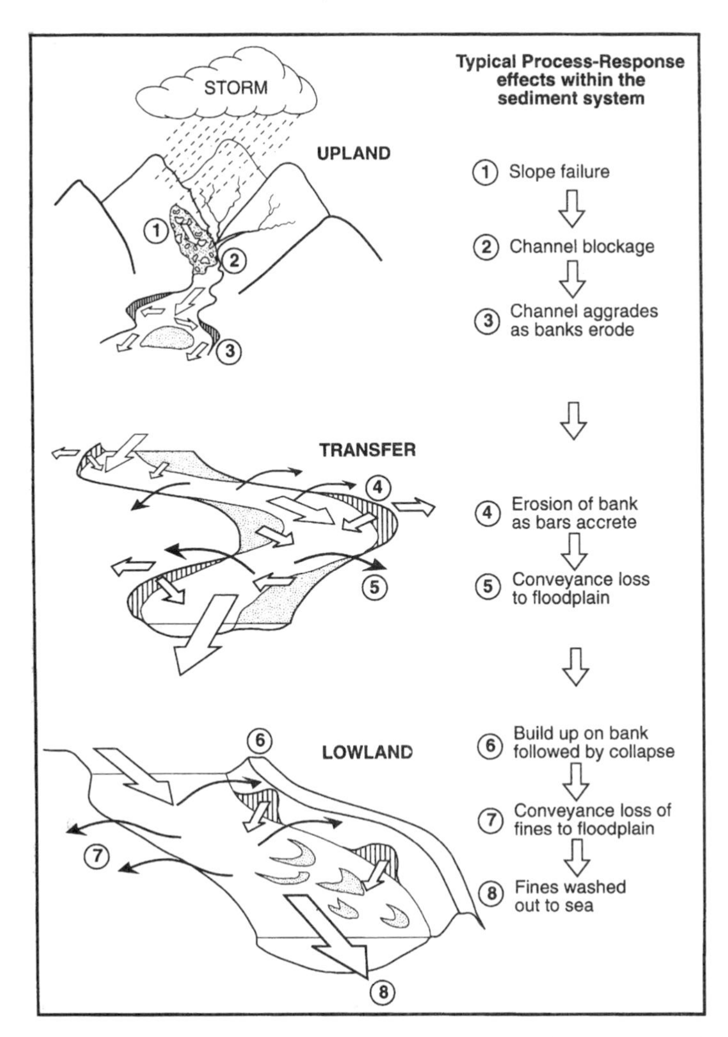

Figure 11.1 Typical process–response effects within the sediment system (after Sear and Newson, 1993)

sediment management is a valid long-term objective. This should identify sediment sources, transmission zones and storage locations.

Such an approach can lead to maintenance which is based on more effective control of sediment through sediment trapping, controlling erosion by a variety of techniques within a reach, and the use of buffer strips to reduce the suspended sediment load supplied by runoff from agricultural land. Sear and Newson (1993) document a procedure termed 'fluvial audit' which is recommended as a standard procedure in England and Wales for evaluating sediment-related problems, potentially linking cause and effect. This procedure was documented in Sear et al. (1995) and is now receiving wider application for solving maintenance problems.

11.3.3 Geomorphological Principles in Channel Design and Maintenance

The significance of geomorphological understanding lies in the ability to design channels which remain stable after construction, thereby requiring little maintenance. Traditionally, there has been little geormorphological evaluation of maintenance solutions. However, in many countries legislative and best practice approaches adopted by various agencies now dictate that environmental issues, such as the conservation of flora and fauna, should be an integral part of channel maintenance. Geomorphological principles can lead to more natural and environmentally preferred designs (e.g. Hey et al. 1991). Increasingly, environmental mitigations and enhancements are being incorporated as part of maintenance strategies.

Accommodation of natural morphological features can be undertaken by extending the floodway as appropriate. Where necessary, multistage channels or flood berms can be constructed to provide additional flood storage capacity without enlarging the natural channel. Flood embankments can also be set back from the channel to accommodate meandering, bank erosion and the in-channel storage of sediments.

11.3.4 Partial Dredging

An alternative option for a channel which *must* be dredged is to conserve a significant proportion of the aquatic plant and animal community by only dredging part of the channel (Lewis and Williams, 1984; Ward et al., 1994). By this means, the variety of habitat conditions and their associated species can be conserved. However, it is recommended that geomorphological principles be followed in this option. A more natural low-flow width should be determined by observing the function of the affected reach as well as adjacent natural reaches, both upstream and downstream. In general, at low flow the pool–riffle sequence may be a possible cause of width fluctuations. On average, riffles at low flow may be 15% wider than pools (Richards, 1976). Figure 11.2 shows the widths for a pool and a riffle for a channel to be partially dredged. The planform of the channel is also important. The width for a pool can be used for the entire bend, although it should be noted that width naturally varies through a bend.

A further principle is that flow in a channel is characteristically unsteady and non-uniform, and nowhere are the streamlines parallel to each other or to the stream banks. Leliavsky (1955) proposed a convergence–divergence criterion whereby at high discharges convergence of flow is associated with scour, and divergence of stream flow corresponds to deposition of sediment. Hence, the inner side of a bend is typically an area of deposition in contrast to erosion at the outer bank. In an unmanaged, natural channel, deposition of sediment forms a point bar on the inner side of a bend adjacent to a pool. It is recommended that these principles be followed in determining where partial dredging should occur to ensure stability across a wide range of flows. Figure 11.2 depicts the areas of sediment which should remain untouched during a partial dredge.

Equally, if there is an inadequate depth of water at low flow then the channel can be narrowed to increase the depth. This can be achieved by redistributing some of the accumulated sediments, or by the use of deflectors. Narrowing the channel provides a more adequate low-flow depth for aquatic life and discourages the deposition of sand and silts within pools and on riffles, by locally increasing low-flow velocities.

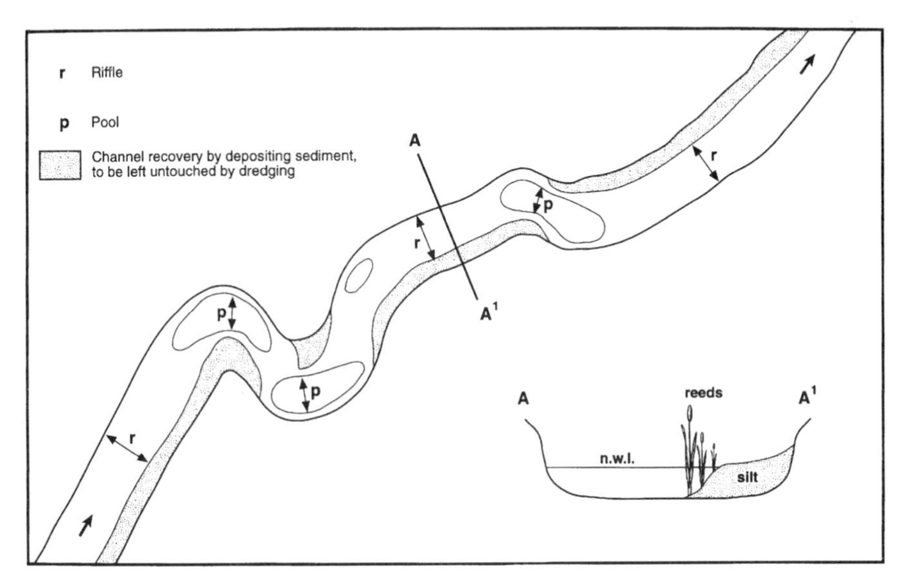

Figure 11.2 Geomorphological principles for partial dredging

Determining the natural low-flow width for a channel is important for a variety of other mitigation or enhancement techniques which may be carried out as part of modern river maintenance. These techniques include the construction of shallow underwater berms or shelves to increase the variety of water depths to support greater biodiversity.

11.3.5 Conserving Pools and Riffles

Maintenance should conserve pools and riffles since a uniform channel bed tends to have very low species diversity. If desilting of the low-flow channel has to be included in the maintenance regime, then pools and riffles should either be preserved or recreated. The location and functioning of existing pools should be determined from a field survey. Relatively deep pools tend to form at intervals of between three and 10 channel widths in natural channels. Each pool is separated by a topographically high area which extends partly or completely across the channel, with a steep downstream face over which the water spills as a riffle at low flows. Pool–riffle sequences are particularly marked where the bed material is gravel to cobbles (2 to 256 mm). In meandering channels pools form adjacent to the outside of a bend, where the flow lines naturally converge and tend to scour sediments.

There is an increasing number of documented instances where the channel bed has been lowered and the pool–riffle sequence recreated at the new lower elevation (Brookes, 1990, 1991). Machine operators need to be experienced at such work and particular care must be taken when matching sediments to the previous conditions. If riffles and pools are not sited according to the convergence and divergence of flow in natural streams, then adverse erosion or deposition may occur, especially during high flows. Perhaps one of the more difficult challenges for geomorphology is to recreate the more natural water surface and bed slopes associated with pool–riffle sequences. Excess slope may cause localised erosion whilst deposition on slopes which are too low may be typical. Mimicking nature and accurately specifying this level of detail in a contract document is not an easy task.

11.3.6 Placement of Stone

The placement of substrate and stone in channels may be undertaken as part of operational maintenance to mitigate the impacts of previous works. This technique needs careful consideration. Calculation of the shear stress exerted on the bed of a channel at the design flow is essential to determine the calibre and size distribution of sediment to be placed. The maximum permissible tractive force is one approach that can be used.

Du Boys' (1879) equation for the bed shear stress (τ_o) (N/m^2) is:

$$\tau_o = \omega R s$$

where ω = specific weight of water (kN/m^3), R = hydraulic radius (m) and s = energy slope (approximated by the bed slope).

Laboratory and field data provide maximum permissible values of bed shear stress as a function of bed material size. Typical values for the movement of coarse gravels and cobbles in water transporting colloidal silts are 32 N/m^2 and 53 N/m^2, respectively (see Webber, 1971). If the shear stress is not too high then appropriately sized materials can be placed. Table 11.2 shows a recommended size distribution of gravels for a medium sized (4–8 m wide) clay-bed channel in lowland south-central England. This is for the upper 150–200 mm of the bed. Experience has shown that very angular or angular washed gravels and cobbles are more likely to form a permanent, stable armour layer than rounded materials. There are numerous indices available for assessing the roundness or angularity of materials (e.g. Briggs, 1977). Typical size distributions have now been identified for a range of different river types in lowland England. Figure 11.3 shows a riffle reinstated in a river in Surrey in 1995. Experience shows that it is necessary to locally raise the bed such that the gravel 'breaks the surface' at low flow. This ensures that the flow is varied, leading to habitat diversity.

Table 11.2 Recommended grain-size distribution for reinstated substrate. Definition and classification of substrate materials: 'gravel-bed material' should consist of angular washed gravel in the proportions indicated

British Standard sieve size (mm)	Weight percentage passing
75	100
37.5	50–85
20	30–55
5	0–5

This is an early general specification used during the period 1990–1993 in Thames Region by the National Rivers Authority. Since that time research into grain sizes for several river types suggests that different gravel sizes should be used. A challenge is to find a supplier who can provide the exact size specified at a reasonable cost.

Figure 11.3 A riffle reinstated in a river in south-central England (courtesy of S. Reed)

Maximum bed shear stress values can also be used to determine the suitability of sizes of stone to be placed in maintained channels for other reasons. These reasons include low stone weirs placed on the channel bed to improve channel habitat, stone revetments for bank protection, and current deflectors (groynes) used to narrow the channel.

11.3.7 Siting of Deflectors

Current deflectors may be installed in bendways to deflect the main force of flow away from an outer eroding bank. These deflectors may be angled upstream so that bank erosion is not exacerbated at moderate flows. When deflectors are most needed, during bankfull flow, they tend to drown out and are ineffective.

Deflectors are also specifically constructed as habitat structures that protrude from one bank but do not extend entirely across a channel (Brookes, 1995b). The primary use of these devices is to generate scour pools by constricting the channel, and accelerating the flow. They should be located far enough downstream from riffle sites and be low enough to avoid backwater effects which would tend to drown out a riffle. Deflectors may be angled upstream, downstream or perpendicular to the bank. They can also be used to narrow over-wide channels and encourage local deposition. Figure 11.4 shows the siting of limestone groynes in an over-wide, lowland flood control channel in south-central England.

For habitat improvement purposes deflectors should be placed typically in straight reaches, or at the inside of a bend rather than the outer bank, although it should be noted that deflectors placed at the outer bank can also enhance habitat in certain circumstances. Deflectors in series are sometimes installed on alternate banks to produce a meandering thalweg and associated physical diversity (Nunnally and Shields 1985).

11.3.8 Bank Protection

Protection of the outside of a bend may be necessary where flows tend to converge and erode. This may be important where the velocities are high, and a suitable material may be

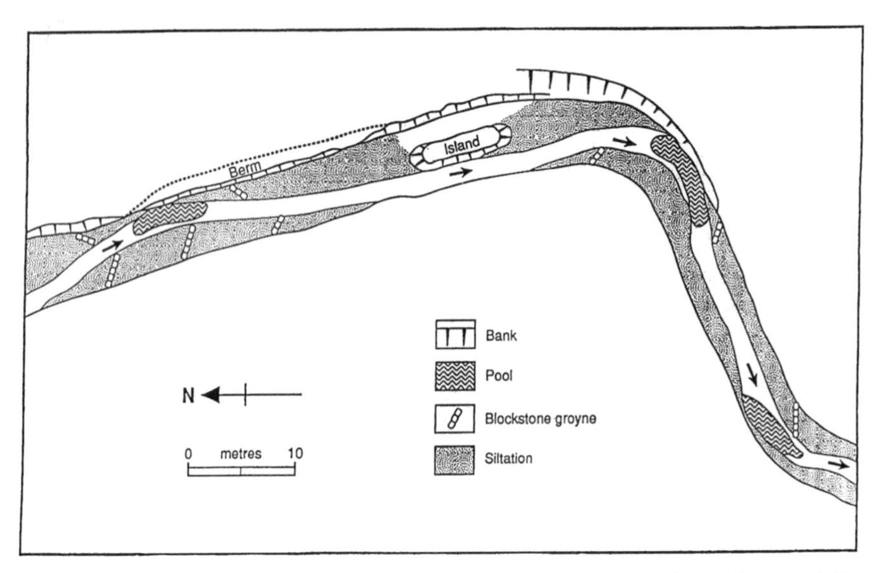

Figure 11.4 Use of deflectors for habitat improvement: River Colne in south-central England (after Brookes, 1995b)

large block limestone riprap. For earth channels in rural areas with low-energy rivers it may be necessary to ensure stability in the short term before natural vegetation becomes established. In these situations a variety of more natural treatments can be tried such as willow spiling, which involves the weaving of willow branches around stakes placed at the foot of a bank. The length of bank requiring treatment in both low- and high-energy environments can be determined from Leliavsky's (1955) principle of velocity reflection (Figure 11.5).

11.4 POST-PROJECT APPRAISAL

Comprehensive post-completion appraisals are rarely carried out, although recently in the UK there has been a slowly increasing number looking at various aspects, including project management, engineering, economics and environmental aspects such as river morphology. Gradually a picture of potential impacts and adjustments is being built up which will allow information to be fed back into the decision-making process. Documented case studies show that operational maintenance based on an understanding of form and process is successful where channel design is tailored to the local conditions (Brookes, 1990; Brookes and Shields, 1996). A sufficient number of traditional and geomorphological maintenance regimes should be evaluated over longer time periods to determine hydraulic and environmental performances, and to devise revised maintenance strategies if appropriate. It is recommended that geomorphological post-project appraisal be undertaken on selected completed projects and associated operational maintenance regimes, in order to identify the strengths and weaknesses of various approaches.

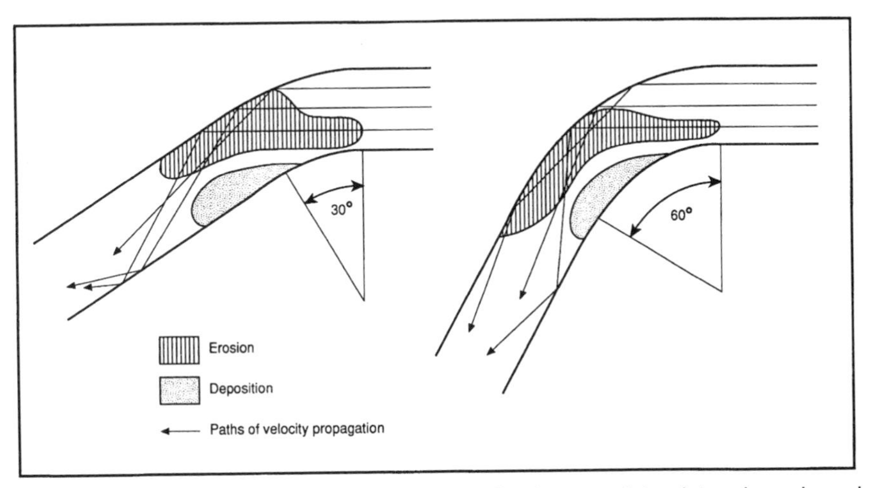

Figure 11.5 Leliavsky's principles of velocity reflection, used to determine where bank protection is required

11.5 CONCLUSION

Research and development work in the UK has demonstrated how geomorphology can be effectively applied to solve operational maintenance problems. There are now many instances where geomorphology has been used to develop guidance, procedures and training methods for river managers. In several countries a framework now exists for a geomorphological approach to river management, leading to more sustainable solutions. However, to ensure that projects are geomorphologically led, and even more extensively managed by geomorphologists, the way forward is to convince decision-makers of the cost-effectiveness and benefits of these alternative solutions. In the UK steps are being taken to evaluate the cost of the geomorphological approach against the benefits and cost savings that may accrue from adopting such an approach.

NOTE

The views expressed in this paper are the author's own and not necessarily those of his employer, the Environment Agency. The majority of this chapter was originally written and approved in 1992 when the author worked for a predecessor to the Environment Agency, the National Rivers Authority. The Environment Agency is currently (1996–1999) funding two PhD studentships which are examining morphological adjustments in managed channels.

11.6 REFERENCES

Brandon, T.W. (Ed.), 1987. *River Engineering – Part I, Design Principles*. Institute of Water and Environmental Management, London.

Briggs, D. 1977. *Sediments*. Butterworths, London, 192pp.

Brookes, A. 1987. The distribution and management of channelized streams in Denmark. *Regulated Rivers: Research and Management*, **1**, 3–16.

Brookes, A. 1988. *Channelized Rivers: Perspectives for Environmental Management.* John Wiley and Sons, Chichester, 326pp.

Brookes, A. 1990. Restoration and enhancement of engineered river channels: some European experiences. *Regulated Rivers: Research and Management*, **5**, 45–56.

Brookes, A. 1992. Recovery and restoration of some engineered British river channels. In: Boon, P.J., Calow, P. and Petts, G.E. (Eds), *River Conservation and Management.* John Wiley and Sons, Chichester, 337–352.

Brookes, A. 1995a. Challenges and objectives for geomorphology in UK river management. *Earth Surface Processes and Landforms*, **20**, 593–610.

Brookes, A. 1995b. The importance of high flows for riverine environments. In: Harper, D. and Ferguson, A. (Eds), *The Ecological Basis for River Management.* John Wiley and Sons, Chichester, 33–45.

Brookes, A. and Shields, F.D. Jr (Eds), 1996. *River Channel Restoration: Guiding Principles for Sustainable Projects*. John Wiley and Sons, Chichester, 433pp.

Dawson, F.H. 1981. The reduction of light as a technique for the control of aquatic plants – an assessment. In: *Proceedings of Conference on Aquatic Weeds*. Association of Applied Biologists, Oxford, 157–164.

Dawson, F.H. and Kern-Hansen, U. 1978. Aquatic weed management in natural streams: the effect of shade by marginal vegetation. *Verhandlungen der internationalen Vereinigung für theorestische and angewandte Limnologie*, **20**, 1440–1445.

Du Boys, P. 1879. Études due régime de Rhone et l'achan excercée par les eaux sur un lit à fond de graviers indéfiniment affoillable. *Annales des ponts et chaissées ser.* 5, **18**, 141–195.

Haslam, S.M. 1978. *River Plants*. Cambridge University Press, Cambridge, 396pp.

Herbkersman, C.N. 1981. *A Guide to the George Palmiter River Restoration Techniques*. Institute of Environmental Science, Miami University, 52pp.

Hey, R.D. et al. 1991. *Streambank Protection in England and Wales*. R&D Note 22, National Rivers Authority, Bristol.

HMSO. 1991. *Water Resources Act 1991, Chapter 57*. Her Majesty's Stationery Office, London, 259pp.

Hydraulics Research. 1987. *Morphological Effects of River Works: a Review of Current Practice*. Report for the Ministry of Agriculture, Fisheries and Food, Report No. SR116, Hydraulics Research Ltd, Wallingford.

Hydraulics Research. 1992a. *Morphological Effects of River Works: Recommended Procedures*. Report for the Ministry of Agriculture, Fisheries and Food, Report No. SR300, Hydraulics Research Ltd, Wallingford, 28pp.

Hydraulics Research. 1992b. *Morphological Effects of River Improvement Works: Case Studies*. Report for the Ministry of Agriculture, Fisheries and Food, Report No. SR151, Hydraulics Research Ltd, Wallingford, 11pp.

Hydraulics Research. 1992c. *Standards of Service Reach Specification Methodology.* Report prepared jointly with the University of Nottingham for National Rivers Authority, Thames Region.

Joglekar, D.V. 1971. *Manual on River Behaviour Control and Training*. Publication No. 60, Central Board of Irrigation and Power, New Delhi, India, 432pp.

Leliavsky, S. 1955. *Irrigation and Hydraulic Design*. Chapman and Hall, London, 492pp.

Lewis, G. and Williams, G. (Eds), 1984. *Rivers and Wildlife Handbook: A Guide to Practices which Further the Conservation of Wildlife on Rivers*. Royal Society for the Protection of Birds, and the Royal Society for Nature Conservation, Lincoln, 296pp.

Little, A.D. 1973. *Channel Modification: an Environmental, Economic and Financial Assessment.* Report to the Council on Environmental Quality, Executive Office of the President, Washington, DC, 37pp.

Nielsen, M.B. 1996. Lowland stream restoration in Denmark. In: Brookes, A. and Shields, F.D. Jr (Eds), *River Channel Restoration: Guiding Principles for Sustainable Projects*. John Wiley and Sons, Chichester, 269–289.

Nixon, M. 1966. Flood regulation and river training. In: Thorn, R.B. (Ed.), *River Enginee*, *Water Conservation Works*. Butterworths, London, 293–297.

Nunnally, N.R. and Shields, F.D. 1985. *Incorporation of Environmental Features in Flood Control Channel Projects*. Technical Report E-85-3, Environmental and Water Quality Operational Studies, US Army Engineer Waterways Experiment Station, Vicksburg, 232pp.

Patric, R. 1971. *The Effects of Channelization on the Aquatic Life of Streams*. Academy of Natural Science of Philadelphia, 12pp.

Richards, K.S. 1976. Channel width and the riffle–pool sequence. *Geological Society of America Bulletin*, **87**, 883–890.

Sear, D.A. 1996. The sediment system and channel stability. In: Brookes, A. and Shields, F.D. Jr (Eds), *River Channel Restoration: Guiding Principles for Sustainable Projects*. John Wiley and Sons, Chichester, 145–177.

Sear, D.A. and Newson, M.D. 1991. *Sediment and Gravel Transportation in Rivers, Including the Use of Gravel Traps*. Preliminary Report No. 232/1/T, National Rivers Authority, Bristol, 93pp.

Sear, D.A. and Newson, M.D. 1993. *Sediment and Gravel Transportation in Rivers, Including the Use of Gravel Traps*. Final Report No. C5/384/2, National Rivers Authority, 50pp.

Sear, D.A., Newson, M.D. and Brookes, A. 1995. Sediment-related river maintenance: the role of fluvial geomorphology. *Earth Surface Processes and Landforms*, **20**, 629–664.

Ward, D., Holmes, N. and José, P. 1994. *The New Rivers and Wildlife Handbook*. The Royal Society for the Protection of Birds, Sandy, Bedfordshire, 426pp.

Webber, N.B. 1971. *Fluid Mechanics for Civil Engineers*. Chapman and Hall, London, 340pp.

Section V

Case Studies and Applications

SECTION CO-ORDINATOR: M. D. NEWSON

12 Case Studies in the Application of Geomorphology to River Management

MALCOLM D. NEWSON[1], RICHARD D. HEY[2], JAMES C. BATHURST[3], ANDREW BROOKES[4], PAUL A. CARLING[5], GEOFF E. PETTS[6] AND DAVID A. SEAR[7]

[1] *Department of Geography, University of Newcastle, UK*
[2] *School of Environmental Sciences, University of East Anglia, UK*
[3] *Department of Civil Engineering, University of Newcastle, UK*
[4] *Environment Agency, Reading, UK*
[5] *Department of Geography, University of Lancaster, UK*
[6] *School of Geography, University of Birmingham, UK*
[7] *Department of Geography, University of Southampton, UK*

12.1 INTRODUCTION: THE GEOMORPHOLOGICAL CONTRIBUTION

During the 1980s and early 1990s a number of trends in public policy converged to produce a reorientation of environmental research towards practical applications. The almost universal acknowledgement that processes of economic development had to be modified either to avoid degrading valuable natural ecosystems or to intervene in their management led to precautionary investigations such as Environmental Impact Assessment. Developers began to seek the services of researchers with field experience, capable of rapid evaluation of a problem or, at the very least, the rapid isolation of those general or site-specific features requiring longer-term research or monitoring. Environmental protection agencies also sought such advice from a regulatory perspective and, at a time of falling financial support for 'strategic' research, environmental scientists were pleased to respond. There are always critics of applied studies but it is not just the survival instinct which has sustained those scientists who have become involved; they have mostly discovered that new perspectives and insights have resulted and that if 'strategic' and 'applied' science are separate entities they are symbiotic or synergistic in the right hands.

Geomorphologists have been in constant demand over this period, their familiarity with environmental process studies within a broader context having made them particularly appropriate to Environmental Impact Assessments and work in interdisciplinary teams; many also have a holistic appreciation of policy matters (see, for example, the papers gathered by Hooke (1987)). Owing to the domination of geomorphological

Applied Fluvial Geomorphology for River Engineering and Management, Edited by C. R. Thorne, R. D. Hey and M. D. Newson.

Table 12.1 Talents brought to geomorphology/engineering integration

From engineering	From geomorphology
Design experience	Field experience
Hydraulics	Sediment supply/transport
Project timescales	Longer timescales
Specialist function	Generalist breadth
Protection techniques	Erosion/deposition processes
Simple channel forms	Complex/sinuous channel dynamics
Reach scale	Basin scale
Accredited standards	Personal insights
Hard hat	Chest waders!

process studies by fluvial scientists during the 1970s there developed a cohort of specialists and a body of knowledge for applications in the 1980s and 1990s. Fluvial geomorphology expanded during this period by making research inroads into difficult topic areas such as gravel-bed rivers in the wild and extreme environments for which protection and/or management was being sought by policy-makers (see Hey et al., 1982; Thorne et al., 1987; Billi et al., 1992).

It has become a feature of the last decade that geomorphologists and engineers have begun working together more closely in the fields of both research and applications. They bring to any problem complementary skills (Table 12.1). The disgruntled civil engineer may claim that society is losing its way by encouraging indirect 'soft' techniques in river management and thereby providing openings for 'non-professional' inputs. Geomorphologists may in response cite the failure of 'hard' engineering and, although they have made some moves to institutionalise their professional approach, have tended to quote the value of truly independent, individual inputs to the field of environmental management.

The case studies reported here are, therefore, examples of the products of a growing bond formed from mutual respect for their mutually supportive approaches, skills and experiences. Several of them are segments of, or encompass, much broader studies with an even greater sweep of expertise. Figure 12.1 therefore categorises studies in which engineers and geomorphologists are beginning to work in tandem and those which are more complex in scope, requiring inputs from biologists, chemists, planners, lawyers and others.

Four main classes of river problem are outlined in Table 12.2 which presents an increasing component of environmentalism down the list. Fluvial geomorphology can, therefore, provide inputs which make direct intervention in river management more efficient, without the need for long-term maintenance to mitigate against environmental damage (though efficient engineering solutions also tend to work with natural processes in the longer term). We may then see geomorphology used in circumstances which are contingent to intervention, for example, in studies of river maintenance. Those applications under category 3 in Table 12.2, for which an understanding is required of the longer-term operation of both natural processes and artificial influences, are much more difficult to simplify for universal practical guidance. Finally, and most challenging of all, are opportunities to be proactively involved in the design of 'enhanced' or 'restored' river habitats (see Table 12.2).

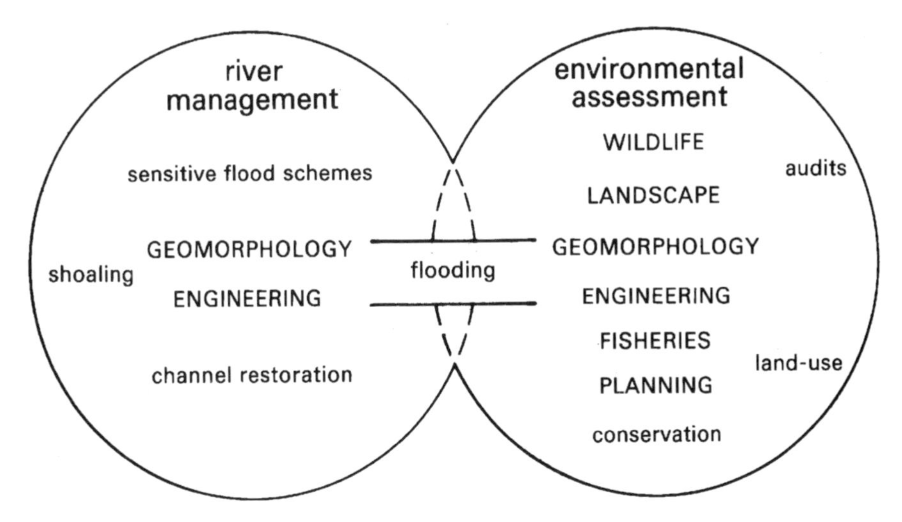

Figure 12.1 Situations in which engineers, geomorphologists and other disciplines work jointly on problems of river management

12.2 DIRECT INTERVENTION IN CHANNEL MANAGEMENT

Modification to a river system, either for capital flood alleviation and land drainage schemes or for heavy maintenance regimes, will precipitate a response by that river to the imposed changes. Alteration to the morphology of the river modifies the flow hydraulics and the sediment transport regime, not only under flood conditions, and can cause erosion and/or sedimentation within the engineered reach and both upstream and downstream from

Table 12.2 Tentative classification of geomorphological advice represented in the case studies

Engineering works	1. Direct intervention in channel management	• Channel design for flood protection • Channel structures/traps/river crossings
	2. Indirect/contingent intervention	• Erosion protection–bed/banks • Channel maintenance
Environmental management	3. Environmental change and mitigation	• Environmental impact of water resource schemes • Understanding/mitigating effects of land use/climate
	4. Proactive intervention	• Habitat improvement/restoration • Geomorphology in support of conservation

it (Brookes, 1987, 1988; Raynov et al., 1986, Hey, 1990). Such changes in the physical character of the river can adversely affect conservation resources and visual amenity.

Traditional methods for increasing the flood capacity of a river channel have included combinations of straightening, widening and deepening, often referred to as 'resectioning'. Such methods, which can lead to major instability problems, have been strongly criticised by conservation and amenity bodies for adversely affecting the river environment and for producing uniform, often straight, trapezoidal channels retaining few natural features and therefore of very limited conservation value (Lewis and Williams, 1984; Newbold et al., 1983; Water Space Amenity Commission, 1980).

Alternative methods are available, some of which have recently been advocated in response to the criticism of traditional approaches, which are considered to be environmentally more acceptable. These include, amongst others, by-pass channels and diversions, two-stage channels, flood banks set back from the edge of the river and partial dredging (Lewis and Williams, 1984; Newbold et al., 1983). Environmentally sensitive designs for flood alleviation and land drainage schemes involving the preservation of meanders, pools, riffles and vertical banks are preferable to traditional approaches, but careful design is required if the channel is to remain stable and the conservation resource is to be preserved (Hey, 1990).

12.2.1 Channel Design for Flood Protection at Environmentally Sensitive Sites

The environmental impact of a particular flood alleviation scheme depends on the nature of the works and the local river environment. Therefore, the performance of a scheme, in terms of stability, maintenance of flood capacity and conservation, fisheries and amenity values, needs to be evaluated under varying river conditions in order to establish its range of application. Distinction has to be made between high-energy (usually upland) rivers which transport bed material load and low-energy (usually lowland) rivers which do not, since the former will quickly respond to any imposed changes in the bed material transport capacity of the river and become unstable. Equally, differences are expected between rural and urban schemes as rural schemes generally have a greater intrinsic environmental value than the latter and, therefore, are more susceptible to adverse changes. However, it should be recognised that in urban areas there may be limited space or other restrictions which can curtail flood control options. In these circumstances it is a question of adopting the scheme which minimises environmental impact.

The new approach to channel design can be described by reference to a research project conducted for the UK Ministry of Agriculture, Fisheries and Food (Box 12.1) which had two main objectives.

1. To evaluate the design performance of a range of flood alleviation and land drainage works which have been implemented in different parts of England and Wales in terms of discharge capacity, channel stability, construction and maintenance costs and conservation value.
2. To develop guidelines for the design of stable flood alleviation and land drainage schemes on mobile-bed channels which preserve the natural stability of the river and their conservation and amenity value on the basis of objective 1.

Box 12.1

Eighteen flood alleviation schemes in England and Wales were surveyed in order to evaluate their engineering performance in terms of their effect on channel stability and maintenance of discharge capacity, together with their impact on the riverine environment. The schemes were chosen to be representative of traditional engineering approaches involving dredging, widening and straightening or the construction of flood banks adjacent to the river, and more environmentally sensitive approaches such as two-stage channels, diversion channels together with those involving the construction of distant flood banks. In order to evaluate the various approaches in different river environments, schemes were chosen in upland and lowland areas with urban and rural locations (Figure 12.2; Table 12.3).

River corridor surveys were carried out at each site with more detailed information on instream and riparian flora for each surveyed cross-section. Regrettably, no pre-scheme river corridor surveys were carried out at any of the sites to enable the environmental impact of the schemes to be directly assessed. To overcome this problem, controls were established in natural sections adjacent to six of the engineered reaches (Figure 12.2). A space–time substitution enables the pre-scheme conditions to be estimated for comparison with the engineered situation.

Various models for predicting the stability of flood alleviation schemes were reviewed. These included regime equations, the permissible velocity method, the critical tractive force method, Griffiths stability index, shear stress analysis, stream power analysis, sediment transport balance and one-dimensional mathematical modelling procedures. The best results, in terms of forecasting observed erosion and deposition at the 18 engineered reaches investigated, were obtained by use of sediment transport equations. Comparisons between predicted transport rates at adjacent sections enables eroding, stable and depositing reaches to

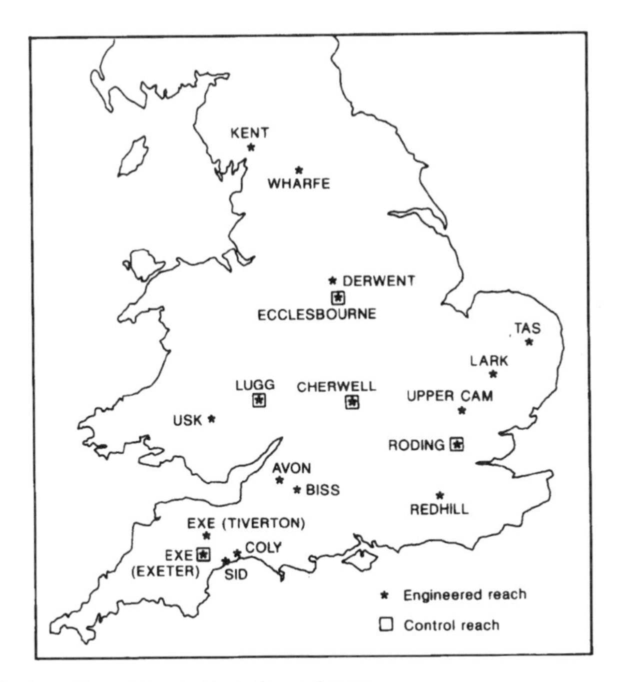

Figure 12.2 Location of flood alleviation schemes

Table 12.3 Classification of flood alleviation schemes (after Hey and Heritage, 1993)

	Urban	Rural
Upland	Usk (Brecon) W,D,B Ecclesbourne (Duffield) W,S,C,R,B Derwent (Matlock) D,B,R Kent (Kendal) W,D,B,R Exe (Exeter) W,D,R,M Exe (Tiverton) W,D	Lugg (Byton) R,F Wharfe (Conistone) F,P
Lowland	Cherwell (Banbury) W,D,S Sid (Sidmouth) D,B,R Avon (Bath) D,B,R Biss (Trowbridge) D,B Lark (Bury St. Edmunds) W,D,C,B Upper Cam (Saffron Walden) W,D,T Coly (Colyton) R,M,F	Redhill Brook (Redhill) D,C,T,W Roding (Abridge) A Coly (Colyford) R,M,F Tes (Newton Flotman) A,M Upper Cam (Saffron Walden) W,D,T

W = widening; D = deepening; S = straightening; T = trapezoidal section; C = concrete-lined channel; B = bank revetment; R = weirs; F = flood banks; M = diversion channel; P = partial dredging; A = two-stage channel

be identified with reasonable accuracy. The Ackers and White transport function (Ackers and White, 1973) performed consistently better than any of the other equations.

The discharge capacities of the various schemes were assessed to determine whether they had maintained their design performance. This presupposes that the channel, when constructed, would have transmitted the design flood. Inevitably, there is no way of establishing whether that was the case. The appraisal, therefore, estimates the current discharge capacity of the scheme on the assumption that the original design capacity was correctly assessed. A range of flow resistance equations is available for estimating discharges. The Hey resistance function (Hey, 1979) was used for this purpose as it has proved to be as accurate as any other for estimating flows under UK conditions. In terms of engineering performance, the only stable schemes were those incorporating distant flood banks, owing to the fact that the sediment transport regime of the river is not affected. All other schemes alter the sediment transport regime and, thereby, cause erosion and/or deposition. The degree of instability varies: those on upland rivers are more unstable than lowland ones because they transport more bed material load and, as a consequence, are more vulnerable to local changes in the sediment transport processes occasioned by the engineering works.

Resectioning creates the greatest problems in upland rivers, causing massive aggradation, while in lowland rivers, maintenance dredging can readily sustain scheme performance (Hey et al., 1990). Natural channels adjust their bankfull flow to the discharge which, in the long term, transports most sediment. This is imposed by catchment conditions. Consequently, any attempt to arbitrarily increase the 'inbank' capacity of the channel above the natural value will promote instability as the river attempts to re-establish its natural bankfull capacity. This, in turn, will destabilise neighbouring reaches upstream and downstream.

In terms of environmental performance, the physical characteristics of the channel under bankfull conditions control aquatic habitats and hence exert a strong influence on instream species in terms of number present, diversity and key species. For some species this is independent of geographical location, suggesting that these particular species may be tolerant of variations in other unmeasured parameters, such as water quality. However, there are some species which are specific to particular sites despite the fact that they have similar engineering treatments and comparable physical characteristics to other rivers. This indicates that either

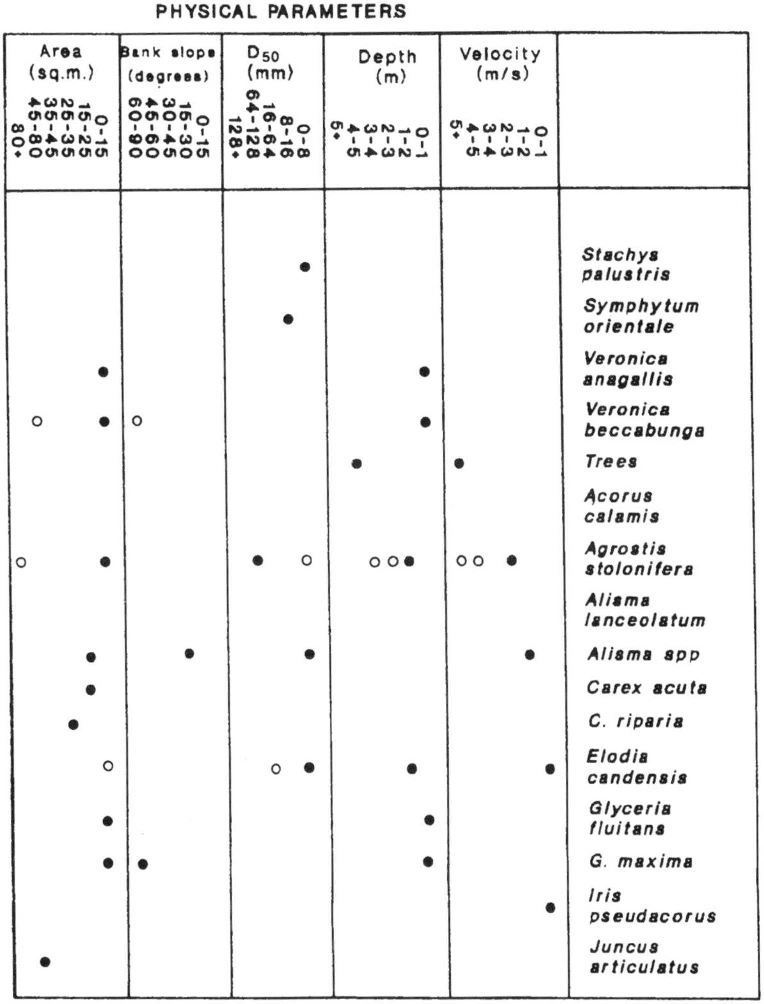

PHYSICAL PARAMETERS

Area (sq.m.): 0-15, 15-25, 25-35, 35-45, 45-80, 80+
Bank slope (degrees): 0-15, 15-30, 30-45, 45-60, 60-90
D_{50} (mm): 0-8, 8-16, 16-64, 64-128, 128+
Depth (m): 0-1, 1-2, 2-3, 3-4, 4-5, 5+
Velocity (m/s): 0-1, 1-2, 2-3, 3-4, 4-5, 5+

Stachys palustris
Symphytum orientale
Veronica anagallis
Veronica beccabunga
Trees
Acorus calamis
Agrostis stolonifera
Alisma lanceolatum
Alisma spp
Carex acuta
C. riparia
Elodia candensis
Glyceria fluitans
G. maxima
Iris pseudacorus
Juncus articulatus

Figure 12.3 Physical preferences of some instream flora: ○, favoured conditions; ●, unfavourable conditions

water quality or the low flow characteristics of the site may be alternative controlling factors. Comparisons between engineered and adjacent control reaches enable water quality effects to be held constant. Paired assessments were carried out for engineered/control reaches on five rivers and this revealed which species and habitats were most affected by the various engineering treatments. The general conclusion was that engineering works had reduced the number of desirable species, such as Celery-leaved Crowfoot and Spiked Water Milfoil, while increases were observed in less desirable species such as Great Willowherb and various algae and weed species.

However, further analyses, on a river by river basis, revealed that some schemes had maintained diversity and preserved and enhanced the occurrence of key species. These changes were explained by the loss or gain of controlling habitats. Two-stage channels and rivers with flood embankments set back from the channel proved to be ecologically superior to more traditional engineering options and, indeed, may actually increase species diversity. Consequently, river engineering works may not necessarily be ecologically detrimental to the river but have the potential to enhance the river environment. Data from every cross-section in both engineered and control reaches enabled a model to be developed relating species occurrence to physical habitat features. Preferred flow depths, velocities, bed material sizes, flow areas and bank slopes were established for all species. As the data were aggregated the engineering treatments were reflected in the predictor variables and the relations clearly have more general application (Figure 12.3).

These relations can be used to predict, at the design stage, the likely response of the river to any particular engineering scheme. Provided that information is available on the existing condition in the river, this approach enables the environmental impact of the scheme, in terms of its effect on river-bed plant species, to be assessed (Hey et al., 1990, 1994).

The geomorphologically based appraisal programme enables comparisons to be made between the engineering and environmental performance of the different flood alleviation works in upland/lowland and urban/rural settings. Set-back levees, where the banks are located along the edge of the meander belt, are the preferred solution as they have no impact on the natural stability of the channel or on the aquatic and riparian environments. Should space restrictions preclude the use of set-back levees, alternative procedures are required, but these can potentially destabilise the river to some degree.

In lowland, urban areas, two-stage channels are a viable alternative, as siltation will be minimal and they can actually increase environmental value. The flood berms can be maintained to provide a linear urban park. If insufficient space is available for this approach, bank-adjacent levees/flood walls are acceptable provided the natural alluvial banks are preserved. Resectioning should only be contemplated as a last resort, as it completely destroys the riverine environment and requires maintenance dredging.

At urban, upland sites, bank-adjacent levees are preferable to relief channels or re-sectioning since they have less impact on channel stability and the river environment. Resectioning can cause serious instability problems and should not be attempted. Two-stage channels are likely to cause similar instability problems owing to the tendency for siltation on the flood berms.

In rural, lowland regions, two-stage channels can be successful provided that the vegetation on the flood berm can be managed to preserve the design discharge capacity of the scheme. Instability need not be a problem, provided that the flood channel is properly aligned and the riverine environment can be enhanced. Bank-adjacent flood banks are preferable to diversion channels and resectioning if space is limited. Resectioning should not be attempted because, even if maintenance requirements are relatively low, there will be a very adverse environmental impact.

Bank-adjacent levees are the only real alternative to set-back levees in upland rural regions. All other procedures will create expensive instability problems and cause unacceptable environmental damage.

12.2.2 Channel Diversions

When designing river diversions, distinction needs to be drawn between upland type rivers, which transport considerable amounts of bed material load, and lowland types where suspended load predominates (bed material transport is often trivial but may occur as sand in suspension). With the former, transport is essentially supply controlled while with the latter it is hydraulically controlled.

In upland rivers it is necessary to maintain bed material load transport continuity through the reach and there is little flexibility in choice of design options as the required channel morphology to maintain stability is prescribed by nature. Only if artificial channel stabilisation is used, for example grade controls and bank revetments, is it possible to vary design procedures. Equally, it is not possible to arbitrarily increase the bankfull discharge in the diverted section as this will certainly promote instability (Hey et al., 1990).

In order to maintain stability in lowland rivers it is only necessary to ensure that inbank flows are below bed material transport thresholds, or that any potential transport is trivial and matches that in adjacent reaches. This allows more flexibility in terms of design procedures and increased channel capacities. See Boxes 12.2 and 12.3.

Box 12.2

West Glamorgan County Council, South Wales, UK, proposed to construct a new dual-carriageway road from Aberdulais to Glyn Neath. The proposed route involved the diversion of the River Neath in several locations. Consultancy services were required to assist them in the design of the river diversions, to promote the scheme and give advice during its construction. The original geomorphological objectives were to collaborate with fisheries and ecology specialists to carry out an environmental impact of the proposed diversions. This was required as part of the planning and consultation process.

During the course of this work it became apparent that environmentally sensitive solutions were required for the diversions in order to obtain consent and, as the geomorphological survey formed the basis for the development of natural design methods, the geomorphological brief was extended to include the design of the diversions. Specific objectives for the study required that the diversions:

1. maintained the natural stability of the river and its sediment transport regime;
2. maintained the natural inbank flood capacity of the channel;
3. incorporated pools, riffles and meanders;
4. maintained the natural size and variability of the bed material;
5. wherever possible, incorporated natural bank features, trees and shrubs and minimised the use of revetments while maintaining bank stability;
6. provided a range of natural instream habitats for fisheries;
7. incorporated a range of instream and bankside features to retain the ecological diversity within the river's riparian corridor.

To design the diversion channels to meet these objectives, a geomorphological survey was required to define existing conditions, natural and engineered, along each reach. Topographic surveys were carried out along the existing channel at each diversion and for seven channel widths upstream and downstream to define approach and exit conditions. This defined riffle/pool and intermediate sections and enabled a contoured plan to be produced. Geomorphological surveys along each reach identified key habitat features, their length and location, and these features included pools, riffles, runs, bar forms, dead zones, vertical banks, protected banks, types of protection, tree cover/shrubs, fish weirs, bridge piers and abutments, fallen trees, etc. (Figure 12.4).

These geomorphological surveys indicated that the river was quite heavily engineered. To prevent bank erosion and meander migration, blockstone revetments had been constructed at numerous locations, while all gravel shoals had been dredged to maintain its flood capacity. To compensate for loss of habitat features a number of fish weirs had been constructed. Sediment surveys, surface (grid-by-number samples) and subsurface (bulk-by-weight sample) were collected at key riffle and pool sections to define bed material size and local variability. Existing geotechnical information defined the nature and shear strength characteristics of the floodplain material both along the banks of the existing channel and along the lines of potential diversion routes.

At the commencement of design the general dimensions of the diversions were defined on the basis of existing river conditions. In order to prevent erosion or deposition within the diversions, given that the bed and bank material in the diversions corresponded to those in the existing channel, the local slope of the river had to be maintained. This, in turn, defined the sinuosity of the realigned river, as the length of each diversion channel had to be equivalent to the length of each cutoff section.

Given the local regime width of the river (both natural in the tree-lined sections and artificial where maintained by revetments), the distance between crossings (and riffles) along

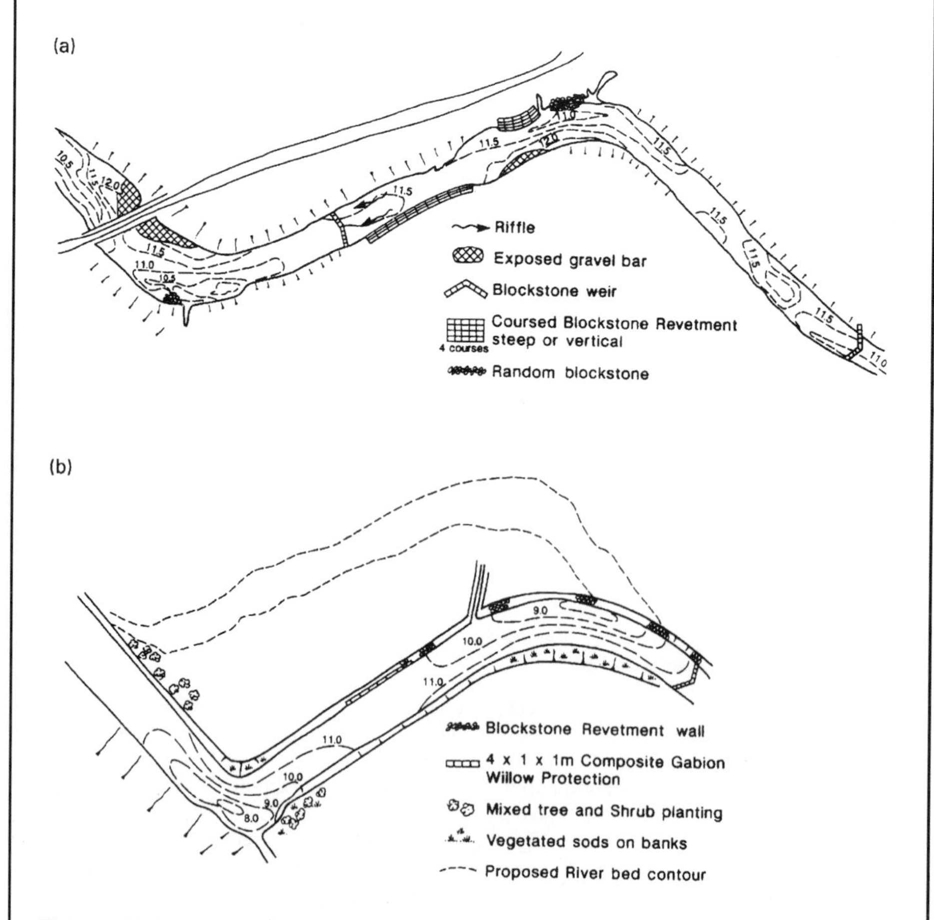

Figure 12.4 Design of diversion, River Neath, south Wales (a) Geomorphological map of original channel; (b) proposed diversion (from Hey, 1992. Reproduced by permission of the Institute of Fisheries Management)

the reach could be specified (2 × pi × channel width (Hey, 1976)). Together with sinuosity, the crossing spacing then defines the planform shape of the river. Care was exercised to ensure that the bends had varying planform geometries within the range exhibited in the adjacent 'natural' reaches. At one site, flow was directed at right angles into the valley side to create a complex 'hammer head' bend with a deep pool, sustained by the main current, and a shallow depression on the upstream side of the bend, maintained by the reverse eddy. A small gravel bar separated the two pool features (Figure 12.4).

Cross-sectional characteristics, in particular mean and maximum depths in pools and riffles, were determined using empirical regime equations for gravel-bed rivers (Hey and Thorne, 1986) given information on dominant discharge, bed material and bank vegetation characteristics. These equations enabled the general channel sizing to be established, while maximum depths in the meander bends were assessed using Apmann's equation, which relates maximum flow depth to bend curvature (Apmann, 1972) (Figure 12.4).

The channel designs were checked to establish that the bankfull capacity of the diversion corresponded to that in adjacent sections using flow resistance equations developed for gravel-bed rivers (Hey, 1979). Sediment transport continuity was confirmed by calculating bankfull bed material load transport rates through the reach (Ackers and White, 1973). The requirement for bank protection was assessed by calculating the stability with respect to bank failure based on the geotechnical properties of the bank and its proposed geometry (Thorne and Tovey, 1981). Where the cutoff channel was to be backfilled, harder revetments involving blockstone were required since this was on the apex of the bend and proximal to the new road embankment. Maximum scour depths in the bend were calculated (Apmann, 1972) to determine foundation depths for the revetment. In general, softer bank protection options were employed, involving willow faggots and spiling, logs or willow fascines, alder and willow trees. On many of the diversions, the establishment of natural vegetation was found to provide sufficient bank protection. The design procedures employed ensured that a natural and environmentally sensitive solution was produced. In view of the engineered nature of the existing river and limited morphological variability, the diversions actually enhanced the habitat value and diversity. The number of riffles was substantially increased, which will further encourage salmon spawning. Shallow, dead-zone areas have been created, which will provide a nursery facility, while larger pools, with associated tree cover, will provide excellent rest areas for the mature fish. The range of habitats incorporated in the design will also encourage colonisation by a natural range of instream macrophytes and riparian flora and fauna.

The scheme was subject to a Public Inquiry in 1989 and obtained planning approval. It was completed in 1996.

Box 12.3

In order fully to exploit the valley gravel resources at Reymerston quarry, Norfolk, UK, it was necessary to divert the River Blackwater from its position in the middle of the floodplain to the valley margin. The planning authority required Atlas Aggregates (the operators) to submit detailed proposals for the diversion, which had to meet certain objectives, prior to giving planning permission. The National Rivers Authority, which sanctions all drainage work on natural watercourses in the UK, required that the proposed diversion should:

1. maintain the drainage standard for the area, in terms of the flood capacity of the river;
2. maintain current bed levels to facilitate land drainage;

3. maintain and, where possible, enhance the conservation, fisheries and recreational amenity value of the river.

The Blackwater is a small, chalk-fed stream which flows roughly west to east to join the River Yare at Hardingham Mill. Prior to land drainage works in the early to middle part of the 19th century, the Blackwater had a sinuous course along the reach in question. Channel improvements to aid land drainage involved the construction of straight, deep, trapezoidal channels. The gradient of the river was increased by this action, in proportion to the reduction in channel length, and the channel deepened by approximately 0.8 m.

A geomorphological and topographic survey of the existing channel was carried out to establish which features, if any, should be recreated, and to assess the hydraulic performance. The survey showed that the river was uniform and trapezoidal with a bed width of 3–4 m, banktop width of 9–10 m and an overall mean depth of 1.7–2.1 m. The bed of the river did not exhibit any pools or riffles and it was composed of compacted gravel overlain in parts by sand, silt and organic material up to a depth of 0.3 m. The vegetated banks were relatively steep, 60–70 degrees, but not vertical.

The river had no natural geomorphological features of note. A companion ecological river corridor survey also indicated that it was devoid of habitat value and interest. The lack of morphological variability and associated hydraulic conditions were responsible for this condition.

The design for the diversion aimed at recreating the river prior to straightening in the mid-1800s, but with a lowered bed elevation to preserve low flow levels. This was achieved by excavating a two-stage channel with the river located within a small floodway 15–20 m wide and 0.8 m below the existing floodplain level.

Within the newly created floodway the river was given a sinuous course, comparable to the pre-drainage condition. The design width for the channel (5.5 m for riffle sections) was based

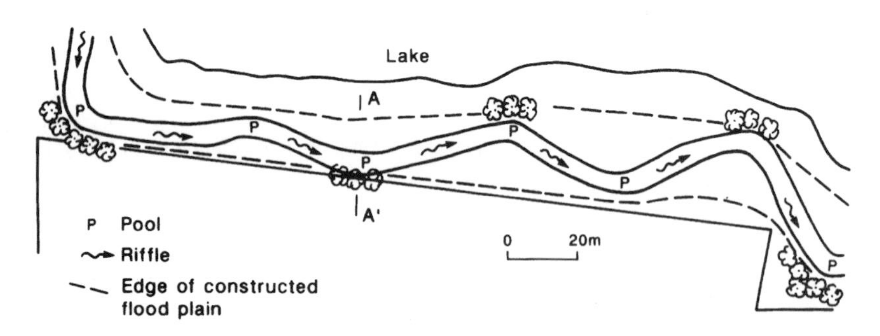

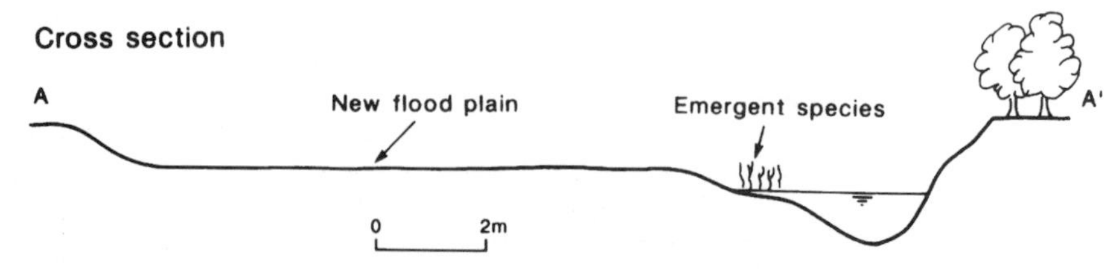

Figure 12.5 Design of diversion channel at Reymerston, River Blackwater (from Hey, 1992. Reproduced by permission of the Institute of Fisheries Management)

on the width of a relatively unaffected reach upstream. This enabled the meander arc length/riffle spacing to be defined (four to 10 times channel width). As this corresponded to the pre-drainage planform geometry, it confirmed the choice of channel width.

A maximum depth of 1.0 m was prescribed for the riffles, this being based on the maximum depth of the existing river and the estimated dredged depth (Figure 12.5). Flow resistance calculations (Hey, 1979) indicated a bankfull discharge for the new channel of about 2.8 $m^3 s^{-1}$. Calculations of channel capacity based on regime equations for stable gravel-bed rivers with between 1 and 5% tree and shrubs lining the banks (Hey and Thorne, 1986) gave a bankfull flow of 2.8 $m^3 s^{-1}$ which confirms the design parameters.

As the overall bankfull capacity of the existing dredged channel is 6.3 $m^3 s^{-1}$, checks were carried out to ensure that the river-plus-floodway could transmit this flow. Flow resistance calculations, based on aggregating discharges on the berms and within and above the channel (Posey, 1967), indicate that flows of 8.9 $m^3 s^{-1}$ could be transmitted with mown berms or about 8 $m^3 s^{-1}$ if they were left uncut.

Calculations were made to assess bed material stability and these indicated that material less than 6 mm in diameter could be transported: a similar size range to that transported in the existing channel. The channel was constructed in 1991 and while vegetation was becoming established on the berm surface, flow through the diversion was restricted by means of a culvert. The range of habitats created in the new channel has led to colonisation by species which were not present in the engineered channel. Fish have also populated the new reach in preference to the existing channel. Flows up to bankfull have been experienced in the new channel without erosion or bank failure. Following a major flood in 1992, the diversion was fully opened. Significantly, neither the floodway nor the channel experienced any erosion, even though the former was not fully vegetated by that time. Within eighteen months of construction, 'kick sampling' of the substrate inidicated that macroinvertebrates had fully recolonised the bed.

12.2.3 River Crossings: Fluvial Geomorphology for a Major Pipeline Project

Under European environmental legislation, development/construction projects of a certain size or in certain categories require an Environmental Assessment (EA) before they can proceed. The environmental impacts of the development can then be mitigated and monitored. Pipeline construction/installation requires EA and the case study presented here represents the efforts of a number of fluvial specialists towards the appraisal of the Shell Chemicals Northwest Ethylene Pipeline from Grangemouth (SE Scotland) to Stanlow (NW England), a route crossing a number of large, mainly gravel-bed rivers (Figure 12.6; Table 12.4).

Environmental Assessment is, by its nature, holistic; thus topics considered at each river crossing were very broad, including effects on adjacent wetlands, 'externalities' such as silt disturbance and associated increased turbidity during channel works and the effects of floodplain flows. However, Shell's agents for the EA and their pipeline engineers also commissioned much more focused assessment of river channel stability around the proposed crossings. Since installation costs rise steeply if the line is bent or buried deeper than 0.9 m these costs must be compared with those of damage to the line or to the environment if the simple 'direct line' philosophy leads to channel destabilisation.

In the contribution by Newcastle University, the entire pipeline route was studied from 1:10 000 maps, using planform indications of reach stability such as abandoned meanders

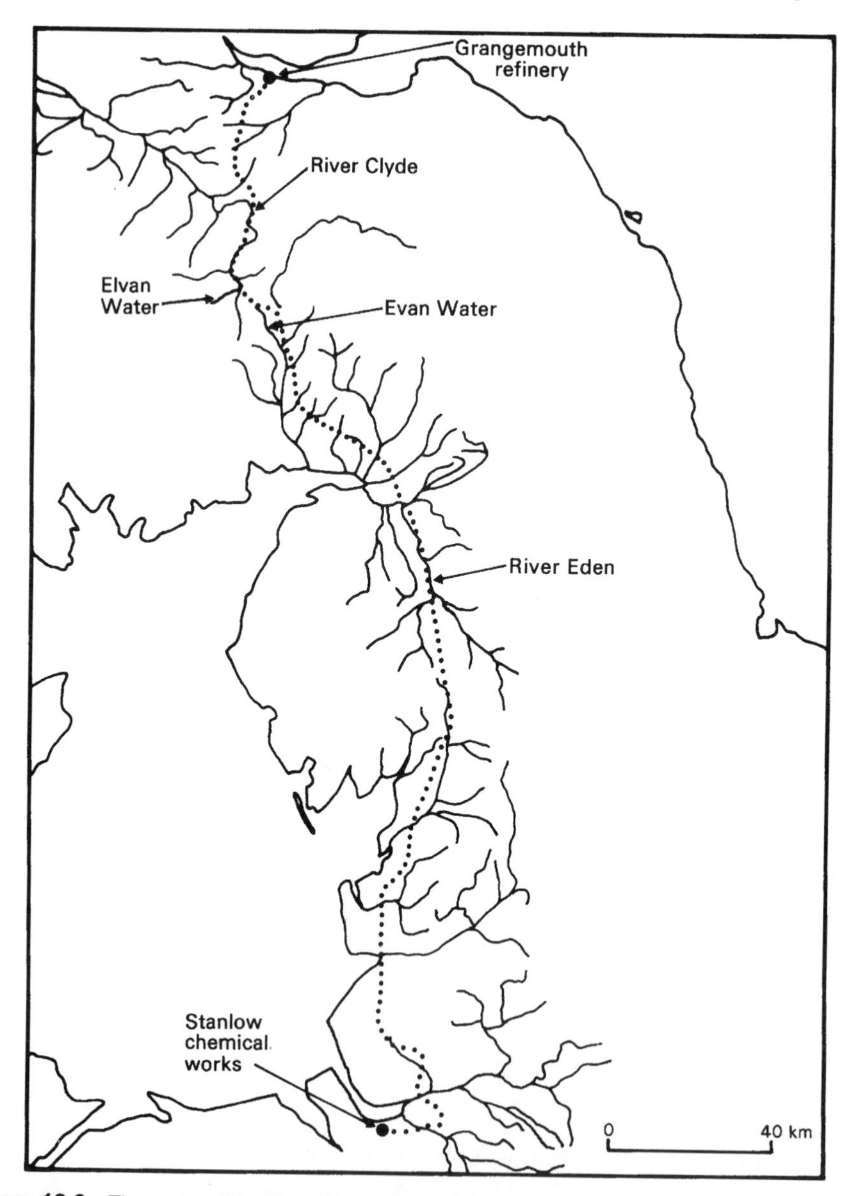

Figure 12.6 The route of the Shell Chemicals Northwest Ethylene Pipeline (shown dotted)

and active river cliffs to select critical sites requiring field visits. A matrix approach was used (Figure 12.7).

During the field inspection of potential 'problem sites', a number of indicators of channel stability (tabulated first by Lewin et al. (1988) and refined by Sear and Newson (1992)) were recorded upstream and downstream of the proposed crossing. A photographic

Table 12.4 Specification for the pipeline

Pipeline specification: 406 km long, Grangemouth–Stanlow; 25 cm diameter
Installation details: Trenched to 0.9 m
Number of river crossings: 137 identified at 1:10 000
Number with potential channel stability problems: 12
Contract advice sought on

- General geomorphology of route: Aberdeen University
- Geomorphology of river crossings: Newcastle University
- Stability calculations for problem crossings: Institute of Freshwater Ecology

Categories of stability prediction

- Entrainment of bed material load
- Transport of bed material load
- 'Regime' channel dimensions
- Bank stability
- Meander migration rates
- River bed scour

record was compiled including scaled photographs of bed sediments for subsequent size analysis.

For those sites where precautionary action by the pipeline engineers was warranted, a separate study provided predictions of the major components of instability in a quantified way suitable as an aid to mitigation measures, including bank protection, deeper emplacement or even choice of a new crossing site.

The work on each site is summarised in Table 12.5 and Figure 12.8. The engineers were provided with a guidance document listing, with bibliographic references, those techniques for prediction which are both practical and validated for the particular conditions of British rivers (Table 12.6). While the origins of the techniques are from engineering science or sedimentology in the main, this review is at pains to point out the importance of the limits to their applicability framed by the site-specific boundary conditions under which they are expected to be valid. Furthermore, this geomorphological bias can be found in the stress laid on historical conditions at the sites. Both sets of information can only be acquired by careful geomorphological fieldwork if they are to be achieved under the short timescale of EA. The details of this stage of the EA are shown in Box 12.4.

12.3 INDIRECT/CONTINGENT INTERVENTION: PROTECTING AND MAINTAINING CHANNEL CAPACITY

Fluvial geomorphologists in the UK are drawn in increasing numbers into work on 'problem sites' where traditional engineering solutions cannot be financed, are environmentally unwelcome or have been tried and have failed. Erosion and deposition problems are the most frequent in this category and involve a political constituency – the riparian owners – which may take independent action at some sites or put pressure on the water management institutions for some form of protection, especially where erosion or deposition leads to a reduced flood protection.

	CROSSING				REACH (downstream)			INTERFLUVE	AQUIFER
SITE	F	I	P	S	F	FP	P	C	P
a	0	0	1	0	0	0	1	0	1
b	0	0	1	0	0	0	1	0	1
c	1	0	2	2	0	0	2	0	0
d	0	0	1	0	0	0	1	0	0
e	2	0	2	2	2	0	2	0	0
f	1	0	1	1	0	0	2	0	0
g	1	0	1	1	0	0	2	0	0
h	2	1	1	1	0	0	2	0	0
i	1	1	1	0	0	0	2	0	0
j	1	1	1	0	0	0	2	0	0
k	1	1	1	0	0	0	2	0	0
l	1	1	1	0	0	0	2	0	0
m	1	0	2	2	0	0	2	0	0
n	1	0	2	2	0	0	2	0	0
o									

F = flooding (affecting construction/maintenance).
FP = flood protection (works requiring reinstatement if breached).
P = pollution – refers mainly to chemical pollution.
S = siltation – as a downstream effect of construction (erosion/siltation of pipeline itself considered under 'instability').
I = instability – liability of crossing or reach to degrade or aggrade its bed or banks during the useful life of the pipeline (25 years in first instance).
C = conservation problems.
0 = impact not applicable.
1 = impact should be considered/checked in addition to normal precautions.
2 = impact likely to require some mitigating action.
3 = impact important enough to require strong mitigating action or re-routing of pipelines.

Figure 12.7 Matrix of impacts by reach and scale of impact for Northwest Ethylene Pipeline

12.3.1 Erosion Protection: Banks

Traditionally, bank erosion is controlled by protecting the eroding bank with some form of revetment ranging from hard riprap, blockstone or gabion structures to softer methods using vegetation. In essence these treat the symptoms, with some form of reinforcement or resistant covering, and they are often prone to failure or outflanking (Hemphill and

Table 12.5 (a) Site details for one crossing (River Clyde) by the Northwest Ethylene Pipeline

River name: Clyde (see Figure 12.6)
Grid reference: NS965442
Distance from source: 48.5 km
Photo: 16
Hydraulic data:
Slope: 0.0010
Bankfull width: 50.0 m
Bankfull depth: 3.5 m
Low-flow width: 30.0 m
Low-flow depth: 1.0 m
Cross-section: trapezoid
Bank inclination: 20 degrees
Vegetation: grass

Notes on morphology and slumping/ instability
This is a large, low gradient gravel/sand-bedded river. The point of crossing is on a straight section of the reach just downstream of a tight meander bend. No dunes were visible but areas of sand and silt deposition were observed. The site is on a large floodplain with an active meander system. Some slumping by undercutting and block movement is present just upstream of the site on the left bank. Banks are composed of reworked gravels, overlain by soil. There is no evidence that scouring will take place although a large-scale lateral shift in channel position could be possible. (This seems very unlikely in the pipeline timescale.) This is a good crossing point with only minimal bank erosion/deposition (relative); however, should meanders shift, pipe will be exposed elsewhere on the floodplain.

Notes on remedial/protection measures
Bank protection needed where trench is sited and immediately upstream. The pipe should be buried at maximum depth (10 m) into the floodplain on both banks.

Table 12.5 (b) Hydraulic/geomorphic data for a problem crossing

Site: Elvan Water (80.9 km)
Hydraulic radius: 0.7500
Average water depth: 1.0000
Water slope: 0.0166
Shear velocity: 0.3490 m s^{-1}
Mean flow velocity: 3.250 m s^{-1}
D_{35}: 0.0500 m
D_{50} (sub): 0.0290 m
D_{84}: 0.1205 m
D_{50} (armour): 0.0720 m

Values of constants in calculations:
g: 9.81 m s^{-2}
Water density: 998.0 kg m^{-3}
Kinematic viscosity: 0.13100 × $10s^{-5}$ ms^2 s^{-1}
Sediment density: 2650.0 kg m^{-3} (at 10°C)

Predicted bed material loads
Meyer-Peter and Muller (1948) formula: 5.35 kg m^{-1} s^{-1}
Ackers and White (1973) formula: 5.38 kg m^{-1} s^{-1}
Parker et al. (1982) formula: 42.00 kg m^{-1} s^{-1}

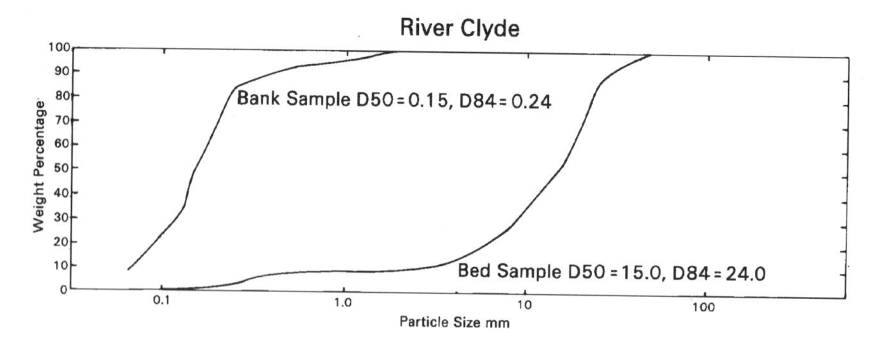

Figure 12.8 Size analyses of bed and bank materials for pipeline crossing of the River Clyde

Table 12.6 Geomorphological inputs to the assessment of river stability at pipeline crossings

1. *Entrainment of gravel into the flow from beds of mixed grain sizes*
   Velocity profile above the bed; driving force and resisting force; Shields diagram; choice of critical particle size to reflect armour/protrusion etc. (Shields, 1936; Komar, 1987; Carling, 1990)

2. *Bedload transport*
   Special problems of gravel size range. Capacity load versus supply limitations from upstream and from reach. Types of grain motion; routing of mixed sediments by size fraction. Shear stress; stream power (Meyer-Peter and Muller, 1948; Ackers and White, 1973; Parker et al., 1982)

3. *'Regime' channel dimensions*
   Stable combination of width, depth and channel slope for imposed discharge and sediment load in straight channels. Degrees of freedom also include hydraulic radius, wetted perimeter, max. flow depth, slope, velocity. Bankfull frequency 1.5 years (Charlton, 1977; Hey, 1982; Hey and Thorne, 1986)

4. *Bank stability*
   Bank retreat controls the aspects of river mechanics, though may result from channel incision. Sediment composition of banks important. Tensile, shear or beam failure (Thorne and Tovey, 1981; Thorne, 1982)

5. *River bed scour*
   Lowering of bed may be due to tectonics, the result of long-term change in sediment supply, an effect of transient movement of bars, or may result from coherent turbulent structures in the flow. Maximum possible depth of scour needs to be known (Laursen, 1962; Jain and Fischer, 1979; Jones, 1984; Colorado State University, 1975)

6. *River meander migration rates*
   Natural rates are 0.1–5.5% of the channel width in UK. Maximum steady rates up to 2.8 m per annum recorded. Forms of meander migration are classified as extension, translation etc. (Hooke, 1980; Hooke and Redmond, 1989)

Box 12.4

A total of 32 river crossings were identified as requiring a site inspection to ascertain if hydraulic calculations would be necessary to determine channel stability. A separate report detailed site reconnaissance information. Following site inspection it was concluded that hydraulic stability calculations might be required at 24 river crossings and that at three sites particular attention needed be given to the possibility of significant channel migration in the future.

The necessary data for performing basic stability calculations were tabulated for the 24 river crossings. Critical shear stress on the bed to induce initial motion of the bed material was calculated, which can be related to a given discharge, depth and average velocity at each location. For three sites, the 10-year and 50-year floods were estimated using the Flood Studies Report (NERC, 1975) procedure and, where possible, using gauging records. Estimates of the riprap size required to stabilise the bed and/or bank were determined for the three potentially unstable sites. Various procedures were used for riprap sizing, obtained from Simons and Senturk (1976). Three bedload transport equations were recommended for use. Two predict low transport rates commensurate with known transport rates in rivers in northern England, while the other predicts somewhat higher rates. These procedures could be used in a mass balance simulation to predict depths of bed scour. It is recommended that the Ackers–White approach is used in any modelling, while the Parker method would indicate the worst possible erosion scenario.

Outline procedures were provided for conducting bank stability assessments where problems might be related to rotational slip or block-fall. In any assessment, site-specific data would be required or gross assumptions made. At none of the existing crossing points is bank stability likely to be a major consideration as long as reasonable protection to the bank toe is provided.

At km 170, the River Eden flows through a tight bend. Calculations indicate a stable river bed for flows in excess of bankfull but this may not actually be the case. Calculations rely to a large extent upon an accurate estimate of the valley slope, which is difficult to assess at this location. Because of the problems noted, no acceptable estimate of the preferred riprap size could be calculated for the River Eden, so instead the size of stable riprap presently in place at the site was given.

Most of the rivers have a ratio of critical depth for motion to bankfull depth close to one. The bankfull depth is taken as producing the maximum shear stress on the bed. Of the exceptions, Dippool Water has much mobile fine sediment at low discharge but overall the channel appears to be very stable. Where the ratio is much larger than unity (e.g. Mouse Water, Duneaton Water, Kirtly Water, River Lyne, River Eden) this is in part owing to engineering works (dredging or embanking) but some bedload transport would normally be expected close to bankfull. With the exception of the upstream crossing on the Evan Water, natural bankfull flow is somewhere between the calculated 2-year and 5-year floods. This is a little less frequent than one would normally expect (i.e. 1–2 year recurrence). With respect to the erosional capacity of the 10- and 25-year floods the conditions are not likely to be more severe than the bankfull case, either because of local dredging to increase channel capacity or because of floodplain storage. Low gradient precludes excessive velocities, i.e. discharge increases in volume but not scouring potential. The upstream crossing on the Evan Water has a much larger capacity than the floods calculated from either the Flood Studies Report or from downstream gauging station data would imply to be necessary. There is evidence of dredging at this site which would explain the over-large capacity, but would also imply problems with flooding in the past.

Box 12.5

Bank erosion at meander bends along a reach of the River Roding at Loughton, Essex, UK, was exacerbated by straightening the adjacent downstream section to enable floodplain gravels to be extracted. Headward erosion resulted and this, in turn, precipitated bank erosion as the river increased its sinuosity in order to reduce its gradient. Bank erosion rates locally exceeded 2 m yr^{-1}.

Thames Water commissioned geomorphologists to evaluate and develop the use of submerged vanes to prevent bank erosion in a meander bend. In order to evaluate the effectiveness of a range of different submerged structures, a fixed-width, mobile-bed meander flume was constructed which enabled tests to be rapidly carried out. Secondary flows and shear stress distributions in a test bend were initially determined, based on measurements with a two-component electromagnetic flow meter, for reference purposes. Subsequently, various shapes and configurations of submerged vane were tested and hydraulic conditions and bed topography compared against the natural condition.

Effectively, the submerged vanes counter the strong skew-induced secondary circulation in the meander bend, which directs faster surface water towards the outer bank, by strengthening the counter-rotating outer-bank flow cell. As a result, downwelling and associated high shear stresses no longer occur adjacent to the outer bank and this reduces toe scour, bank over-steepening and mass failure. The downwelling instead occurs towards centre channel inside the line of the submerged vanes, causing bed scouring in mid-channel. The best results, in terms of scour reduction, were obtained using hydrofoil-shaped vanes.

The number, size and location of the vanes were calculated to ensure that they produced sufficient opposing force to neutralise the centrifugally induced torque, due to the secondary circulation, on an imaginary wall along the proposed hydrofoil installation line (see Paice and Hey, 1989). Flume trials and hydrofoil theory provided the following guidance on the design of hydrofoil schemes.

1. Hydrofoil protection should extend throughout the meander to prevent outflanking. This could take the form of buried hydrofoils which would automatically become effective if bed degradation exposed them.
2. The hydrofoils should be positioned along a line of curvature of the bend, ideally at the base of the point bar face.
3. Hydrofoils should be oriented at an angle of 10–20 degrees to the direction of the primary flow. If this is exceeded, stalling will occur and flow separation from the hydrofoil surface can cause undermining and possible collapse of the structure.
4. The bed topography and bend geometry should be accurately surveyed as close to the time of installation as possible to ensure that the scheme is tailor-made to fit prevailing conditions.

At the field site on the River Roding, scaffolding bridges at the bend apex and adjacent upstream and downstream crossing riffles enabled flow measurements to be obtained with an electromagnetic flow meter at discharges up to and including bankfull flow. This information enabled secondary and primary velocities and shear stresses to be determined. The number, size and location of the hydrofoil vanes to counter the skew-induced secondary circulation were calculated, based on the field data (Figure 12.9). This proved to be half the number calculated using Odgaard and Mosconi's (1987) method. In part this results from the fact that observed secondary velocities, rather than predicted ones, are used in the calculation. More significantly, the procedure recognises that the secondary flow, as it develops in the bend, has to be progressively counteracted and that each vane contributes to this process. In contrast, in Odgaard and Mosconi's method each vane is made to counteract the fully developed secondary flow and, thereby, ignores the contribution of the upstream vanes to this process.

Each vane was 3 m long and 1 m high. They were installed with top elevations such that they remained submerged even during low flow conditions. The angle of attack was fixed at

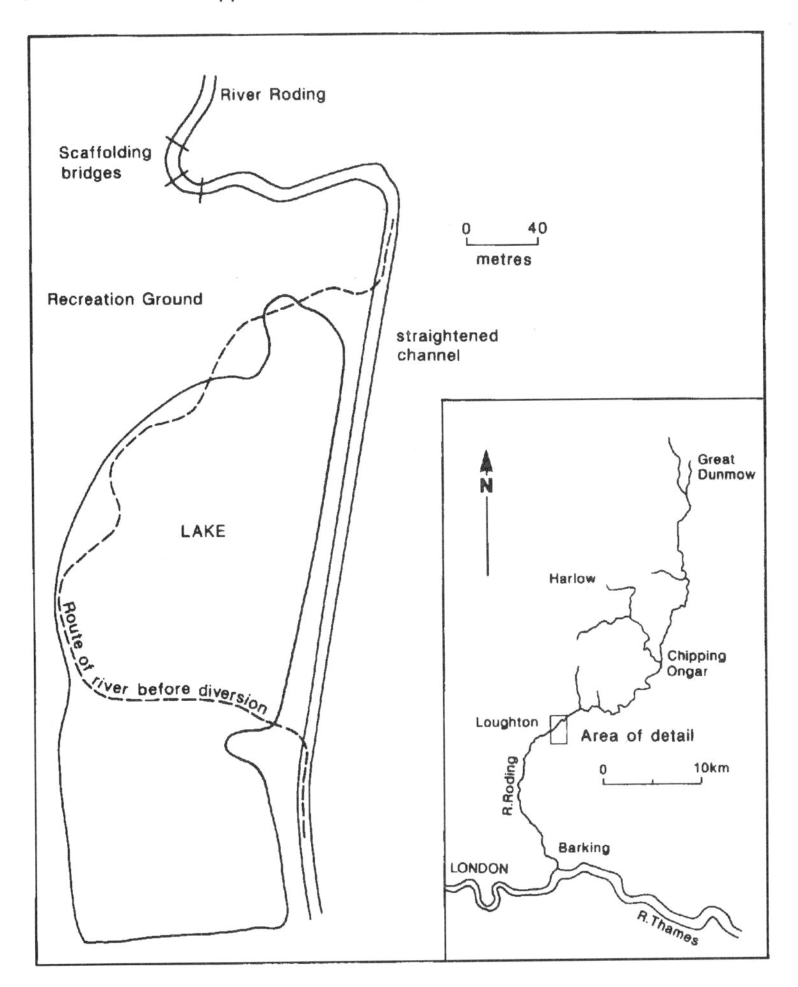

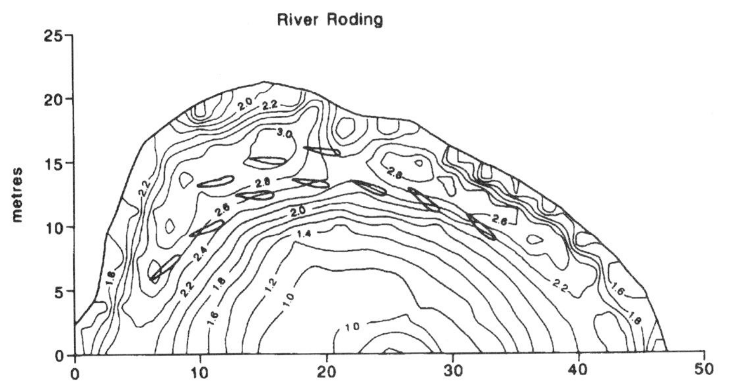

Figure 12.9 Location of submerged vanes (hydrofoil shape), River Roding, Loughton, Essex (after Paice and Hey, 1989)

15 degrees to provide a degree of safety from stalling should approach flow directions change. The vanes were fixed in position with a main, anchoring pile and a smaller secondary pile to prescribe their orientation. Installation was completed in January 1989.

A topographic survey was carried out immediately prior to installation of the hydrofoil vanes and subsequently after flood events. Comparisons indicated that the scour hole adjacent to the outer bank had been infilled and scouring had taken place across the toe of the point bar inside the line of the hydrofoils. Bank erosion rates on the 3 m high outer bank were significantly reduced. The bank was no longer being undercut and over-steepened and mass failures were no longer occurring. Some slight subaerial weathering was taking place towards the top of the bank, reducing upper bank slopes, but this only occurred where the bank was devoid of vegetation.

Bramley, 1989). An alternative method, based on the installation of submerged vanes on the river bed, treats the cause of the erosion by modifying the secondary flows in the meander bend and is environmentally less intrusive as the bank is left in its natural state (Box 12.5) (Odgaard and Mosconi, 1987).

12.3.2 River Channel Maintenance and Sedimentation

In England and Wales there is a strong tradition of structural flood protection whose assets, in terms of channelised reaches, must be maintained. River channel maintenance is a multi-million pound Sterling management function, under engineering direction, but geomorphological insights are proving increasingly valuable, especially for sensitive sites or sites where costs could be cut by controlling sedimentation or erosion.

For example, Mimmshall Brook is a small gravel-bed stream located north of London in southern Hertfordshire. The catchment comprises a varying geology, from London Clay on the valley sides to Chalk in the valley bottom. Nevertheless, much of the sediment supply is derived from alluvium laid down by the stream through lateral migration across the valley floor. Relevant catchment details are listed in Table 12.7 and shown in Figure 12.10.

The hydrology of Mimmshall Brook is dominated by 'flashy' floods due to surface runoff from the impermeable London Clay, augmented by urban and road runoff. At Water End a significant proportion of the catchment runoff is absorbed by a number of swallow holes in the bed of the stream. The swallow holes are of national conservation significance ('Site of Special Scientific Interest', SSSI) but are frequently blocked by fine-grained sediment and trash from urban/motorway sources. Further fine sediment loads are supplied from soil erosion of arable fields bordering the channel upstream.

Table 12.7 Catchment characteristics: Mimmshall Brook, Thames Region NRA

Area: 52 km^2
Main river length: 38 km
Geology: Chalk, London Clay, Reading Beds, alluvium
Land use: urban (18.5%); rural (81.5%); motorway (17 km)
Principal streams: Mimmshall Brook, Potters Bar Brook, Catherine Bourne, Clare Hall Stream, Brookmans Park Stream

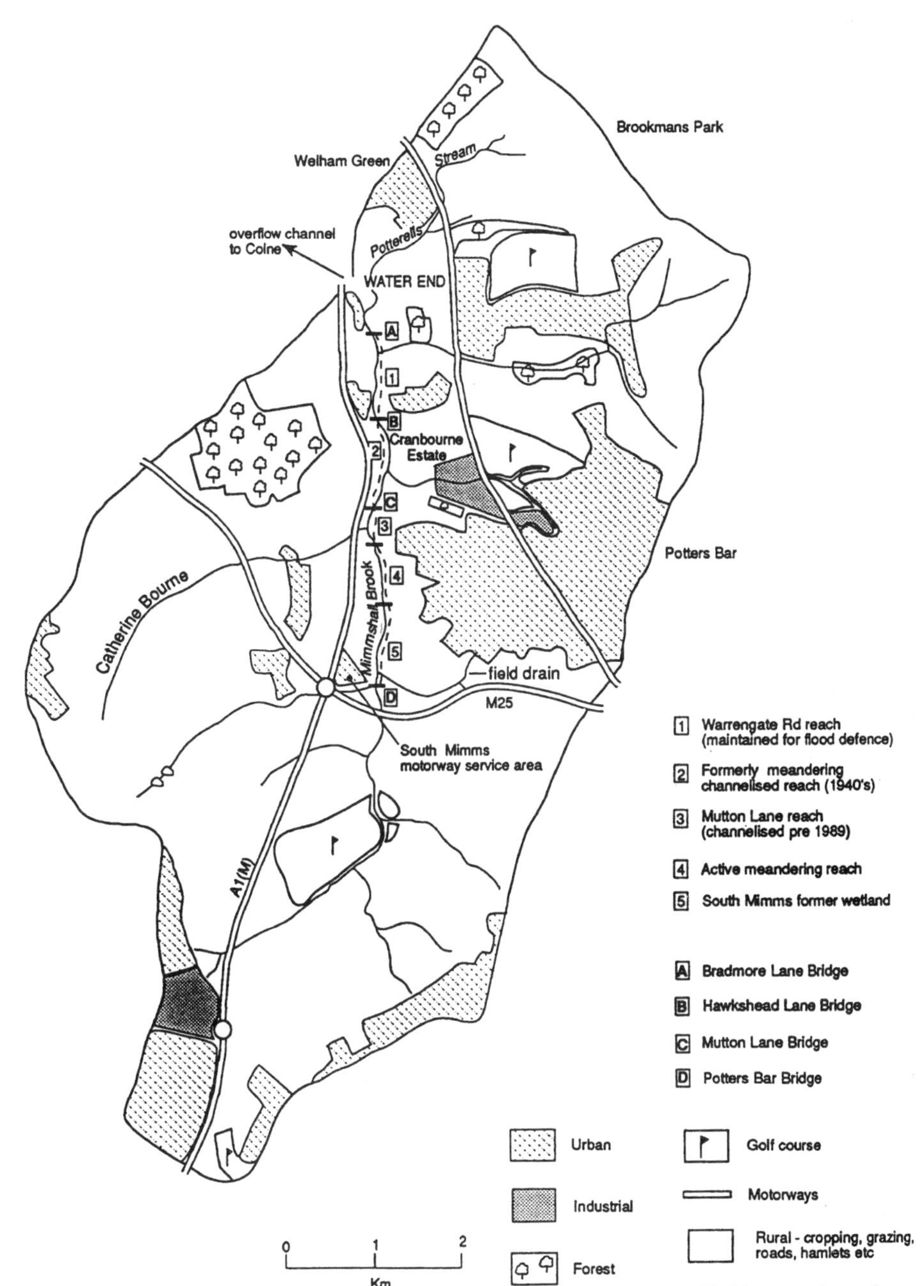

Figure 12.10 Catchment of Mimmshall Brook: land use and location of bridges and reaches

Improved management of the fluvial system of the Mimmshall Brook catchment is required to achieve reduced flooding frequency of properties due to increased ponding in the Water End SSSI and gravel deposition within a maintained reach immediately downstream of Hawkshead Lane Bridge (Figure 12.10). Conservation laws require action to prevent deterioration of the officially designated swallow holes.

The Thames Region of the National Rivers Authority (NRA) became aware of consistent flooding problems in the lower reaches of Mimmshall Brook in 1988. Some flood control measures had been conducted along 400 m of channel in the 1950s. These measures included widening of the channel and raising the left bank with the excavated spoil. Subsequently, the channel has required maintenance dredging due to accumulation of gravel shoals within the widened reach and a loss of channel capacity. Flooding is still a problem along this section of the channel and consequently a study has been initiated to identify the benefit/cost of a flood alleviation scheme. As part of this scheme, a geomorphological assessment was conducted to establish the causes of the sediment build-up, and to provide practical management strategies for reducing the maintenance commitment (Box 12.6) (Sear, 1992).

Box 12.6

River managers are concerned by excess storage of sediments in sensitive locations, or by erosion of the channel bed and/or banks where this affects flood defence assets. In Mimmshall Brook, the problem of sedimentation may be framed in terms of four possible explanations:

1. the problem is due to excessive sediment supply;
2. the problem is due to reduced sediment transport capacity within the channel;
3. the problem is a combination of both factors;
4. the problem is not sediment related.

The sources consulted for assessing the long-term contributions to the present sediment system in the Mimmshall Brook catchment include published maps at various dates, water authority surveys, and records and general documentation on the topography and development of the catchment, as well as the usual climatic and hydrological data.

The historical analysis reveals that significant catchment land-use changes have occurred as a result of urban development. These developments have triggered changes in the sediment supply from the catchment. In addition, considerable channelisation has occurred throughout the Mimmshall Brook and in most of the tributary streams. The effects of channelisation and urbanisation have been shown to be capable of producing major disturbances of the sediment system in a catchment (Brookes, 1988; Roberts, 1989). Four factors are identified as causing major changes in the Mimmshall sediment system:

(a) *Urbanisation* (1900–the present). Urban development is known to increase runoff and decrease time to flood peak. Downstream channel reaction is typically characterised by an increase in channel capacity through bed and bank erosion.

(b) *Agricultural land drainage* (1939–1950). Robinson (1989) identifies land drainage as producing decreased time to flood peak and increased baseflows on clay soils. Fine sediment yields are also increased.

(c) *Channelisation* (1872–1980s). Robinson (1989) identifies channelisation with decreased time to flood peak, and increased flood magnitude. Brookes (1988) has described major upstream and downstream channel changes associated with increased stream power due to increased flood magnitude; upstream erosion is balanced by downstream deposition. Paradoxically, the channelisation upstream of the new properties at Warrengate Road will have exacerbated the flood hazard through increased deposition and flood stage.

(d) *Motorway construction* (1980s). Runoff and fine sediments from the M25 Motorway have led to fine sediment input and burial of the gravel bed. Channel modifications for the construction of the South Mimms Motorway Service Area have removed the storage capacity of the former Mimms Wash Wetland and also increased fine sediment loads to the channel while at the same time decreasing the quality of sediment and water inputs.

The combination of catchment changes effectively increased the flood frequency of Mimmshall Brook and Potters Bar Brook (due to flood wave steepening, loss of floodplain storage and increased runoff) against a climatic trend of reduced winter rainfall (Harold, 1937). This is clearly demonstrated in Figure 12.11.

Morphological and sedimentological surveys were conducted in the Mimmshall catchment. The objectives were to assess the current extent of sediment storage and supply within the channels, the location and severity of bed and bank erosion, the type and development of channel morphology and the stability of the stream system. The surveys involved walking the stream channel and noting the range of geomorphological features which are outlined in Table 12.8.

The results of the survey can be summarised into the following key points relevant to the sediment management of Warrengate Road.

1. Major sources for sediment of the size identified as causing an immediate problem in the Warrengate Road reach are generated from:
   - deformation of the river bed in the Warrengate Road reach;
   - erosion of the bed and banks of Potters Bar Brook;
   - erosion of the bed and banks of the channelised reach of the Mimmshall Brook for 200 m downstream of the confluence with Potters Bar Brook.
2. Additional sources of the fine sediment which is of concern to the conservation and ecological value of the swallow holes SSSI include:
   - erosion in the meander belt upstream of Potters Bar Bridge, although the rate of sediment supply to downstream reaches is regulated by the considerable storage potential in point bars and pools within this reach;
   - soil erosion (during overbank floods) of arable cropped farmland adjacent to the channel.
3. Additional sediment sources for future concern are identified at the reach immediately downstream of and adjacent to the South Mimms Motorway Service Area where fine sediment is generated from the M25. Additional sediment inputs stem from road washings and path erosion around the South Mimms Motorway Service Area. Further bank erosion is associated with destabilisation of channels by upstream progression of knickpoints. The runoff from the Motorway Service Area is also a source of pollution which decreases sediment and water quality.

A geomorphic approach to mitigating the loss of flood capacity by sedimentation can adopt one of three options.

The first option is to *reduce the sediment supply* to the Warrengate Road reach by methods such as sediment trapping, management of upstream sediment supply at source or by reducing sediment transport rate upstream and increasing sediment transport storage potential upstream.

Gravel traps act as bars/pools, storing excess sediment in specific (advantageous) locations. Problems associated with gravel traps involve too much loss of downstream sediment supply through over-design which can initiate bed and bank erosion together with channel armouring/compaction. Location of gravel traps upstream of sensitive structures such as bridges should be avoided for these reasons. In addition, the upstream lip of gravel traps should be protected against knickpoint progression caused by the local increase of bed slope upon entry to the trap.

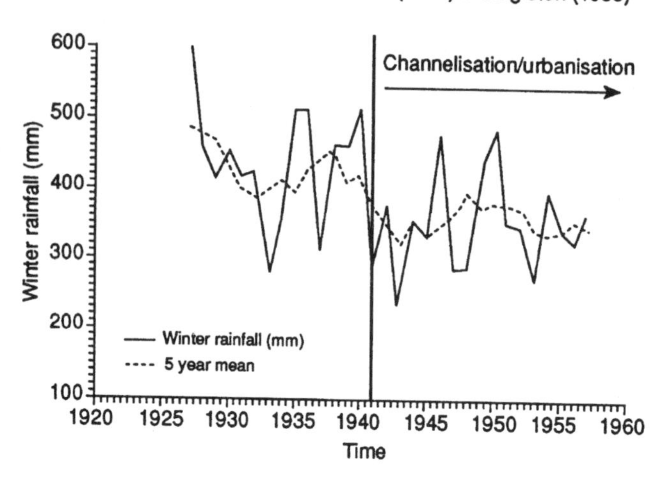

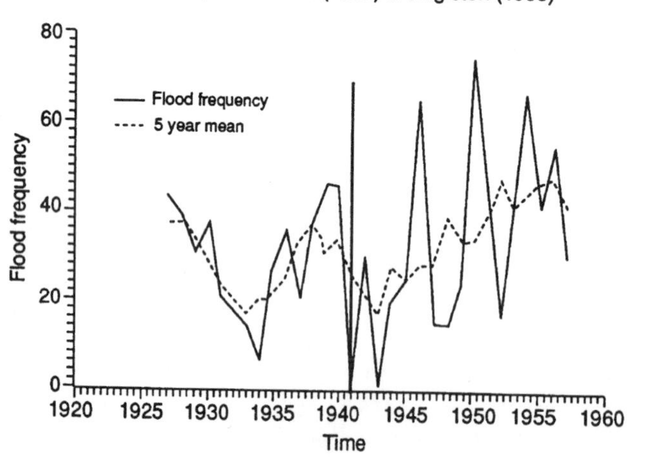

Figure 12.11 Rainfall and flooding trends for Mimmshall Brook, 1920–1960

In conclusion, the study recognised that gravel traps are not suitable as a mitigation option due to sensitivity of the Warrengate Road reach, and the accentuated stream power of the channelised reach upstream.

The management of sediment sources seeks to reduce the excess gravel and fine sediment loads derived from the erosion in Potters Bar Brook in particular and upstream destabilised banks, and represents a necessary goal for long-term maintenance reduction since at present the channel is still adjusting to accentuated stream power. This can be achieved by controlling unauthuorised dredging of the bed (which effectively increases bank instability) and the

Table 12.8 Geomorphological features noted from surveys

		Date:	Location
Region:	Officer:	Catchment/river:	Problem type:

Meandering:		Straight:		Floodplain (m):	
Landuse %:		Valley slope:		Old channels:	

Meandering:		Straight:		Braided:		Islands:	
HiSiN		Artificial:		High		No.	
ModSiN		Natural:		Moderate		Natural	
LowSiN				Low			

Riffle/pool (spacing):		Uniform:		Artificial:		Other:	
Gradient:		% Riffle:		% Pool:		% Bar/berm	

Banks	Height	Steep	Berm	Cliff	Artificial	Stratified	Embanked
Right bank:							
Left bank:							

Banks	Block	Slip	Flake	% Erosion	Artificial	Stratified	Embanked
Right bank:							
Left bank:							

DATA	RIFFLE	POOL	UNIFORM	ARTIFICIAL
Width: Low Q				
Bank Q				
Embankments				
Depth: Low Q				
Bank Q				
Embankments				
W/D: Low Q				
Bank Q				
Embankments				
SYM/ASYM				

SEDIMENT	Point bar(%)		Medial bar (%)		Alternate bar (%)		Tributary bar (%)		Berm (%)	

STORAGE	(%)	Bedrock(%)	Boulder(%)	Gravel(%)	Sand(%)	Silt/clay	D50(mm)	D16(mm)	Armour ratio	Structure(%)	Strength	Fresh/old
Bars:												
Riffles:												
Pools:												
Berms:												
Banks: Right												
Banks: Left												

SEDIMENT SOURCES	Type supplied (eg sand)	Rank order (increasing)
River cliff		
Banks		
Bed		
Urban runoff		
Field drains		
Ditches		
Tributaries		
Other		

Vegetation	Aquatic (%)	Riparian (%)	Trees (%)	Landuse
Left bank				
Right bank				

STRUCTURES Inspected	Stable	Unstable	Buried	Exposed
Bridge footings				
Outfalls no.				
Buildings				
Bank protection % RB				
Bank protection % LB				

installation of drop structures in Potters Bar Brook (maximum height per structure 350 mm as determined by Fisheries requirement) designed to reduce channel slope and to decrease bank height with respect to bed levels.

The fine sediment and pollution problem associated with the South Mimms Motorway Service Area is best addressed by the restoration of the Mimms Wash, a former wetland site, drained and infilled over the past 50 years.

Reducing sediment supply from bed and bank erosion in Potters Bar Brook is a viable long-term goal for reducing maintenance commitment and maintaining flood capacity; however, it should not be carried out without reducing the transport capacity of the downstream channelised reach. Reducing accentuated bank retreat upstream is a viable option for fine sediment management (Darby and Thorne, 1992).

Modification of sediment transport and storage capacities mimics the direction in which the stream system will develop naturally in the longer term, principally via a reduction in stream power and sediment deposition via meandering. Meander restoration could be used to decrease stream slope, increase sediment storage (by increasing the opportunity for bar development and increased pool depths) and reduce flood velocities through this reach by increasing form roughness. Long-term maintenance will be reduced but at an initial expense involving land acquisition and channel restoration. Maintenance will still be required under existing channel dimensions in the Warrengate Road section. The sediment storage capacity of the meandering channel will effectively provide a buffer for enhanced sediment yields from upstream catchment instabilities. The flood storage upstream of the maintained Warrengate Road reach will be beneficial for the flood defence obligations associated with this reach.

The second option consists of increasing the sediment transport capacity through the Warrengate Road reach by increasing stream power and decreasing sediment storage potential.

This can only be achieved by channel modification involving a reduction in channel width or an increase of channel slope. A multistage channel design to introduce a low-flow, active-bed channel with riffles and pools would be required in order to maintain the required flood capacity. A problem is faced downstream where the levels of bedload to the SSSI would be increased. This could be arrested by constructing a gravel trap immediately downstream of Bradmore Lane Bridge. This option would involve major road realignment and reconstruction works to three bridges, a gauging station and a grade control structure. The costs are considered to be prohibitive.

The third option is to *do nothing*.

In the do nothing option, the present rate of sediment transport will continue, and in time will increase as sediment from Potters Bar Brook initiates the first stages of meander development in the channelised reach downstream. In the longer term there is a problem of increasing maintenance as a result of fine sediment supply from the upstream South Mimms Motorway Service Area. At present, the Potters Bar Brook and the channelised reach of Mimmshall Brook are susceptible to destabilisation from high-magnitude, low-frequency flood events resulting in stream powers above the empirically derived 35 W m^{-2} stability threshold.

The recommended solution involves sediment supply management together with the reduction of sediment transport capacity in the channelised reach upstream of Hawkshead Lane Bridge. A bank stability survey should be conducted to identify unstable banks. The Mimms Wash wetland should be restored to function as a silt trap as well as a reedbed purification system. To reduce sediment transport capacity to Warrengate Road, the meandering planform of Mimmshall Brook should be restored and a floodway created to accommodate overbank floodplain flows. The combination of these options effectively seeks to restabilise the stream network and to minimise sediment flux divergence between reaches. Further benefits of these recommendations include flood storage upstream of the flood-prone Warrengate Road, and an improvement in ecological and recreational integrity of the Mimmshall Brook and Water End SSSI.

The application of geomorphological assessments to catalogue the sediment system of a catchment over timescales of 100+ years, provides the information to integrate the objectives of flood defence, customer demand for environmentally acceptable channels and economic efficiency in the long term. This can be achieved by the adoption of out-of-channel works such as appropriate morphological restoration of the stream network upstream of problem reaches, land-use control at sensitive sites (e.g. riparian corridors and headwater streams), controls on catchment development and/or recognition of the geomorphic implications of catchment development in the planning process.

12.4 THE GEOMORPHOLOGY OF CHANNELS AND WATER RESOURCE MANAGEMENT

Three case studies are presented here to exemplify the applied value of understanding the relationship between sediment supplies from points upstream of water abstraction points, and between channels and the flows they carry after modifications occur through water transfer or abstraction.

12.4.1 Regulating Reservoirs and Channel Stability

In the 1970s a feasibility study was carried out to investigate the engineering, hydrological and environmental implications of enlarging the Craig Goch Reservoir (situated at the head of the Elan Valley, a tributary of the River Wye in mid-Wales, UK) for water resource development.

It was proposed to establish Craig Goch Reservoir as a major pumped storage reservoir with water being abstracted from the River Wye in winter and releases being made to the River Wye and across the watershed to the River Severn during periods of low river flow. The rivers were to be used as an aqueduct to supply water principally to Birmingham and the West Midlands (Box 12.7).

To date (1997) the Craig Goch dam scheme remains strategic, but regulation of flow in the reaches described in Box 12.7 occurs by releases from the Clywedog Reservoir. Geomorphology has played a continuing role in enabling river managers to operate a release strategy which, even in the case of occasional maximum scour-valve releases, reduces the impact of high flows on the upper reaches of the Severn (see Leeks and Newson, 1989).

12.4.2 Sedimentation Problems at Intakes and Outfalls

Water resource engineering can also encounter problems at riverside intakes from both regulated and unregulated channels. The case study described in Box 12.8 arises as the result of sedimentation of both diversion weirs in the headwater zone and of a large offtake chamber downstream. Geomorphological assessment was employed to establish a 'fluvial audit' (rapid assessment by field observation of the sources and sinks of sediment and the influence of flood history).

Because river channel capacity and form adjust, under natural conditions, to the imposed water and sediment regime, rivers whose flow becomes modified as a result of abstraction or of water transfers may modify their channels in response. The impacts of impoundments, via modifications to flood flows, are likely to be more rapid and pronounced because of the formative role of flows near bankfull, but other flow changes may have an impact on the details of channel capacity and form, e.g. those related to bank exposure, fine sediment load and vegetative cover (both submerged, emergent and riparian).

Geomorphologists are, therefore, employed by water managers to investigate the channel changes which have occurred, or may occur, in relation to modifications to the whole flow regime. This interest is expanding, parallel to the expansion of interest in ecological effects of river management. In terms of habitat, the degree to which the flow

Box 12.7

Severn–Trent and Welsh Water Authorities required advice on the maximum levels of release that could be made to the rivers Wye and Severn which would enable their natural stability to be maintained. Alternatively, if higher release levels were required, they needed to know the sections of river which would require operational maintenance or capital improvement works.

The natural stability of a river downstream from an intake or outfall can be maintained provided that abstractions or releases do not alter the frequency of flows which transport bed material load or cause bank erosion further downstream. Attempts to make larger abstractions and releases by seasonally balancing erosion and deposition will not be successful as bank erosion, unlike bed elevation changes, is not reversible (Hey, 1986).

The basic requirement was to establish threshold discharges for bed material entrainment downstream from the proposed intake/outfalls. Not only will this vary downstream as the catchment area increases, but locally it can vary between pools and riffles and between reaches which are naturally eroding, stable and depositing. It is also dependent on geology as this can influence the local calibre of the bed material. In addition, it was also necessary to ascertain critical discharges for bank erosion and failure.

In view of these factors, the choice of sampling location was important. Sites were selected after carrying out a geomorphological assessment of the river downstream from the proposed intake/outfall. First, historical maps were consulted to identify reaches where sinuosity had changed over the previous hundred years. Increased sinuosity implies net erosion, and a reduction in channel gradient, and vice versa with decreased sinuosity. Extreme deposition will lead to the development of a braided condition. Second, a walk-over reconnaissance survey was carried out to confirm the results obtained from the map analysis and determine the degree of natural instability. This involved the assessment of current river conditions and the evaluation of local morphological and sedimentary features which are indicative of erosion and deposition.

This reconnaissance identified a sequence of eroding, stable and depositing reaches downstream from the outfall. In each reach, bed material transport thresholds were assessed at a representative pool and riffle. For this purpose the size frequency characteristics of the surface gravel-bed material were determined at each cross-section and 50 representative clasts were removed, painted and replaced in line across the section. Regular observation was maintained at each site after different flow events to establish the critical discharge for surface bed material entrainment.

Threshold discharges for bank erosion were determined at a number of sites where erosion was active, as evidenced by the presence of failed bank blocks in the river. The bank material was characteristically fine cohesive material overlying gravel, and different techniques were employed to ascertain erosion in each layer. With the former material, a marker peg at a known distance back from the edge of the bank provided a survey datum, while with the gravel a reinforcing rod was driven horizontally into the bank as an erosion pin until the end was flush with the surface and any subsequent exposure indicated erosion. In addition, paint lines were sprayed down the face of the bank to identify any surface erosion. Results showed that transport of the gravel-bed material adjacent to the bank was a prerequisite for failure, indicating that bed material transport at the toe is the limiting condition for bank retreat. In order to determine whether the proposed releases would cause additional instability it is necessary to compare the threshold discharges for transport and erosion with the maximum regulated flows (namely, the release plus natural flow) in the river. On the River Severn below the Llanidloes outfall this was 18.5 $m^3\ s^{-1}$, rising to 22.3 $m^3\ s^{-1}$ at Abermule due to the downstream increases in base flow when releases were being made. Comparisons can then be made between threshold discharges for bed material movement and maximum regulated flows (Table 12.9) for various points downstream from

Table 12.9 Maximum regulated flows (17.4 $m^3 s^{-1}$ maximum release) and threshold discharges for bed material movement, Rivers Dulas and Severn (data from Hey, 1986. Reproduced by permission of *Journal of Hydrology*)

Site	Estimated maximum regulated flow ($m^3 s^{-1}$)	Probable threshold discharge for bed material movement ($m^3 s^{-1}$)
Dulas		
A Railway Rank	2.3	2.0
B Pen-y-bont	2.3	1.5
C Pen-y-bont	2.3	1.5
D Upper Glandulas	2.4	2.2
E Lower Glandulas	2.5	2.5
Severn		
F Dulas confluence	3.7	9.8
Clywedog confluence		
G Llanidloes	18.5	18.1
H Morfodion	18.7	11.0
I Dolwen	18.7	11.0
J Lower Penrhuddlan	19.0	16.1
K Craig Fryn	19.0	30.1
L Llandinam	19.2	11.0
M Maes-Mawr	20.7	16.2
N Ty-Mawr	20.9	17.1
O Penstrowed	21.2	17.1
P Llanllwchaian	21.9	18.5
Q Abermule	22.3	18.5

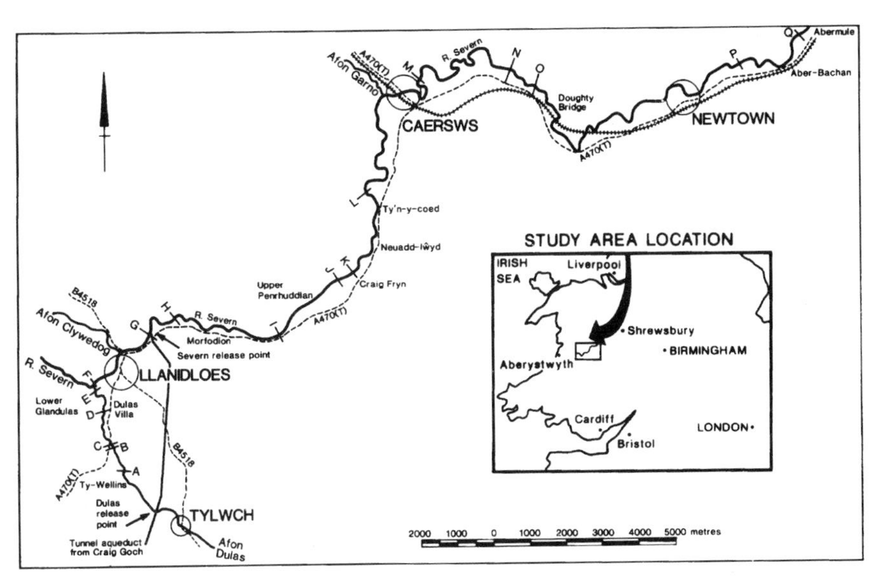

Figure 12.12 Bed material movement: monitoring sites Rivers Dulas and Severn, mid-Wales (from Hey, 1986. Reproduced by permission of *Journal of Hydrology*)

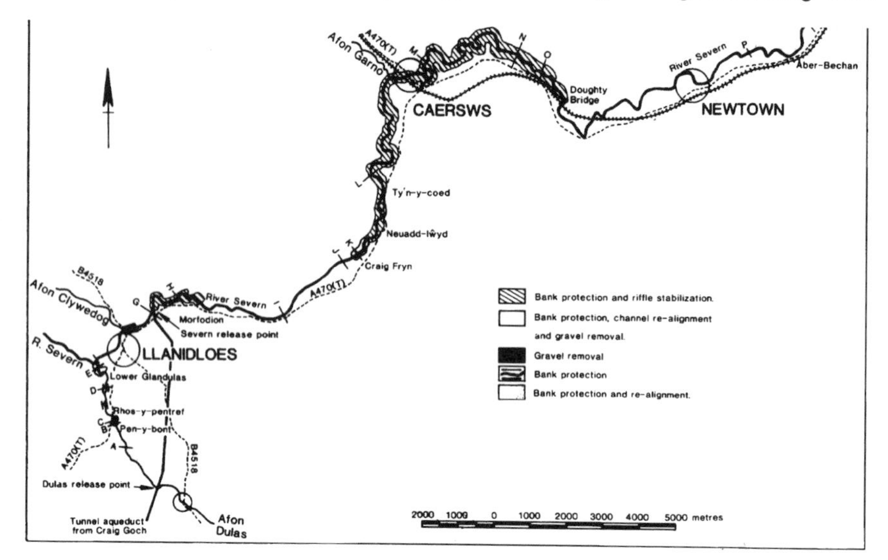

Figure 12.13 Channel maintenance requirements to allow river regulation at full potential (from Hey, 1986. Reproduced by permission of *Journal of Hydrology*)

the outfall (Figure 12.12). If maximum regulated flows exceed transport thresholds then erosion is expected, with deposition further downstream where transport thresholds exceed maximum regulated flows. Channel maintenance requirements to enable such releases to be made can then be identified (Figure 12.13).

fills the channel determines the gross biotic properties of a reach, feeding back to velocity distributions. Together, these properties determine the way in which stream vegetation grows and the use of the channel by animals as substrate, food source, shelter and breeding ground.

The Instream Flow Incremental Methodology (IFIM) allows users to model changes in habitat quality as flow varies in regulated rivers. IFIM represents a major field and computer modelling exercise, dominated by hydrologists and biologists. For the purposes of this chapter we include two simple case studies of the impact of flows on channel capacity (see Boxes 12.9 and 12.11).

12.4.3 Monitoring Environmental Impacts of River Regulation

It is part of the spirit of Environmental Assessment that the impacts of developments are monitored after completion, both as an 'insurance' against failure to predict them correctly (and therefore failure to prescribe suitable mitigation) and as a method of 'accounting' to other users of the affected environment. In the case of impacts on rivers it is essential to maintain monitoring systems as part of good management but, for example, in the cases of

Box 12.8

The Dunsop catchment in the Bowland area of Lancashire, UK, has (since 1884) provided water supplies to the town of Blackburn. Major stream intakes are located at the head of the Brennand and Whitendale valleys, tributaries of the River Dunsop. In addition, some 12 minor water intakes intercept small streams before they join the Brennand and Whitendale rivers.

With the increasing emphasis on the cost-effective operation of water resource systems, every effort is being made to maximise the use of the cheaper sources such as the Dunsop system, and in particular the gravity intakes within it. However, operational experience combined with hydrological analysis has shown that full advantage cannot be taken of the available supplies from the Dunsop system because of geomorphological processes that are taking place in the catchment. Erosion has made many of the intakes and aqueducts susceptible to collapse, and problems with sediment movement reduce the efficiency of the intakes. Such is the remote location of many of the intakes that they are difficult and expensive to maintain. The eroded materials accumulate not only in the small gravity intakes but in the stilling basin for the Footholme pumps, rendering this supply unusable after flood periods. The regional water authority therefore commissioned a geomorphological study.

During a one-week study the following topics were addressed.

1. The state of the gravity intakes in both the Brennand and Whitendale valleys; this included the present rate of supply from their small catchments and the effect of sediment deposition on the intake capacity and associated engineering works such as aqueducts.
2. The channel stability of the main-stem Brennand and Whitendale streams which has a bearing on, among other things, the maintenance of the supply pipeline and the stability of the intake aqueducts. It also clearly influences the rate of sedimentation at Footholme.
3. The pattern of sedimentation in the abstraction basin at Footholme, its estimated rate, and simple hydraulic methods of reducing its impact on the pump screens at the works.

The results of the geomorphological investigations and the recommendations in relation to both operational management and the capital investment necessary to maintain the intakes can be costed. Then a judgement can be made as to whether these would be cost-effective in comparison to the results of the yield and average supply (and hence operating costs) assessments.

The reconnaissance showed that relatively few eroding sites contribute sediment directly to the lower and middle Whitendale. Only in the steeper and narrower reaches above the Whitendale intake is there strong evidence of sediment freshly deposited in the main stream from erosion scars (Figure 12.14a and b). In contrast, the Brennand receives a much larger volume of direct sediment input. Its headwaters tap an eroding peat area while eroding river bluffs are a feature both in the headwaters and the bedrock-lined gorge section lower down. Much of the upper main stream is also bedrock-lined, presenting few opportunities for storage and thus enabling a rapid supply of eroded material to the downstream area critical to the gravity and Dunsop (pumped) intake schemes.

In the middle reach of the Brennand below the intake, sediment sources abound, both from river cliffs yielding sediment directly to the channel and from tributaries producing prominent gravel fans towards the main stream. Further, the channel exhibits signs of instability in the form of braiding (caused by a major flood on the catchments in August 1968) through which it is able to tap the large volume of sediment forming the valley-floor floodplain. At its downstream end the increased load of the Brennand is demonstrated by the size of the deposit of material in the Footholme intake stilling basin. Only a comparatively minor deposit lies at the Whitendale exit.

Finally, the channel beds contain a considerable volume of fine material (sand, silt and particulate organic load). While protected beneath the armour layer of coarser material at low

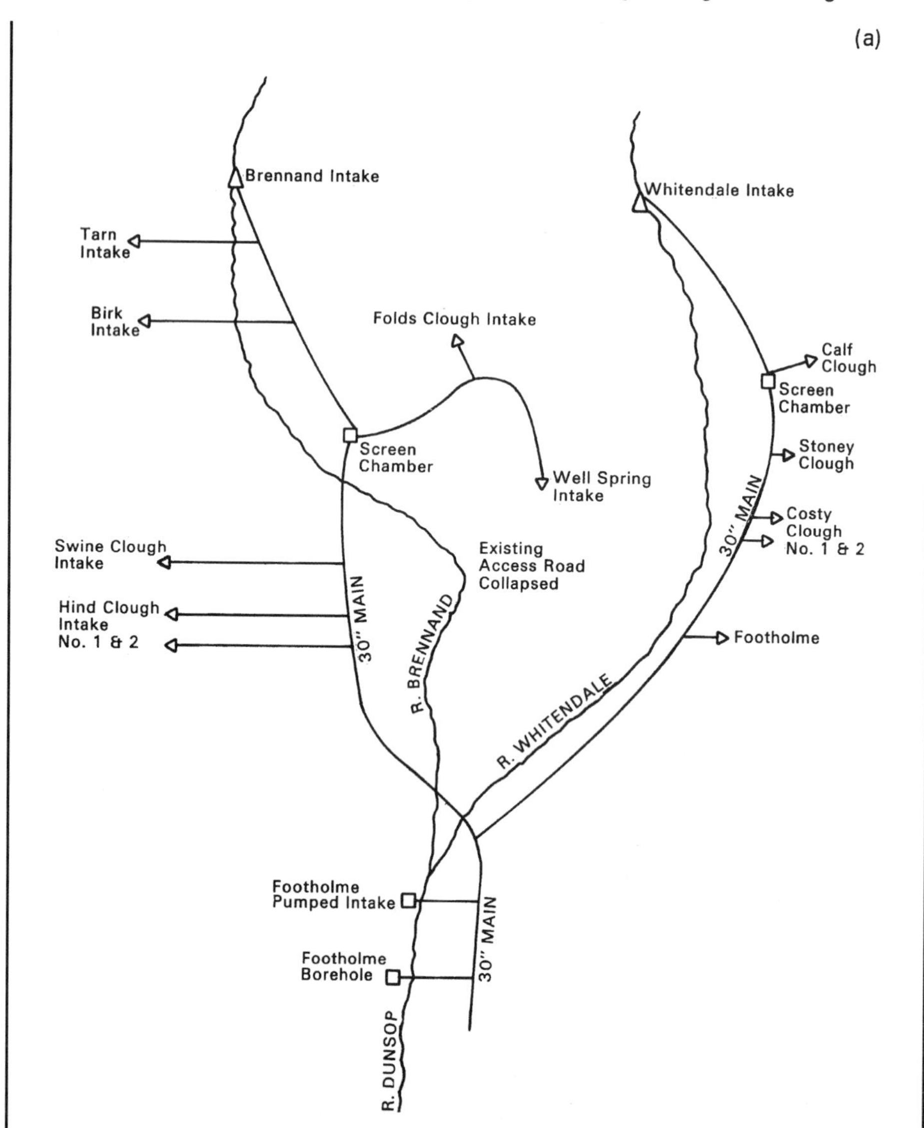

Figure 12.14a Fluvial audits of the Forest of Bowland intakes: engineering design of the intakes and offtakes

flows, these fines are likely to make a significant contribution to total load during floods which disrupt the armour layer.

Two small flood peaks were monitored during the survey, giving some indication of conditions for initiation of motion of the finer gravel sizes at the Brennand and Whitendale exits. Additional information on the flows required to move different sizes of gravel from within the overall size distribution was obtained from two sources. First, a survey was made of the fresh shoal material which had accumulated in the Footholme stilling basin since it was

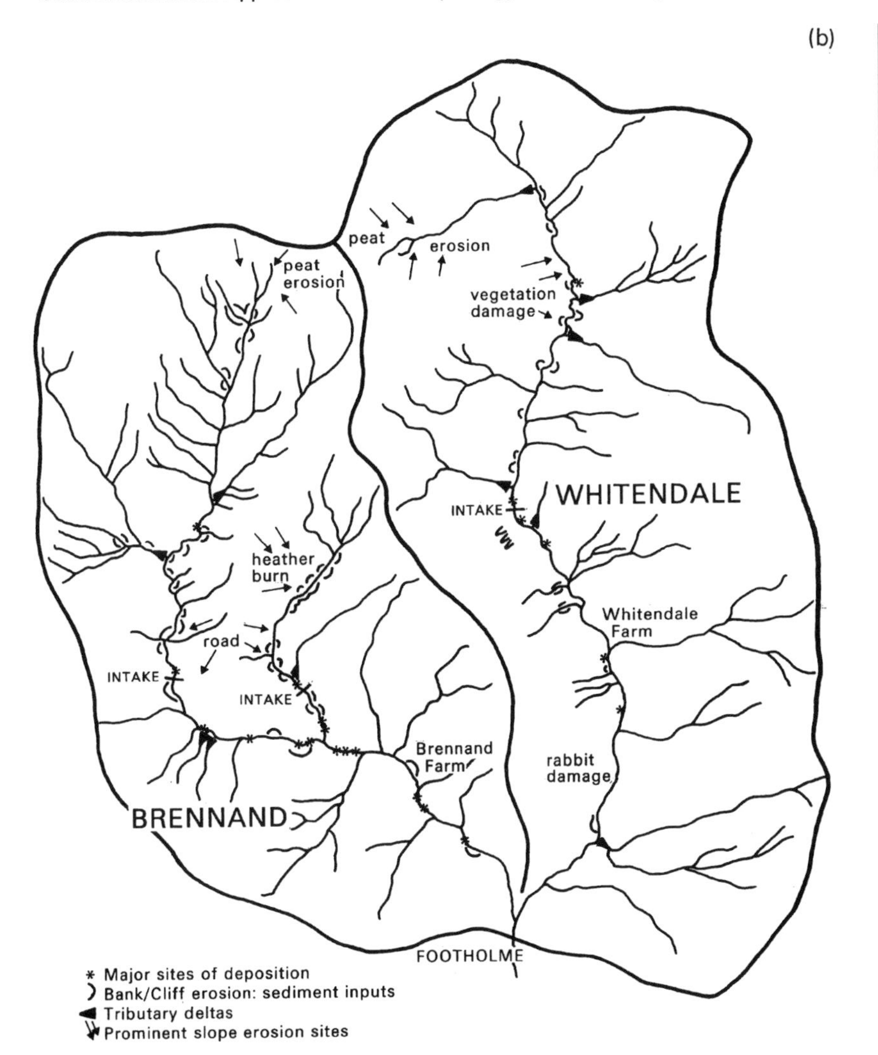

Figure 12.14b Fluvial audits of the Forest of Bowland intakes: sediment supplies and sinks, Whitendale and Brennand catchments

emptied five months before the field visit in October 1987. Second, painted tracer pebbles covering the size range 22.5–125 mm were placed in the Brennand channel above Footholme and checked for movement seven months (October 1987) and 13 months (April 1989) later. The results indicate movement of the D_{50} size (80 mm) at discharges of around 8 to 9 m^3 s^{-1} and the D_{84} size (210 mm) at around 10 to 11 m^3 s^{-1}. In the Brennand these discharges correspond to total discharges at Footholme (Brennand and Whitendale) of about 18 to 20 m^3 s^{-1} and 23 to 25 m^3 s^{-1} respectively.

Analysis of these data supports an approach to calculations of critical discharge given by Bathurst (1987). Application of this approach to the Brennand and Whitendale intakes gives critical discharges for initiation of transport of the finer gravel sizes in the ranges 1.68 to

3.35 $m^3 s^{-1}$ and 1.0 to 2.0 $m^3 s^{-1}$ respectively. The D_{84} sizes for the two sites are estimated to move at discharges of about 6.5 $m^3 s^{-1}$ for the Brennand ($D_{84} = 240$ mm) and 3.5 $m^3 s^{-1}$ for the Whitendale ($D_{84} = 255$ mm).

The relatively large transport of material along the river systems (particularly the Brennand) regularly creates a problem of deposition where the two channels meet in the Footholme stilling basin. Convergence of the inflowing streams, aided by the creation of separation eddies off the exit shoals, leads to the formation of two counter-rotating secondary flow cells and the generation of a shear layer between the two streams. A strong scouring action is produced, with the flow plunging to the bed and returning to the surface along the walls of the scour hole. The scouring action and upward flow component tend to maintain the avalanche faces of the exit shoals at steep angles.

The minor intakes clearly experience regular problems of sedimentation. It is difficult to see how these problems can be mitigated by any form of management other than regular mechanical emptying. This would require costly improvements to access. Any alternative measures aimed at curtailing sediment supply upstream are doomed to failure. For example, revegetation of scars is not feasible and gravel traps placed further upstream also need emptying, raising even more difficult problems of access.

Bracken control on the steeper slopes has apparently removed an important soil protection, and slopewash has clearly occurred before heather can take over as the vegetative cover; on such slopes a major rabbit infestation has clearly exacerbated the situation.

In two small forestry plantations recently established in the Whitendale catchment, drainage practice has resulted in incision and erosion. In the upper Brennand, dense networks of 'moor grips' (open drains) have eroded severely.

Vehicular access to shooting butts, particularly in the Brennand, together with low-cost road construction (without proper culverting) has led to incipient gullying.

Since supplies of further sediment cannot be curtailed it is difficult to suggest an efficient future for these intakes; clearly, had it not been for the 1967 flood these intakes might have lasted under manual care for a further century.

dam construction and operation impacts considered here, the geomorphological contribution needs to be regulated and updated by competent specialists who are often outside the responsible river management institutions.

Our case study (Box 12.9) reports on an ongoing monitoring programme designed to assess the impacts of the Roadford Dam, Devon, UK, on the channel-bed sediment characteristics and channel cross-profiles in the River Wolf below the dam, by comparison with control sites on neighbouring rivers (Figure 12.15).

12.5 CHANNEL RESTORATION AND HABITAT CREATION

River engineering works involving, for example, dredging, widening and straightening have, over the centuries, adversely affected hundreds of kilometres of rivers. The destruction of instream and riparian habitats and the creation of uniform conditions with little morphological and hydraulic diversity has homogenised the river environment and adversely affected conservation and amenity values.

Box 12.9

The monitoring programme uses nine sites in the Roadford area (seven downstream from the dam construction works and two control sites unaffected by the works), selected during a pilot survey of the spatial variation of channel-bed sediments. Data were derived for the armour layer and the underlying substrate, including the upper substrate to a depth of 15 cm and the lower substrate (15–30 cm). Four key sites were divided into three cells across the channel. From each of these cells and at the five other salmon spawning grounds, the following samples are obtained during each annual survey (undertaken at the start of the hydrological year):

(a) 100 armour layer particles from a Wolman grid-by-number sample;
(b) 20–25 kg substrate sample by freeze-coring, derived from five randomly chosen point samples.

Surveys of the channel cross-section were undertaken at the four key sites.

The channel-bed sediments are dominated by gravel with at least 20%, and usually 35%, coarser than 37.5 mm. The median particle size is 27 mm (see Table 12.10).

Table 12.10 Diagnostic sediment sizes below Roadford Dam

	Armour layer D_{50} (mm)			Substrate D_{50} (mm)		
Site	1987	1988	1989	1987	1988	1989
1	55	47	47	38	22	26
2	60	54	53	29	22.5	31
3	50	33	47	28	13	20
4	71	58	66	14	38	21
5	71	63	62	44	29	25
6	75	61	73	28	46	34
9	59	48	55	23	37	26
12	73	53	61	25	26	31

The cross-sections showed little change, with only minor aggradation at most sites. The exception was at site 2 which had degraded by a maximum depth of 20 cm across one-third of the upstream section.

At the end of the phase of bed clearance and dam construction, five sites on the River Wolf, extending downstream for 5 km below the dam site, showed elevated concentrations of fines with about 14% finer than 1 mm. Maximum concentrations of fines (18% < 1 mm) occurred at a site located 3 km downstream of the dam. More than a quarter of the sediment in the lower (15–30 cm) layer was less than 1 mm.

Over a period of two years a depositional front was shown to have progressed downstream from the original site 1 (located 300 m below the dam construction works). Elevated concentrations of fines were identified at site 1 in 1987, at sites 1, 2 and 3 in 1988, and at sites 1 to 4 in the 1989 survey. This represents an average rate of advance of 2.5 km yr^{-1} – a sedimentation rate of about 17 500 m^2 yr^{-1}. Control sites and sites remote from the works (sites 5, 6, 9 and 12) showed variations over the period, but no trends were apparent in the data.

The results suggest that basic clearance and the construction works had a marked impact on the sediment characteristics of salmon spawning grounds for 5 km downstream. Within this reach, fine sediment infiltrated into the gravel bed, leading to partial substrate clogging/

compaction. Estimates suggest that the increase in fines (<1 mm) represents the addition of about 1300 tonnes of sediment over a channel bed area of 35 000 m^2.

Prior to the initiation of construction works, natural levels of fines (<1 mm) at salmon spawning grounds was less than 10%. This compares well with data for similar streams where gravels have a median content of sediment less than 1 mm of 7.9%. These data suggest that in the affected reach the proportion of fines is about 5% higher than the median level for natural salmon spawning gravels in this type of stream. Although there is no simple relationship between substrate characteristics and salmon spawning success, sedimentation is known to have adverse effects.

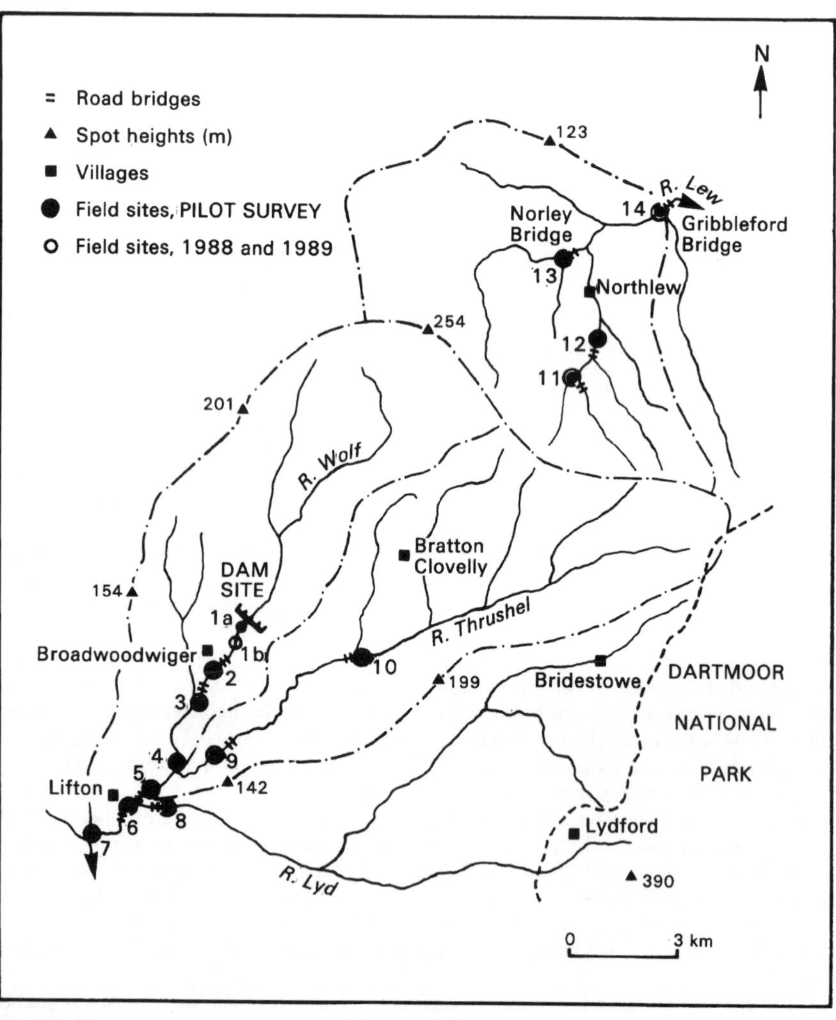

Figure 12.15 Roadford Dam impact monitoring: sediment sampling sites on the Wolf and tributaries

12.5.1 Restoration and Channel Stability

Lowland rivers have low regional slopes and small rainfall surplus over evaporation, which combine to give low stream power. The rivers are, therefore, insufficiently powerful to overcome the erosion resistance of the banks, particularly where the river bank is derived from soft clay bedrock or drift, making the sediment fine-grained and cohesive. Where non-cohesive glacio-fluvial sands and gravels form the river bank, such as in glaciated areas to the north of London, then there may have been considerable change in recent times. In general terms, the channels of the River Thames and tributaries may be described as geomorphologically inactive; that is, they have not changed their courses measurably since the first accurate surveys of the late 19th century (Ferguson, 1981), with the exception of those channels which have been relocated or straightened. Generally, these channels have a stable gravel bed and sediment movement is limited to the transport of finer sediments derived from building construction and urban runoff.

Box 12.10

The low-flow widths for restored channels in the Thames basin are determined by observing the functioning of the affected reach as well as adjacent natural reaches, both upstream and downstream. If there is an inadequate depth of water at low flow then the channel can be narrowed to increase depth. This can be achieved by the construction of a multistage earth channel, whereby the lowest stage accommodates the low flow, or by the use of deflectors. Narrowing the channel provides a more adequate low-flow depth for aquatic life and discourages the deposition of sands and silts within pools and riffles by locally increasing the velocities during lowflows. In a previously over-wide channel, narrowing may also improve water quality by increasing velocities and, therefore, aeration. Figure 12.16a depicts the establishment of a suitable low-flow width for an earth river channel in Essex, England. At 5.5 m width the channel is currently over-wide and has a natural low-flow width of 3.2 m at bends and 3.7 m in straight reaches.

Restored pools are usually adjacent to the outside of a bend, where the flow lines naturally converge. Locating pools here also tends to keep them scoured of silts derived from the catchment upstream. In several instances, where the channel bed has been lowered, the pool–riffle sequence has been recreated at the new lower depth (Brookes, 1990). Figure 12.16b shows part of a chalk stream proposed for restoration in Berkshire. Flows are concentrated in the newly excavated pools by installing deflectors on the inside of the bends to establish the desired low-flow width. These deflectors also trap sediment and help to build up a point bar on the inside of the bend, thereby controlling the sediment deposition patterns in the stream.

The instatement of a stable substrate for aesthetic or ecological reasons is an important component of restoration. Calculation of the bed shear stress exerted on a channel at the design flow is essential to determine the nature and size of substrate to be reinstated. The maximum permissible tractive force approach can be used. The Du Boys equation for the unit tractive force (τ) is:

$$\tau = \gamma RS$$

where γ is the specific weight of water; R is the hydraulic radius; S is the bed slope.

Laboratory and field data provide maximum permissible values of unit tractive force. A typical value for the movement of coarse gravels in water transporting colloidal silts is 32 N m^2 (or a velocity of 1.8 m s^{-1}). For cobbles and shingles the value is 53 N m^2, or a

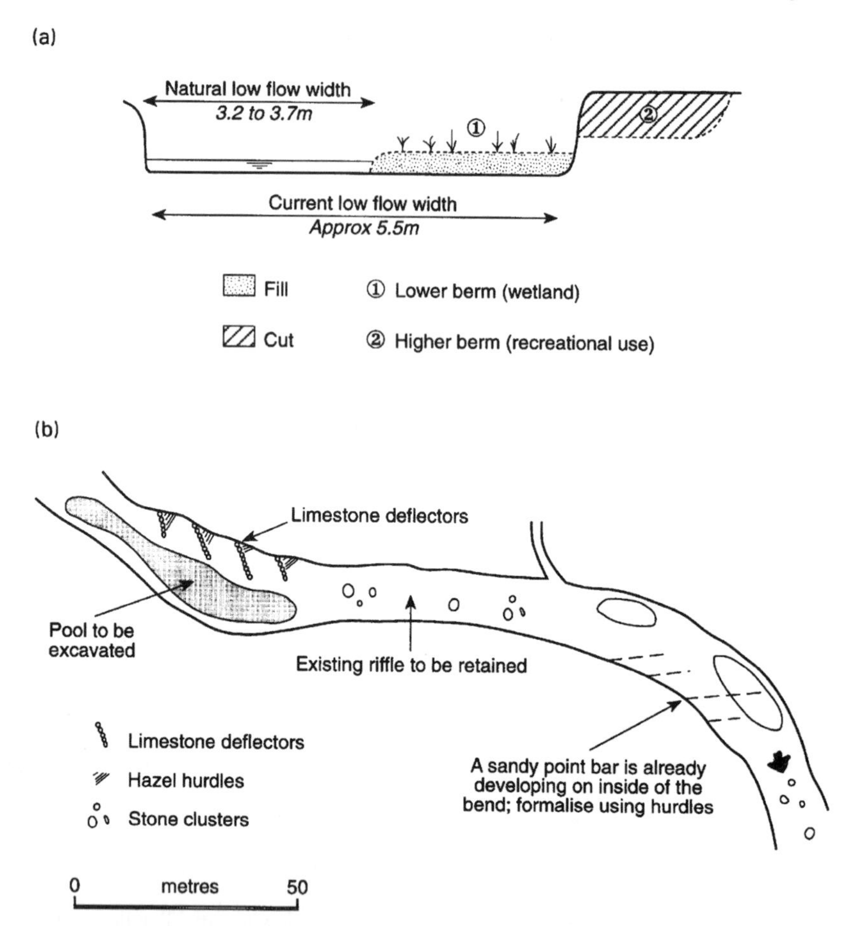

Figure 12.16 Restored river sections and plans, Thames Region NRA: (a) cross-section of lowland clay stream, Essex; (b) planform of chalk stream (gravel-bed), Berkshire

velocity of 1.7 m s^{-1}. If the shear stress is not too high then appropriately sized materials can be reinstated. Table 12.2 shows a recommended size distribution of gravels for a medium-sized (4–8 m wide) clay bed channel in the Thames region. This is for the top 150–200 mm of the bed. Experience with a number of projects in the Thames region has shown that very angular, washed gravels and cobbles are more likely to form a permanent, stable armoured layer than rounded materials. There are numerous indices available for assessing the roundness or angularity of materials (e.g. Briggs, 1977).

If the shear stress is too high, as for example in urban areas, then alternative methods of placing the substrate have been tried such as forming the bed from rolled angular cobbles locally embedded in concrete. The interstices can be filled with gravels, either imported or from an adjacent natural reach. Brookes describes various methods of restoring the substrate to urban river channels.

The significance of a geomorphological understanding in such systems is to allow the design of channels which recover forms and features lost to past engineering and which remain stable after restoration. Design principles include the selection of an appropriate low-flow width, the construction of pools and riffles, choice of an appropriate substrate, bank profiles which emulate more natural channels, and the nature and siting of more natural bank protection measures. Apart from the restoration of river channel cross-sections and sedimentary bedforms, advice is also given on the recreation of natural river planform patterns and extension of the natural drainage network by removing culverts and concrete pipes. Perhaps one of the greatest challenges is the restoration of earth channels in urban areas, where the increased frequency and magnitude of flood flows arising from paved surfaces enhances the potential for instability (see Box 12.10).

12.5.2 Restoration and Fluvial Features: Habitat Creation

Major restoration schemes should involve the naturalisation of long reaches of river by reinstating such features as meander bends, pools and riffles, and they need to be designed to be compatible with local constraints. Minor restoration projects are essentially local and involve the restoration of a small number of pools and riffles and the creation of overhanging or vertical banks, dead zones, etc. These can be created non-structurally by the installation of gravel/quarry rock bars, submerged vanes or deflectors.

Many lowland rivers in the UK are heavily managed. Long dredged and ponded reaches are created to provide the necessary head and volume of water for milling purposes. On many rivers a series of mills have been constructed and only short reaches remain unaffected by backwater conditions. Even these sections have been dredged, widened or straightened for flood alleviation. Although none of the mills are now operational, the ponded reaches are retained (Boxes 12.11 and 12.12).

12.6 RETROSPECT AND SUMMARY

To conclude a collection of case studies such as this it is appropriate, because of the interdisciplinary nature of river management, to be more analytical and critical of the material we have included so as to draw out general themes of technology transfer or of technical improvement.

12.6.1 The Importance of Technique and Field Measurement

It has been a tradition in the UK to build up a body of simple, highly practical river engineering techniques (see IWEM Handbook). While these have been standardised through professional bodies, there has been considerable variation in 'custom and practice'. Such variation results partly from such socio-economic factors as land tenure,

Box 12.11

The Norfolk Anglers Conservation Association have fishing rights along a 3 km stretch of the River Wensum, immediately downstream from Lyng Mill. Advice was required on measures to improve fisheries and habitats within a reach of the river together with plans for the restoration of instream and bankside habitats. In particular, they were keen to improve habitat for chub.

A geomorphological survey was carried out along the reach in question, to ascertain the size and extent of existing pools and riffles, bank profiles, instream and riparian flora, the size characteristics of the bed material and the upstream limit of ponded conditions (above Lenwade Mill). This information, which was plotted on a 1:1000 base map (Figure 12.17), provided the basis for the design of the restoration scheme. Essentially, it was a question of incorporating natural features in the proper, natural locations if they are to be self-sustaining.

Chub prefer faster flowing reaches and bankside cover and this type of habitat was currently lacking (Smith, 1989). There were, therefore, three main elements in the design: creation of two large riffles, construction of two pools and the provision of bankside tree cover.

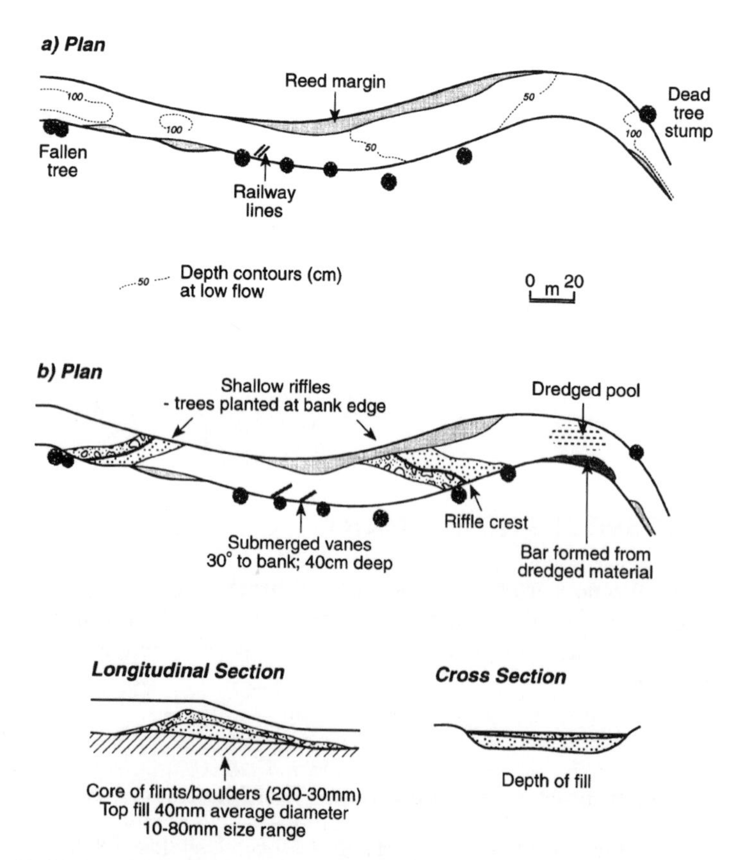

Figure 12.17 Recreation of pools and riffles at Lyng, River Wensum: (a) geomorphological map of original channel; (b) proposed riffle–pool sequence (from Hey, 1992. Reproduced by permission of the Institute of Fisheries Management)

Riffles were proposed at two inflection points where the river had been over-widened. The restoration reach was upstream from mill backwater effects so that riffles would be able to create both ponded sections and an increased gradient over the riffle itself (Figure 12.17). Flint blocks were proposed for the base of the riffles overlain by graded material in the 5–80 cm range. Each riffle was angled across the channel, directing flow towards the outer bank in the adjacent pool downstream, and was slightly parabolic in cross-profile.

Two pools were plannned. The downstream one enlarged an existing pool by dredging the bed adjacent to the left bank and creating a shelving point bar on the opposite bank with the spoil. This narrowed the channel considerably and provided scouring conditions at high flow (Figure 12.17). The other pool, between the two riffles and adjacent to the right bank, was designed to be formed and sustained by the river. To aid the process, two submerged vanes were proposed, each 3 m long and 40 cm deep and angled downstream at 30 degrees to the bank, which would direct faster, surface water towards the bank where the associated downwelling would scour the bed (Figure 12.17). A tree planting programme, using willows and alders, was proposed at key locations to prevent bank erosion adjacent to the pools and to provide cover for fish.

The scheme was implemented in 1990 and has proved to be successful. The riffles created ponded reaches upstream and shallow, faster flowing water across the downstream apron. The gravel substrate was colonised by a range of invertebrates, including freshwater crayfish and stone loach, even during the construction phase. The submerged vanes, which were installed to create the second pool plus some bank erosion, had to be removed owing to their success. The flow off the upstream riffle is now maintaining this pool and cut bank. The pool that was dredged suffered some siltation immediately after construction, but following a series of high flows it is being maintained in a silt-free state. The tree planting programme has been completed, but more time is required before its full benefits materialise.

Box 12.12

The River Wensum at Fakenham, Norfolk, UK, is a lowland river which has been extensively straightened and dredged, resulting in a lack of habitat diversity. The National Rivers Authority, Anglian Region, required advice on the deployment of instream structures over a 1000 m length of the Wensum below Fakenham Mill to improve fisheries habitats. This was subsequently to serve as a demonstration project for riparian landowners.

A standard geomorphological survey was carried out along the river to identify existing river conditions. This confirmed the uniform nature of the reach, but also identified a number of silted pools. Traditional structural methods for habitat improvement involve the construction of weirs or deflectors. The former cause extensive ponding, adversely affect flood capacities and trap floating debris. Deflectors, which restrict channel width up to and above bankfull flow, also limit the discharge capacity of the river and trap debris. For this scheme, unconventional structural measures were advocated to create pools. Paired submerged vanes, both symmetrical and asymmetrical, angled upstream at approximately 30 degrees, were designed to desilt existing pools (Figure 12.18). The vanes direct near-bed water towards the bank, while faster flowing surface water was made to converge at the centre of the channel. This causes downwelling at the convergence zone and an associated increase in near-bed velocities. Alternatively, a single, straight vane angled downstream at 45 degrees can be installed extending from the bank, and this will create a pool adjacent to the bank. Bank erosion and meander initiation could occur if scouring is significant. As the

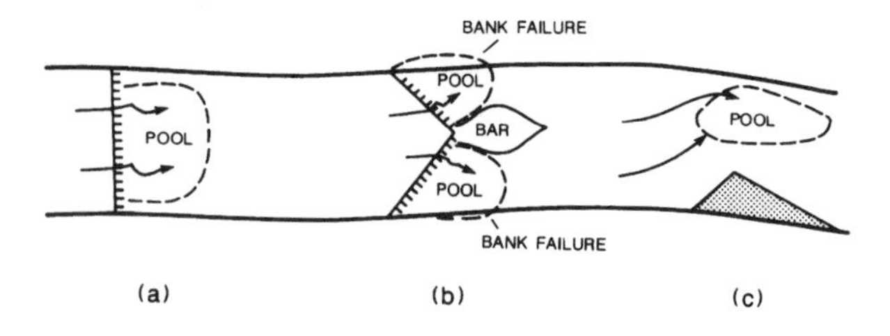

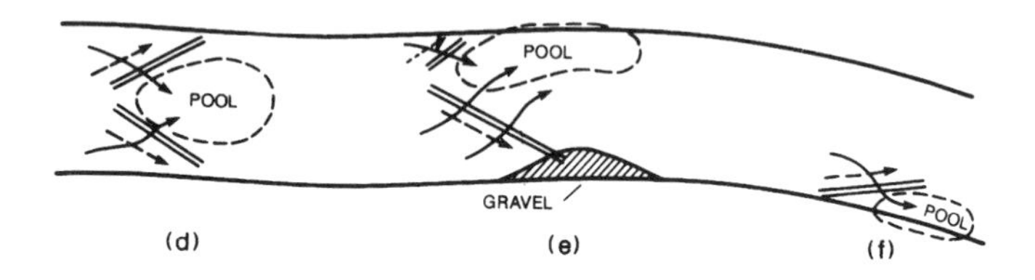

a) Full width weir – perpendicular
b) Full width weir – V-ing downstream
c) Wing deflector
d) Submerged vanes – symmetrical V-ing upstream
e) Submerged vanes – asymmetrical V -ing upstream
f) Wing vane

Figure 12.18 River restoration using structural measures (from Hey, 1992. Reproduced by permission of the Institute of Fisheries Management)

vanes are submerged at low flow they have minimum impact on channel capacity and do not trap debris (Hey, 1990).

Given the geomorphological characteristics of reach and the location of existing pools, a number of submerged structures were designed for installation along the reach. Vane sizes had to reflect local low-flow depths and widths and were up to 3 m long and 20 cm deep. Wooden boards, secured by reinforcing rods, were sufficiently robust for use as vanes on this particular river, as gravel-bed material was not transported. The submerged vanes were installed by hand in 1989 without the need for mechanical aids. The wooden vanes were protected by flint cobbles, principally to naturalise and camouflage the structures.

Subsequent site surveys indicated that the vanes provided sufficient downwelling and increased bed scour to desilt pools. Single straight vanes were seen to cause local bank erosion and, as these had been located on alternate banks on the straightened section, slight meandering was initiated. In terms of fisheries improvement, the few brown trout which colonised the existing river were seen to congregate adjacent to the submerged vanes in the faster flowing water. The riffled water surface also provides cover from predators. According to the angling club the modifications to the habitat along this reach have improved the fisheries and the biotic carrying capacity of the reach has been considerably improved.

hazard rating and personal preference, but also reflects an innate sympathy for variability in the force and resistance factors in the rivers themselves. Thus, the regional tradition for river bank protection in Wales has been for quarry waste revetment, whereas in Yorkshire it is much more common to see willow spiling. Part of the contribution made by geomorphology is to further refine this sensitivity to environmental conditions at scales as local as within-the-reach.

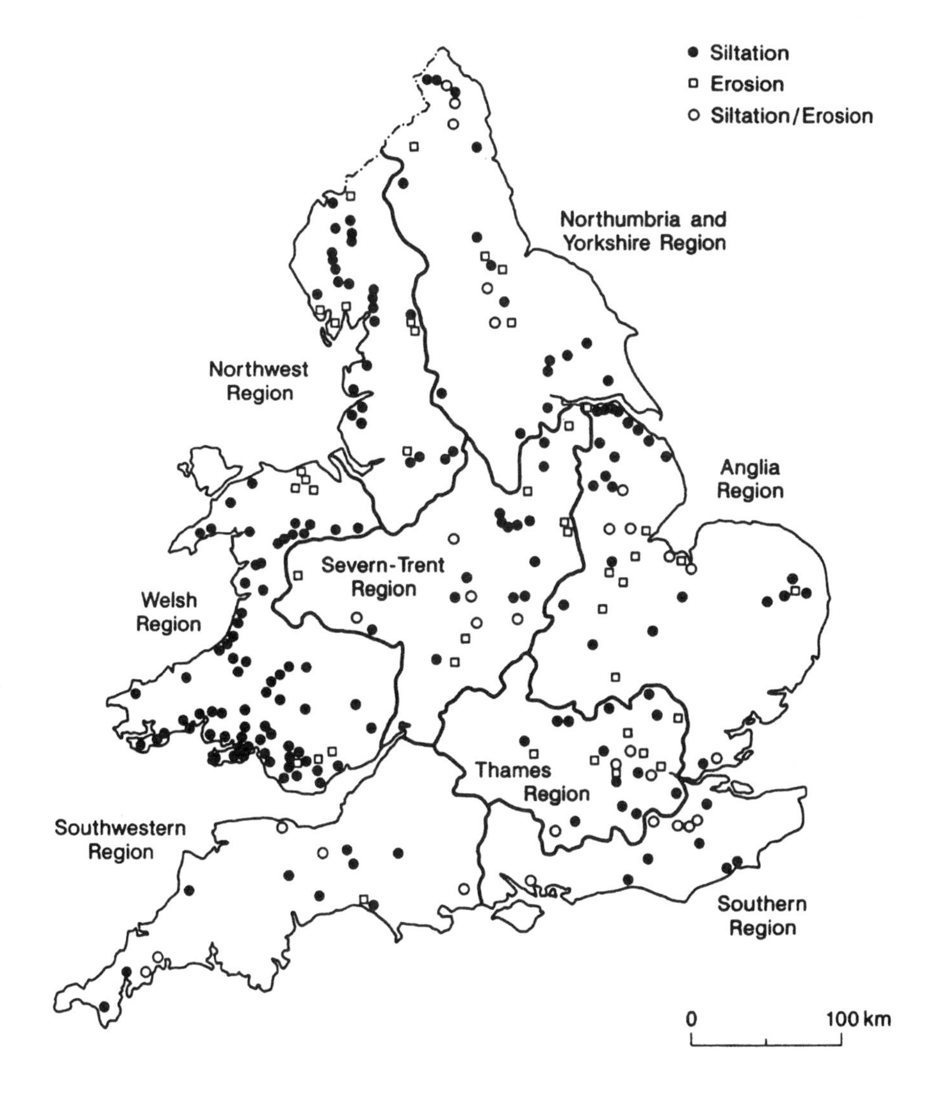

Figure 12.19 Siltation, erosion and joint problems of river channel maintenance in England and Wales

In order to live up to this claim it is necessary to have at hand a body of field techniques with which to research the boundary conditions to any imposed changes to a reach, be it natural or artificial.

At present the parameters which are measured by geomorphologists as a basis for channel design or habitat protection, for example, are the normal preserve of research institutions and academics; they have, by diffusion, become fairly standard, but are not uniform. Certainly, with the exception of flow variables collected by the rivers authorities at gauging stations, none is routinely monitored. It is particularly problematic that, in the UK, there is no monitoring system for sediment yields, yet we increasingly realise that the 'stable bed, zero flux' assumptions often made for design in the relatively quiescent channel conditions produced by British geology and climate will not hold. Furthermore, UK geomorphologists have begun to reveal that quiescence is but relative; Figure 12.19 reveals how widespread routine sediment-related river maintenance has become in the area covered by the National Rivers Authority. Solutions must be sought which, sadly, incorporate data that we lack!

From the case studies presented here it would appear that the fluvial geomorphologist joining our interdisciplinary river management team will need a 'shopping list' of measurements which might include some or all of those shown in Figure 12.20.

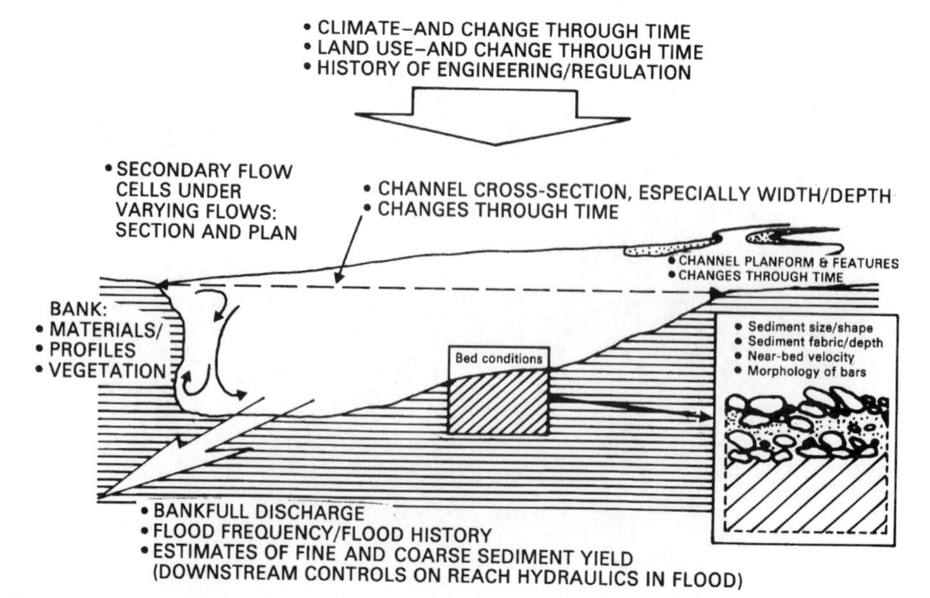

Figure 12.20 Elementary information needs for geomorphological treatment of river management problems, seen in both cross-section and upstream/downstream

12.6.2 Bridging Disciplinary Gaps: Explaining 'Whole River' Geomorphology

One outcome of the case studies reported here is the realisation by engineers, biologists, anglers and others that fluvial geomorphology has both a legitimate broad technical role (e.g. stability analyses), utilising numerical or statistical predictions, and a qualitative, observational and field measurement role which is much harder to codify and to access. In some ways fluvial geomorphology's practitioners work as natural historians, basing some of their expertise on an experience built up from observations in the field. Until recently it could be claimed that they further restricted access to this usable knowledge by employing a terminology which confused those who were unaccustomed to 'riffle', 'clast', 'planform' etc.

Fortunately, one outcome of the exposure gained by geomorphologists in applied studies such as those reported here has been a codification, tabulation and clarification of hitherto unorganised knowledge from the field of river studies. This section attempts to rationalise this transferable information but itself is largely the outcome of a case study in which Sear and Newson (1991,1994) attempted to inform river engineers working with the National Rivers Authority about the importance of geomorphological processes in channel maintenance operations. The section also comes with the clear warning that, although the purpose is to transfer knowledge to an interdisciplinary field (and a lay audience), there

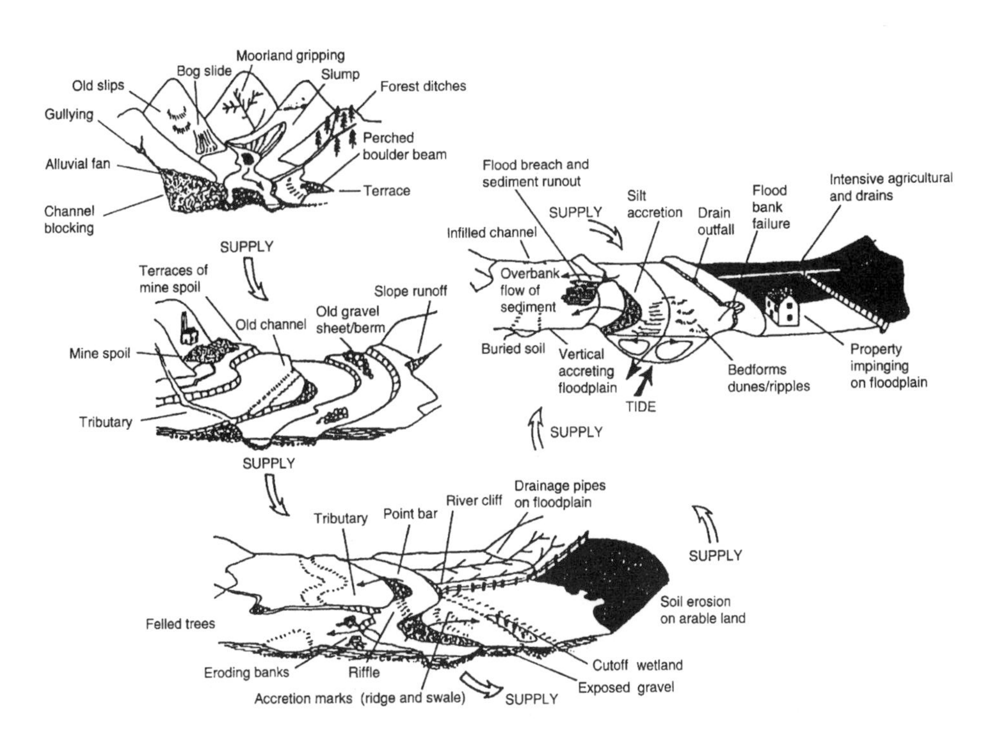

Figure 12.21 Use of upstream/downstream 'cartoons' to illustrate field indications of river instability, both in planform and section (from Sear and Newson, 1992)

Table 12.11 Identifying sediment sources; indicators of stability

	Upland zone	Transfer zone	Lowland zone
Potential sediment sources	Slope debris Peat slides Alluvial fans Boulder berms Channel bars Bank erosion Forest ditches Forest roads Moorland 'grips' Mining	River cliffs Terraces Bank erosion Channel bars Field drains Field runoff Urban runoff Tributaries/upstream Mining	Upstream Bank erosion Bed movement Field drains Field runoff Wind-blown material Tributaries Urban runoff Tidal sediments Mining
Evidence of erosion	Perched boulder berms Terraces Old channels Old slope failures Undermined structures Exposed tree roots Narrow/deep channel Bank failures both banks Armoured/compacted bed Gravel exposed in banks	Terraces Old channels Narrow/deep channels Undermined structures Exposed tree roots Bank failures Armoured/compacted bed Deep gravel exposed	Old channels Undermined structures Exposed tree roots Narrow/deep channel Bank failures Deep gravel exposure
Evidence of aggradation	Buried structures Buried soils Large uncompacted bars Eroding banks at shallows Contracting bridge space Deep fine sediment in bank Many unvegetated bars	Buried structures Buried soils Large uncompacted bars Eroding banks at shallows Contracting bridge space Deep fine sediment in bank Many unvegetated bars	Buried structures Buried soils Large silt/clay banks Eroding banks at shallows Contracting bridge space Deep fine sediment in bank Many unvegetated bars
Evidence of stability	Vegetated bars and banks Compacted weed-covered bed Bank erosion rare Old structures in position	Vegetated bars and banks Compacted weed-covered bed Bank erosion rare Old structures in position	Vegetated bars and banks Weed-covered bed Bank erosion rare Old structures in position

Table 12.12 Engineering feedbacks to channel process/morphology

Engineering work	Upland bounder-bed, steep channel	Transfer gravel-bed, moderate channel	Lowland sand-silt/clay channel
Straightening and deepening	Destabilises channel, especially in large floods. High maintenance frequency	Upstream incision and downstream aggradation. Shoaling reforms. High maintenance frequency	Encourages downstream aggradation and upstream incision. Bank erosion increases siltation and maintenance
Dredging to grade	Difficult to maintain	Riffles/pools/bars are natural features; high frequency of upstream and downstream adjustments	Upstream incision, downstream aggradation. Channel silts up. Banks need protection
Close-set embankments	Increased stream power	Increased erosion, deposition, migration	Bank and bed erosion. High frequency maintenance
Set-back embankments	Limited value in these channels	Accommodates flood, allowing migration	Leaves low-flow hydraulics natural
Two-stage channel	Limited value	Berm unstable	Must accommodate meanders
Culverting	High strength required. Costly	Dredging required. High cost	Rapid siltation. Bed can aggrade downstream
Trapping	Protection required downstream. Fills on flood	Site on riffles. Regular maintenance	Large trap required. Bed mobility problem
'Hard' bank protection	Environmentally unacceptable. High cost/short life	High cost. Limited success. Frequent maintenance	High cost. Limited success
'Soft' bank protection	Encourages vegetation, providing bank protection	More successful–install before winter floods	Needs time to colonise
Grade control (rock dams/piling)	Used to stop incision. Can cause bank erosion	Can increase instability. Mobile bed/bank problems	Flood scour problems
Underpinning bed/banks	High costs; steel and concrete	High cost to be successful. Bank protection required	Deep working required + bank protection

is no substitute for the recruitment or hiring of a professional geomorphologist for important applications.

The recent attempt to transfer geomorphological approaches has stressed three aspects.

1. An emphasis on the river basin context for the treatment of problems at the reach scale. Geomorphologists, therefore, stress upstream–downstream linkages (especially over longer impact or design timescales) but they also take a valley-floor view beyond the channel cross-section (see Figure 12.21).
2. A classification of river features and their practical utility as indicators of the status and trend of channel dynamics at a reach scale.
3. Utilisation of a field knowledge of the boundary conditions for the major physical processes to advise on:

   (a) the choice of appropriate predictive techniques from river mechanics;
   (b) the likely success of engineering intervention, especially of river structures.

Figure 12.21 and Tables 12.11 and 12.12 are taken from National Rivers Authority reports (Sear and Newson, 1991, 1994). Further simplified guidance is to be found in Newson (1986), Hey (1990) and Newson and Brookes (1994).

It is clear that a promising new dialogue is opening between engineering science, geomorphology and practical engineering, which will allow river managers to use detailed knowledge of natural processes as a less expensive and more sustainable way of protecting both human communities and ecosystems.

12.7 REFERENCES

Ackers, P. and White, W.R. 1973. Sediment transport: new approach and analysis. *Journal of the Hydraulics Division, ASCE*, **99**(HY11), 2041–2060.

Apmann, R.P. 1972. Flow processes in open channel bends. *Journal of the Hydraulics Division, ASCE*, **98**(HY5), 795–809.

Bathurst, J. 1987. Critical conditions for bed material movement in steep boulder-bed streams. In: Beschta, R.L., Blinn, T., Grant, G.E., Ice, G.G. and Swanson, F.J. (Eds), *Erosion and Sedimentation in the Pacific Rim.* IAHS Publication 165, 309–318.

Billi P., Hey, R.D., Thorne, C.R. and Tacconi, P. 1992. *Dynamics of Gravel-bed Rivers.* Wiley, Chichester.

Bishop A.W. and Morgenstern N. 1960. Stability coefficients for earth slopes. *Geotechnique*, **10**, 129–150.

Briggs, D. 1977. *Sediments.* Butterworth, London.

Brookes, A.B. 1987. River Channel adjustments downstream from channelisation works in England and Wales. *Earth Surface Processes and Landforms*, **12**, 337–351.

Brookes, A.B. 1988. *Channelised Rivers: Perspectives for Environmental Management.* Wiley, Chichester.

Brookes, A.B. 1990. Restoration and enhancement of engineered river channels: some European experiences. *Regulated Rivers: Research and Management*, **5**, 45–56.

Carling P.A. 1990. Particle over-passing on depth-limited gravel bars. *Sedimentology*, **37**, 345–355.

Charlton F.G. 1977. *An Appraisal of Available Data on Gravel Rivers*. Report INT 151, Hydraulics Research Station, Wallingford.

Colorado State University. 1975, *Highways in the River Environment: Hydraulic and Environmental Design Considerations*. Report to Federal Highway Administration, US Department of Transportation.

Darby, S.E. and Thorne, C.R. 1992. Impact of channelisation on the Mimmshall Brook, Hertfordshire. *Regulated Rivers: Research and Management*, **7**, 193–204.

Ferguson, R.I. 1981. Channel form and channel changes. In: Lewin, J. (Ed.), *British Rivers*. George, Allen and Unwin, London, 90–125.

Harold, C.H.H. 1937. The flow and bacteriology of underground water in the Lee Valley. In: *The Results of the Chemical and Bacteriological Examination of the London Waters for the 12 Months ended 31st December 1937*. 32nd Annual Report to the Metropolitan Water Board, Chapter 17, 89–99.

Hemphill, R.W. and Bramley, M. (Eds) 1989. *Protection of River and Canal Banks*. Butterworths, London, 200pp.

Hey, R.D. 1976. Geometry of river meanders. *Nature*, **262**, 482–484.

Hey, R.D. 1979. Flow resistance in gravel-bed rivers. *Journal of the Hydraulics Division, ASCE*, **105**(HY4), 365–379.

Hey, R.D. 1982. Design equations for mobile gravel-bed rivers. In: Hey R.D., Bathurst, J.C. and Thorne, C.R. (Eds), *Gravel-bed Rivers*. Wiley, Chichester, 553–580.

Hey, R.D. 1986. River response to inter-basin water transfers: Craig Goch Feasibility Study. *Journal of Hydrology*, **85**, 407–421.

Hey, R.D. 1990. Environmental river engineering. *Journal of the Institute of Water and Environmental Management*, **4**(4), 535–540.

Hey, R.D. 1992. River mechanics and habitat creation. In: O'Grady, K.T., Butterworth, A.J.B., Spillet, P.B. and Domaniewski, J.C., (Eds), *Fisheries in the Year 2000*. Institute of Fisheries Management, Nottingham, 271–285.

Hey, R.D. and Heritage, G.L. 1993. *Draft Guidelines for the Design and Restoration of Flood Alleviation Schemes*. R&D Note 154, National Rivers Authority, Welsh Region, 98pp.

Hey, R.D. and Thorne, C.R. 1986. Stable channels with mobile gravel-beds. *Journal of Hydraulic Engineering, ASCE*, **112**(8), 671–689.

Hey, R.D., Bathurst, J.C. and Thorne, C.R. 1982. *Gravel-bed Rivers: Fluvial Processes, Engineering and Management*. Wiley, Chichester.

Hey, R.D., Heritage, G.L. and Patteson, M.T. 1990. *Design of Flood Alleviation Schemes: Engineering and the Environment*, Ministry of Agriculture, Fisheries and Food, London, 176pp.

Hey, R.D., Heritage, G.L. and Patteson, M.T. 1994. Impact of flood alleviation schemes on aquatic macrophytes. *Regulated Rivers: Research and Management*. **9**, 103–119.

Hooke, J.M. 1980. Magnitude and distribution of rates of river bank erosion. *Earth Surface Processes*, **5**, 143–157.

Hooke, J.M. and Redmond, C.E. 1989. River channel changes in England and Wales. *Journal of the Institute of Water and Environmental Management*, **3**, 328–335.

Jain, S.C. and Fischer, E.E. 1979. *Scour Round Circular Bridge Piers at High Froude Numbers*. Report FHWA-RD-79-104, Federal Highways Administration, US Deptment of Transportation.

Jones, J.S. 1984. Comparisons of prediction equations for bridge pier and abutment scour. *Transport Research Record*, **950**, 202–209.

Komar, P.D. 1987. Selective grain entrainment by a current from a bed of mixed sizes: a reanalysis. *Journal of Sedimentary Petrology*, **57**, 203–211.

Laursen, E.M. 1962. Scour at bridge crossings. *Transactions of ASCE*, **127**, 166–180.

Leeks G.J.L. and Newson M.D. 1989. Responses of the sediment system of a regulated river to a scour valve release: Llyn Clywedog, Mid-Wales, UK. *Regulated Rivers: Research and Management*, **3**, 93–106.

Lewin, J., Macklin, M.G. and Newson, M.D. 1988. Regime theory and environmental change – irreconcilable concepts? In: White, W.R. (Ed.), *International Conference on River Regime*. Wiley, Chichester, 431–445.

Lewis, G. and Williams, G. 1984. *River and Wildlife Handbook*, RSPB and RSNC, 295pp.

Meyer-Peter, E. and Muller, R. 1948. Formulas for bedload transport. *Proceedings of the 3rd Annual Conference of the International Association of Hydraulics Research*, Stockholm, 39–64.

Natural Environment Research Council (NERC). 1975. *Flood Studies Report* (five volumes). Wallingford UK.

Newbold, C., Purseglove, J. and Holmes, N. 1983. *Nature Conservation and River Engineering*. Nature Conservancy Council, 36pp.

Newson, M.D. 1986. River basin engineering – fluvial geomorphology. *Journal of the Institution of Water Engineers and Scientists*, **40**(4), 307–324.

Newson, M.D. 1992a. River conservation and catchment management: a UK perspective. In: Boon, P., Petts, G. and Calow, P. (Eds), *River Conservation and Management*. Wiley, Chichester.

Newson, M.D. 1992b. *Land, Water and Development*. Routledge, UK.

Newson, M.D. and Brookes, A.B. 1994. Channel geomorphology and habitat. In: Hose, P. and Holmes, N. (eds), *Rivers and Wildlife Handbook*. RSPB, Sandy, Bedfordshire.

Newson, M.D. and Sear, D.A. 1994. *Sediment and Gravel Transportation in Rivers*. National Rivers Authority R&D Note C5/384/2, Bristol.

Odgaard, A.J. and Mosconi, C.E. 1987. Streambank protection by submerged vanes. *Journal Hydraulic Engineering, ASCE*, **113**(4), 520–536.

Paice, C. and Hey, R.D. 1989. Hydraulic control of secondary circulation in meander bend to reduce outer bank erosion. In: Albertson, M.L. and Kia, R.A. (Eds), *Design of Hydraulic Structures*. Balkema, Rotterdam, 249–254.

Parker, G., Klingeman, P.C. and McClean, D.G. 1982. Bedload and size distribution in paved gravel-bed streams. *Journal of Hydraulic Research, ASCE*, **108**, 544–571.

Posey, C.J. 1967. Computation of discharge including overbank flow. *Journal of Civil Engineering, ASCE*, 62–63.

Raynov, S., Pechinov, D., Kopaliany, Z. and Hey, R.D. 1986. *River Response to Hydraulic Structures*. UNESCO, 115pp.

Roberts, C.R. 1989. Flood frequency and urban-induced channel change: some British examples. In: Beven, K. and Carling, P.A. (Eds), *Floods: Hydrological, Sedimentological and Geomorphological implications*. Wiley, Chichester.

Robinson, M. 1990. *Impact of Improved Land Drainage on River Flows*. Institute of Hydrology Report No. 113, 225pp.

Sear, D.A. 1992. *Mimmshall Brook: Geomorphological Assessment*. Unpublished report to Thames Region NRA.

Sear, D.A. and Newson, M.D. 1992. *Sediment and Gravel Transportation in Rivers Including the Use of Gravel Traps*. National Rivers Authority Project Report 232/1/T, 93pp.

Sear, D.A. and Newson, M.D. 1994. *Sediment and Gravel Transportation in Rivers: a Geomorphological Approach to River Maintenance*. National Rivers Authority R&D Note 315, Bristol.

Shields, A. 1936. *Anwendung der ahnlichkeitsmechanil und der turbulenzforschung suf die geschiebebewegung*. Milleilungen der Preussischen Versuchsanstlt fur Wasserbau, Berlin.

Simons, D.B. and Senturk, F. 1977. *Sediment Transport Technology*, Water Resources Publications, Fort Collins, Colorado, 807pp.

Smith, P.R. 1989. *The Distribution and Habitat Requirements of Chub* (Leuciscus cephalus L) *in Several Lowland Rivers in Eastern England*. Unpublished PhD thesis, University of East Anglia.

Thorne C.R. 1982. Proceses and mechanisms of river bank erosion. In: Hey R.D., Bathurst, J.C. and Thorne, C.R. (Eds), *Gravel-bed Rivers*, Wiley, Chichester, 227–259.

Thorne, C.R. and Tovey, N.K. 1981. Stability of composite river banks. *Earth Surface Processes and Landforms*, **6**, 469–484.

Thorne, C.R., Bathurst, J.C. and Hey, R.D. 1987. *Sediment Transport in Gravel-bed Rivers*. Wiley, Chichester.

Water Space Amenity Commission 1980. *Conservation and Land Drainage Guidelines*. WSAC, 121pp.

13 Application of Applied Fluvial Geomorphology: Problems and Potential

C. R. THORNE[1], M. D. NEWSON[2] AND R. D. HEY[3]
[1] *Department of Geography, University of Nottingham, UK*
[2] *Department of Geography, University of Newcastle, UK*
[3] *School of Environmental Sciences, University of East Anglia, UK*

13.1 INTRODUCTION

In this closing chapter the editors address current issues in the application of fluvial geomorphology to river engineering and management. Sections deal in turn with the professional, practical and broad contextual questions.

13.2 PROBLEMS IN APPLYING FLUVIAL GEOMORPHOLOGY IN PRACTICE

This book presents an overview of fluvial geomorphology intended to provide river engineers and managers with useful insights and understanding of natural channel forms and fluvial processes. However, there remain four main areas of difficulty that limit the widespread adoption of geomorphological approaches to improved river management and, in particular, engineering.

1. Geomorphology is 'new' in that it has not been incorporated into the institutional and professional frameworks which circumscribe all public works. For example, neither Bachelor nor Master level courses in geomorphology are accredited by any institution and there is, in fact, no professional qualification or registration in geomorphology nationally or internationally.
2. As an analytical science, geomorphology is different from related disciplines, such as open channel hydraulics or geotechnical engineering, in that it involves the collection of a very wide range of field information that includes not only hard, quantitative data, but also material that is descriptive, narrative and historical. Though it is essentially a field-based science, extensive use may also be made of archive documents, paper maps and geographical information systems (GIS). These

Applied Fluvial Geomorphology for River Engineering and Management, Edited by C. R. Thorne, R. D. Hey and M. D. Newson.

are media that are outside the experience of many river engineers and managers. Much of the data are spatially distributed, making the use of conventional statistical treatments at best, inappropriate, and at worst, misleading.

These complicating factors remove applied geomorphology from the remit of the project team, making it the reserve of a relatively small cadre of experienced individuals who are often added to the project team on a short-term or consultancy basis as 'subject specialists'. This type of arrangement is often efficient and, despite high daily rates, cost-effective, but it tends to perpetuate the application of geomorphology as something of a mysterious 'black art'.

3. Geomorphological input to a project is often associated with the advice given by conservationists and 'single-issue' environmental pressure groups. It is true that geomorphological input is usually environmentally aligned; however, this is not because the geomorphologist is an environmentalist or a conservationist *per se*, but because approaches to river engineering and management that follow a geomorphologically sound approach will almost always produce benefits to ecological habitat value and biodiversity. These are often collateral benefits in that the primary purpose of engineering designs and management practices that preserve, create or reinstate natural channel forms and processes is usually to increase engineering efficiency and reduce post-project maintenance requirements. Many policy-makers and some operations staff remain unconvinced of the validity of an environmentally led, 'green' approach to river engineering and management and, perhaps inevitably, some of their scepticism also extends to geomorphologically based approaches.
4. A more tangible and practical extension of the problem outlined in point 3 is a difficulty currently facing many applied sciences – that of adequately assessing the current status of natural systems in a short, project or political timeframe, and then extrapolating the results to predict future developments in the system over decades and centuries with or without the proposed engineering intervention or management regime. In the case of issues related to global environmental, climate and sea-level changes, the scientific community, governments and society are being driven closer together as they try to cope with considerable uncertainty in the spatial and temporal scales of change. Geomorphology has always had to deal with changes at a wide variety of timescales that are nested within one another and which relate to a cascade of spatial scales. Hence, the need to span a wide range of scales in time and space is not a new challenge for applied geomorphologists. What perhaps is less familiar to geomorphologists is the need to assimilate the role of geomorphology fully with the needs and perspectives of project planners and team leaders, who operate outside academia and who must be convinced that spatially diffuse and long-term impacts are relevant.

This is fundamentally an introductory, factual and methodological book and it has not attempted to address these problems directly. Nevertheless, elements of the answers to some of the questions raised in points 1 to 4 may be found here as a substantial component of the UK fluvial geomorphological community 'networks' with river engineers and managers in an attempt to bridge the gap between basic geomorphology and its application.

Point 1 concerns professionalism in applied geomorphology and, while there is no suggestion that the contributors to this book have a monopoly in this area, they do perhaps demonstrate the existence of a body of expertise that is both sound and accessible, even though it is not codified through a professional institute or association.

Point 2 concerns the fact that geomorphological databases and analytical approaches are different from those routinely used in river engineering and management. The book has set out the methodological basis for the collection of geomorphological data and information and shown how quantitative and qualitative geomorphic analyses and classifications may be applied to gain useful knowledge and insight regarding the past, present and future behaviour of the system. The thematic chapters and, particularly, the case studies demonstrate that, while there remain considerable elements of judgement, major contributions to geomorphological understanding usually come from the careful assemblage and objective analysis of all available data and information. In this regard, the book should have gone some way to demystifying applied fluvial geomorphology.

Point 3 concerns the delineation of approaches to river engineering and management incorporating applied fluvial geomorphology from those driven by conservation or environmental issues. The case studies presented in this book leave no room for doubt concerning the inherent superiority of geomorphologically sound approaches over those that fail to account for fluvial forms and processes. The case for geomorphologically sound engineering is also implicit in 'the four Es' identified as evidence of best-practice by General Hatch when he was the Chief of the US Army Corps of Engineers:

Engineering
Economy
Efficiency
Environment

General Hatch pointed out that schemes could only be considered to be successful if they met attainable levels of excellence in all four categories. Applied fluvial geomorphology is vital to each category in the following ways.

(i) Concepts of 'taming' rivers have now been abandoned and the adoption of works that harmonise with the natural forms and processes of the river are accepted to represent best-practice in engineering.
(ii) Cost-benefit analyses of geomorphologically sound schemes and management regimes show them to be highly cost-effective. This is not withstanding the difficulties of properly accounting for the value of intangible benefits such as protecting environmental capital, sustaining the transfer of landscape and heritage features from this generation to the next, and increasing biodiversity. It appears likely that as methods of accounting for intangible benefits improve, the economic advantage of geomorphologically aligned approaches will further increase.
(iii) Geomorphologically aligned channels have been shown to operate more efficiently over the entire range of flows in conveying water and sediment, and to have lower maintenance requirements than conventionally engineered channels.
(iv) The environmental benefits of river engineering and management based on geomorphological principles are now firmly established.

Point 4 addresses problems of time and space scales, particularly in the prediction of future evolution of fluvial systems with and without human interference. The book has illustrated that fluvial geomorphologists are accustomed to working at a variety of temporal and spatial scales. The chapters and case studies demonstrate that geomorphologists appreciate the potential for present-day, local problems to have wider implications that are often neglected by other specialists. While not underestimating the difficulties associated with individual situations and geographical locations, geomorphological analyses will at least help the engineer or manager to view local problems and particular parts of the fluvial system with the appropriate perspectives of time and space. These perspectives are the essential prerequisites for the formulation of sustainable solutions.

13.3 USING THE BOOK IN PRACTICE

Figure 13.1 presents a conceptual diagram that attempts to illustrate where each chapter in the book fits into a framework dimensioned by time and space. This is a broad 'map' of the book and should assist engineers and managers in identifying which chapters are relevant to a particular application or project.

However, a more structured guide is required if the book is to be used in project studies at pre-feasibility and feasibility levels. Project-related studies are usually complex and multifaceted, requiring an ordered or modular system that allows several subunits of study to proceed semi-independently and in parallel to one another.

A blueprint for the geomorphological aspects of the project should be developed. The aim of a blueprint is to support a raft of geomorphological studies that are comprehensive in their coverage of the issues, but within which problems or hold-ups in one topic do not

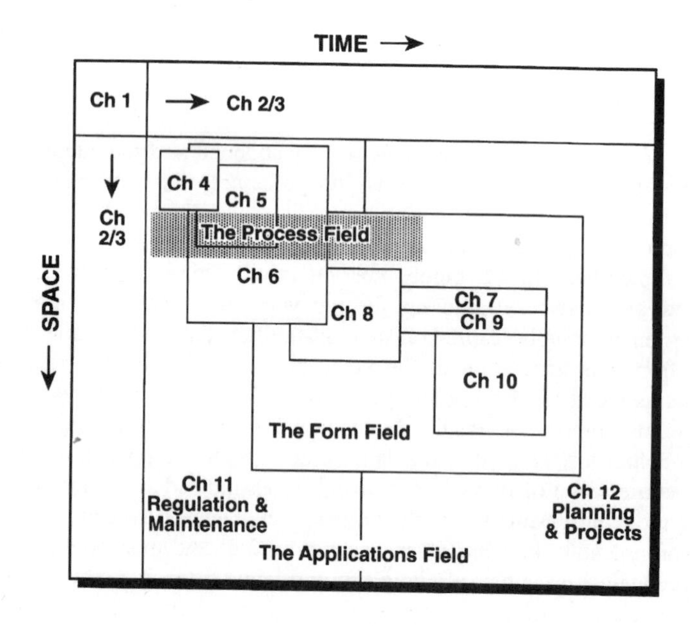

Figure 13.1 Schematic plan for chapters in the book to assist end-users in applying geomorphology to river engineering and management projects over a variety of time and space scales

delay progress in the others. To this end, the complex and interrelated field of applied geomorphology may be divided into subtopics, each of which can be pursued in a single, definable study with a clearly identifiable end product. At the conclusion of the study, the products of the topic-based studies must then be reassembled into an integrated geomorphological report which is both broad and deep. The details of a geomorphological blueprint will to some extent be dictated by the scale of the system and the approaches to data acquisition appropriate to the resources available. A generic example of a geomorphological blueprint may be found in a recent paper by Thorne and Baghirathan (1993).

Clearly, each blueprint must be customised according to the particular application at hand and in light of the resources available to support geomorphological studies. It may be possible to ignore some of the aspects of fluvial geomorphology covered in this book in studies with restricted geomorphological relevance. However, the concept of the blueprint does provide a basis from which to plan the geomorphological components of a study; care should be taken to avoid eliminating any geomorphological topic which has a relevance that perhaps is not apparent at the outset, but which may become important later. Conversely, customising may involve the addition of topics not dealt with directly in this book as it is impossible to foresee or allow for each and every contingency.

13.4 APPLIED GEOMORPHOLOGY, UNCERTAINTY AND PUBLIC POLICY

The public now realise that 'traditional' approaches to river engineering and management involving 'taming the river' or 'improving' the channel by replacing natural variability with simple geometric shapes and planforms (trapezoidal cross-sections, straight alignments) represent inappropriate engineering, involve the inefficient use of resources, produce poor value for money and, inevitably, lead to adverse environmental and aesthetic impacts. This realisation has coincided with the development of catchment-scale, co-ordinated management plans and improved methods of project appraisal in water and river agencies in the developed nations. This coincidence is important because society's desire to see more sustainable and sympathetic treatment of river systems, the adoption of catchment-wide planning, and the measurement of 'success' by a range of criteria in post-project appraisal, all require approaches to river engineering and management that are broader, smarter and multifunctional.

Applied fluvial geomorphology lies at the heart of the synthesis of improved river engineering and management because it is the point of contact of a range of fluvial processes, channel attributes and agency functions. For example, a sound understanding of the fluvial geomorphology of a system can help to inform scientists, engineers, planners and managers concerned with channel stabilisation, navigation, flood defence, recreation, fisheries and conservation.

In applying modern principles to protect or enhance the river environment and achieve multiple objectives, some compromises between conflicting goals are inevitable. For example, questions arise concerning the requirement to balance correctly the needs of nature against those of people and to protect the environment while not exposing river users and the inhabitants of floodplains to undue risks. In an era where 'accountability' has never had a higher profile, it is vital that the sound, scientific basis for decision-making is both rigorous and transparent. This is necessary not only to guarantee that the outcome is the best that can be achieved using the available time and resources, but also to protect the project team and decision-makers from charges that it acted unreasonably or recklessly if, as must occasionally occur, something goes wrong.

In terms of larger and more difficult questions regarding the development of sustainable solutions to river problems that are capable of solving today's crises while still being sufficiently flexible to cope with the evolution of unstable systems and the impacts of environmental change, applied geomorphology has some, but not all, of the answers. If goals for sustainable management, increased biodiversity and the intergenerational transfer of environmental capital are to be achieved in the case of our rivers and riparian corridors, then the potential contribution from applied fluvial geomorphology cannot be ignored.

13.5 REFERENCES

Thorne, C.R. and Baghirathan, V.R. 1993. Blueprint for morphological studies. In: Schumm, S.A. and Winkley, B.R. (Eds), *The Variability of Large Alluvial Rivers*. ASCE, New York, 441–454.

Rivers Index

Subject Index